Butterflies East of the Great Plains

Butterflies

East of the Great Plains

An Illustrated Natural History

Paul A. Opler
George O. Krizek

THE JOHNS HOPKINS UNIVERSITY PRESS
BALTIMORE AND LONDON

PAUL A. OPLER
is director of the editorial offices of the Fish and
Wildlife Service, U.S. Department of the Interior. He is also the
author of one book and more than seventy articles on entomology
and lepidoptery.

GEORGE O. KRIZEK
is a psychiatrist in private practice in Washington, D.C.
He has published numerous articles on both medicine
and entomology; his photographs of butterflies have appeared in
*The Audubon Society Field Guide to North American Butterflies, De-
fenders Magazine,* and *Endangered Species.*

The Johns Hopkins University Press, Baltimore, Maryland 21218
The Johns Hopkins Press Ltd., London

Library of Congress Cataloging in Publication Data
Opler, Paul A.
Butterflies east of the Great Plains.
Bibliography: p. 273
Includes indexes.
1. Butterflies—United States.   2. Insects—United States.
I. Krizek, George O.   II. Title.
QL549.0'64 1983   595.78'90974   83–6197
ISBN 0–8018–2938–0

This book was composed in Galliard text type
by Brushwood Graphics Studio
and Delphin No. 1 display type
by the TypeWorks from a design
by Cynthia W. Hotvedt.
It was printed on 70-lb. Glatfelter Offset
and bound in Holliston Sail Cloth
by the Maple Press Company.

Contents

List of Illustrations and Tables

Figures

Tables

Plates

Foreword

Butterflies occupy a unique position in the study of natural history. Among all invertebrates, no group is so conspicuous to laypersons and historically has received so much attention from amateur collectors, taxonomists, and evolutionary biologists alike. Characterized in art from prehistoric times to the present, butterflies have filled a diversity of roles, from representing colorful beauty in advertisements to serving as the primary subjects of population biology studies.

Butterflies traditionally have been important in education, as in demonstration of insect metamorphosis to schoolchildren, and a great many biologists, if pressed, will admit that their first lessons in taxonomy, nomenclature, species relationships, geographical distribution and variation, and ecological relationships came from Lepidoptera.

Butterflies have been important study animals in evolutionary biology because of the conspicuousness of their characters. They have also served as models for studies of geographical and environmentally induced variation, mimicry, genetics of polymorphism, and migration, as well as for research on population parameters, such as individual dispersal, longevity, and courtship behavior. The association of butterflies with ants, with larval food plants, and with one another, as in mimicry, have all been subjects of intensive investigations, quite apart from the few species that have received attention in agricultural entomology. Often they are the invertebrate group first used in analyses of geographical distribution patterns (for example, studies of island faunas) and in recent years they have served as important indicators of disappearing communities. There are twelve insects on the U.S. Department of Interior's list of endangered and threatened species: nine are butterflies.

Why do butterflies command such attention? One important reason seems to be their easily comprehended diversity. There are relatively few kinds of butterflies, compared to most insects, and, like birds or snakes, all the species in a given region can be learned, surveyed, and dealt with in an identification manual. Anthropologists have shown that primitive peoples in different parts of the world develop communication systems about plants or animals that generally contain 250–800 items, irrespective of community richness. The items included are those for which recognition and the ability to communicate names are important to users of the language. There are about 750 species of butterflies on the North American continent north of Mexico, a number of names that can be and is comprehended by many members of the butterfly cult. This means that any youth can take up the study of butterflies and can easily learn the 50–100 kinds in the neighborhood, thereby gaining a feeling of orderliness and accomplishment, which in turn enhances the intrinsic attractiveness of the specimens.

Morever, field guides and monographs set the stage for and invite more detailed study of such aspects as host specificity and mating behavior and longevity, which are documented in mark-release tests. The remaining Lepidoptera—moths—by contrast, comprise fifteen times as many species at any given site, and neither the regional nor the local fauna is completely known for any place in North America. Even if it were, no lepidopterist could comprehend the 11,000 species names and communicate about them.

Thus it is not surprising that there are more students of butterflies in the United States than there are species of butterflies, in marked contradistinction to other insects, for which there is one specialist to every 100 or 1,000 species. In terms of faunal monographs or field guides, butterflies incomparably outdistance other insects. In North America, for example, comprehensive books have been published every decade or two since the late nineteenth century, with some volumes dedicated solely to selected regions, such as California (1927), the District of Columbia and Virginia (1932, 1935), Kansas (1940), the eastern United States (1951), and Colorado (1957). In the past two decades, almost every year has seen the arrival of a major new publication, including treatises on butterflies of America north of Mexico (1961, identification; 1975, monographic; 1981, field guide) and of Florida (1965), the San Francisco Bay region (1965), Georgia (1972), southern California (1973), Washington (1974), Oregon (1980), and the Rocky Mountain states (1981).

What sets this book apart from its predecessors is its emphasis on biological relationships. Whereas the authors of the treatments mentioned above have dwelt primarily on identification, geographical distribution, and variation, and have limited their coverage of biology to dutiful lists of larval host plants (often repeating errors from preceding books, passing on misconception as if it were folklore), Opler and

Krizek have paid particular attention to behavior, population structure, and ecological associations, such as nectar resources in relation to tongue length. Paul Opler has been an enthusiastic butterfly collector since his boyhood days in Pleasant Hill, California. In contrast to most biologists, however, he went on to study other organisms, yet maintained a productive interest in Lepidoptera. Most butterfly enthusiasts have never studied other animals and possess skewed taxonomic and biological viewpoints, while biologists working with other insects usually have lost or disdain admitting any interest in butterflies.

Opler has researched the taxonomy of butterflies and small moths, as well as the host-plant relationships of microlepidoptera, especially leaf-mining species and their fossils. He spent four years working on plant phenology in Costa Rica, where he studied dioecism and spatial relationships in tropical trees, bat activity and pollination, foraging of solitary bees, taxonomic isolation and accumulation of herbivorous insects, and the mimicry of wasps by a polymorphic mantispid. During the past eight years he has been a primary motivator for insect conservation and the establishment of endangered species in the United States. His diverse experience and in-depth understanding of biology, combined with his extensive knowledge and enthusiasm for butterflies, make him uniquely qualified to undertake a book like this one.

George Krizek has had a long-term interest in nature photography, especially of butterflies. His efforts have resulted in most of this volume's excellent photographs, and they stimulated initiation of the project. Krizek's research into entomological etymology provides the first comprehensive account of the origins of the scientific names of butterflies in North America.

More than thirty years have elapsed since the appearance of Klots's *Field Guide to the Butterflies of North America*. Published when the fledgling Lepidopterists' Society was just beginning to have its now overwhelming influence on communication among American lepidopterists, Klots's accurate and detailed manual, together with increasingly better means of transportation and the growing population, had enormous effect on the growth of interest in and knowledge of butterflies in this region. A summary of traditional kinds of information alone would have warranted a new treatment, and, in compiling this volume, Opler and Krizek have recruited unpublished data from more than 100 collectors and have eloquently summarized biological and geographical information, along with aspects of variation.

It is to be hoped that this book will encourage butterfly enthusiasts to turn their attention to the fascinating world of biological studies in the manner that Klots's *Field Guide* encouraged attention to identification and geographical distribution three decades ago. Longevity and mating behavior, nectar resource partitioning among sympatric species, oviposition, larval food preferences, and seasonal variation in behavior and form are all fertile fields for investigation that can be explored readily, without sophisticated equipment, in yards and woodlots throughout the eastern United States.

JERRY A. POWELL
UNIVERSITY OF CALIFORNIA, BERKELEY

Acknowledgments

Many persons contributed their time, knowledge, and expertise to this book.

The professional staff at Image, Inc., prepared the color photographs. Our own pictures are augmented by those graciously provided by Andrew Beck, John and Norma Riggenbach, Clark Shiffer, and Ray Stanford.

The personnel at several parks and refuges kindly allowed us to study or to photograph butterflies on lands under their care; we are particularly indebted to the personnel at the Great Dismal Swamp National Wildlife Refuge and Biscayne National Monument.

George Steyskal and F. Martin Brown were of great assistance in providing derivations of scientific names.

Deane Bowers, John M. Burns, Frances Chew, Thomas C. Emmel, J. Richard Heitzman, William McGuire, Charles Oliver, Jerry A. Powell, Robert Robbins, James A. Scott, Jon Sheppard, and Jason Weintraub reviewed portions of the manuscript or advised us on technical matters.

Many people made available unpublished data on the distribution, host plants, and other biological aspects of eastern butterflies. These persons are listed in alphabetical order, with the names of those who provided particularly extensive data shown in italics:

W. A. Andersen, R. A. Anderson, R. T. Arbogast, *H. D. Baggett,* G. J. Balogh, H. E. Barton, V. Board, R. W. Boscoe, W. F. Boscoe, D. Bowers, R. W. Bracher, J. Brewer, J. Brill, A. E. Brower, *C. T. Bryson,* C. Burkhart, *R. W. Cavanaugh, Jr.,* O. Chermock, F. S. Chew, J. M. Coffman, P. J. Conway, *C. V. Covell, Jr.,* D. A. Currutt, R. P. Dana, J. M. DeWind, H. Doss, *J. C. Downey,* J. A. Ebner, G. Ehle, *J. H. Fales,* F. D. Fee, L. A. Ferge, I. L. Finkelstein, H. A. Flaska, C. W. Freeman, Jr., M. Furr, L. L. Gaddy, L. F. Gall, C. A. and S. M. Gifford, T. A. Greager, L. P. Grey, W. R. Grooms, S. W. Hamilton, P. H. Hammond, W. D. Hartgroves, *J. R. Heitzman, D. F. Hess,* G. Holbach, J. A. Hyatt, J. Ingraham, R. R. Irwin, *M. L. Israel, F. H. Karpuleon,* P. J. Kean, R. O. Kendall, W. J. Kiel, A. B. Klots, *L. C. Koehn, R. M. Kuehn,* J. L. Lavasseur, Jr., C. Leahy, R. E. Leary, B. Lenczewski, V. P. Lucas, *R. Mancke,* L. L. Martin, *B. Mather,* T. L. McCabe, *J. W. McCord,* G. M. McWilliams, L. D. Miller, P. F. Milner, M. Minno, R. T. Mitchell, M. Monica, J. C. Morse, S. J. Mueller, J. Muller, G. B. Murray, *M. C. Nielsen,* P. Nugent, D. K. Parshall, *L. J. Paulissen,* R. Pine, L. L. Pechuman, R. S. Peigler, A. P. Platt, *J. M. Prescott,* R. E. Price, Jr., E. L. Quinter, J. and N. Riggenbach, R. K. Robbins, M. Roth, C. E. Schildknecht, *D. F. Schweitzer,* Y. Sedman, A. H. Showalter, J. Shuler, R. and J. Simpson, R. H. Smith, Jr., R. E. Stanford, J. B. Sullivan, M. D. Taylor, S. Temple, A. A. Towers, E. Tryon, W. H. Wagner, Jr., C. N. Watson, Jr., M. L. Wenger, D. A. West, N. A. White, B. D. Williams III, T. S. Williams, G. D. Willis, W. D. Winter, Jr., E. N. Woodbury, W. B. Wright, and J. D. Zeligs. There are many others unmentioned who contributed directly to various state compilations.

This book has been brought to publication with the generous assistance of Nathaniel P. Reed and Byron Swift.

We thank the staff at the Johns Hopkins University Press, particularly Cynthia Hotvedt, Jim Johnston, Jane Warth, and Anders Richter, who are in large part responsible for the form and design of this book. Maria Coughlin did a superb and sensitive job of copyediting the text.

Finally, our wives, Sandra Opler and Blanca Krizek, assisted us in many ways and with good humor and tolerance sat through many evenings of butterfly slide shows.

INTRODUCTION

The Study of Butterflies

Understanding Butterflies: Misconceptions and Realities

Many people don't even *notice* butterflies, and anyone carrying a butterfly net risks comic characterization as an escapee from, or a likely candidate for, a mental institution. Even lepidopterists and other naturalists have misconceptions about butterflies, and adult butterflies are often viewed in anthropomorphic terms. Their activities have been variously interpreted as fun-loving, playful, warlike, or social. However, perhaps much to our disappointment, adult butterflies engage almost entirely in activities that can best be understood as being directly related to reproduction, maintenance, or dispersal.

Another misconception is that butterflies are static in their occurrence, and that local colonies persist until destroyed by man's activities or that they go extinct forever, never to reappear again. Another persistent view, perhaps engendered by the writings of Rachel Carson, is that butterflies have suffered irreversible declines due to spraying of pesticides. This view may be offered as an explanation for the paucity or absence of butterflies in the eastern forests, but, as we show, while some local populations of some species may persist for decades, the occurrence of most species undergoes tremendous yearly changes, not only on a local basis, but also continent-wide, and, of course, as most collectors know, a mature deciduous woodland is almost the last place to go to find most butterflies. Butterflies are most abundant in diverse, sunny situations. A visit to a hayfield, abandoned pasture, or power line any time from late May through September should quickly dispel the notion that butterflies have disappeared. Even near the centers of our largest cities,

butterflies abound in season: Rock Creek Park in Washington, D.C., and Van Cordtland Park in New York City are examples of urban settings in which butterflies are reasonably abundant and diverse.

To understand what butterflies are, and what they do, if they're not the playful, fun-loving sprites of popular writing, one needs to understand their four life-history stages. The first stage, the egg, is the stage at which, after fertilization by the male's sperm, the embryo is produced, uniting two parent cells and recombining their genetic heritage into a new individual, replete with complex chemicals, tissues, and organs that will serve the individual's needs and enhance its chances for survival. The caterpillar (the technical term is *larva*) is the butterfly's primary feeding and growth stage. The *chrysalis* (or pupa) is the transformation stage, in which the caterpillar's tissues are broken down and reconstituted to form the mobile, winged *adult*. The adult's primary functions are reproduction and dispersal, and adult butterflies engage in mate location, courtship, mating, egg-laying (or oviposition), feeding, temperature regulation, resting, and dispersal, as well as the related, more stereotyped phenomenon, migration. During much of the year, climatic or host-plant conditions are not suitable for adult flight or caterpillar feeding, and at these times any of the four stages, depending on the particular species, may enter a period of metabolic slowdown, called *diapause*. Thus, each stage may also serve the additional function of simply preserving the species until the next stage (or function) is possible.

Historical Roots of Butterfly Study

The first recorded study of butterflies in the eastern United States is the painting of the Tiger Swallowtail made on Roanoke Island in 1587 by John White, who was the commander of Sir Walter Raleigh's third expedition to "Virginia." At that time the local Indians also had a word for "butterfly."

Certainly, there must have been persons in Colonial America who collected butterflies and sent them by ship to

Sweden, England, and France, since Carolus Linnaeus, the Swedish father of animal and plant classification, named seventeen North American species between 1758 and 1771. Other Europeans, such as Fabricius, Drury, and Cramer, described more than fifty other North American butterflies before the work of our first American naturalist, John Abbot, was published. Abbot, who sailed from England to America in 1773, spent most of his life in coastal Georgia, where, until

1841, he collected natural-history specimens and painted the species he found there. His paintings include many butterflies, together with depictions of their early stages, and he made notes on their life histories. Most of Abbot's work was sold abroad. His works on butterflies were published by James Edward Smith of England in 1797 and by Major John LeConte of New York and J. A. Boisduval of Paris in 1833. Nineteen southeastern species designations were based upon his work, and he worked out and painted many life histories. Some of this work has never been repeated. (The only problem in Abbot's butterfly work is an instance of artistic license, which unfortunately has been faithfully repeated in the literature for more than 150 years. Abbot apparently depicted caterpillars on the wrong host in several instances, and in the text we point out some species for which Abbot portrayed the wrong plant.) Abbot must have been a keen observer, since one of his species, the Rare Skipper (*Problema bulenta*), was not recollected for more than 100 years!

The most significant period in the study of eastern butterflies came in the 1860s and 1870s, when William Henry Edwards of Coalburgh, West Virginia, and Samuel Hubbard Scudder of New Britain, Connecticut, studied and described the majority of eastern butterflies that had yet to be named and provided information on the life histories of many species. Edwards, whose *Butterflies of North America* was published between 1868 and 1897, described many species of the barrens, the Midwest, and the mountainous western states. Scudder concentrated on New England butterflies and described many of their life stages in exquisite detail. The quality and thoroughness of his work, as exemplified in his monumental *Butterflies of the Eastern United States and Canada, with Special Reference to New England* (1889), have rarely been repeated.

As part of the growth of large museums that began late in the nineteenth century, Henry Skinner's work at the Philadelphia Academy of Sciences included the description of eight new eastern butterflies. In the first half of the present century, only a few new species were discovered and described, the last being the Missouri Woodland Swallowtail (*Papilio joanae*), described by J. Richard Heitzman in 1974. Subsequently, only a few previously named forms have been discovered to represent separate species. The sibling species

TABLE 1. State and regional treatments of butterflies of the Eastern United States

State/Region	Author(s)	Year of Publication
District of Columbia	A. H. Clark	1932
Florida	C. P. Kimball	1965
	B. Lenczewski	1980
Georgia	L. Harris, Jr.	1972
Illinois	R. R. Irwin and J. C. Downey	1973
Louisiana	E. N. Lambremont	1954
Maine	A. E. Brower	1974
Massachusetts	F. M. Jones and C. P. Kimball	1943
Michigan	S. Moore	1960
	M. C. Nielsen	1970
Minnesota	R. W. Macy and H. H. Shepard	1941
Mississippi	B. and K. Mather	1958, 1960
New Hampshire	W. F. Fiske	1901
New Jersey	W. P. Comstock	1940
New York	A. M. Shapiro	1973, 1974
North Carolina	C. S. Brimley	1938
Pennsylvania	A. M. Shapiro	1966
	H. M. Tietz	1952
Virginia	A. H. and L. F. Clark	1951
	C. V. Covell, Jr., and G. B. Straley	1973
West Virginia	B. M. Drees and L. Butler	1978
Wisconsin	J. A. Ebner	1970

[a] Refer to Bibliography for complete citation.

of the satyrine genera *Satyrodes* (see Cardé, Shapiro, and Clench 1970) and *Enodia* (Heitzman and dos Passos 1974) and of the lycaenid genus *Celastrina* (Clench 1972) were notable among such discoveries.

A number of monographs have clarified the relationships of difficult groups, for example, Burns's (1964) work on the dusky wings (*Erynnis*). A number of states and regions have been the subject of separate books or publications (see Table 1).

Procedures for Studying Butterflies

Collecting. Many biologists, both professional and amateur, have developed an early interest through the collection of insects, including butterflies. Making at least a small collection of local butterflies is an almost indispensible start to understanding the group and the identification of the species found in an area. Many people maintain large collections of particular groups or geographic areas. Most popular books present extensive descriptions of the equipment and techniques involved in collecting specimens, and they need not be repeated here. However, a few basic principles should be understood by all collectors and are outlined briefly.

Always know your goal when collecting so that you do not take more than necessary. For a local representative collection, you need only a few of each sex of each species. A single individual of a rare species or of a species difficult to identify is all that is necessary to document its occurrence at a particular location. Sight records of common, easily recognized species are adequate, but should be carefully recorded in your notes. To study geographic variation, a series of ten to twenty of each sex is necessary to permit statistical comparison. Collection of large numbers of a rare species or from small populations of other native species for purposes of sale or exchange is unethical unless you know that the area will soon be destroyed as a suitable habitat.

It is important that all specimens be completely labeled. Label data should include state, county, and place name—accurately, so that the site may be revisited by others—the elevation in mountains, the date, and the name of the collector. Other information, such as township and range, Universal Transverse Mercator (UTM) grid reference, time of collection, and associated biological data (flowers visited, prey records, mating time) will increase the value of your material. When no longer needed, collections should be donated to appropriate local, state, or national collections.

Raising Butterflies. As is noted in some individual species accounts, life histories are often incompletely or entirely unknown. Accurate descriptions of life stages may be lacking even for common butterflies. Probably the most satisfactory way to study life stages is to start with eggs, which are obtained by watching females in nature and taking the eggs or by caging females with appropriate host plants and nectar sources. Shining a lamp on the cage may be necessary to encourage females of some species to oviposit. During rearing, careful notes should be taken of all observations. Young caterpillars should be handled carefully, with a small camel's hair brush, and should be provided with fresh food periodically. Always be sure that containers are not too wet or too dry. Avoid overcrowding. Overwintering stages should be kept outside under natural temperature and humidity conditions or may be refrigerated in tight plastic freezer boxes.

Examples of all stages may be preserved. Eggs are best frozen and then kept in small vials or envelopes. Larvae should be distended and hardened in KAAD soluation (7% kerosene, 14% acetic acid, 72% ethyl alcohol, 7% dioxane), then transferred to vials of 95% ethanol. Indelible ink must be used for labels that will be put in alcohol. For drawing larvae or preparing setal maps, specimens are best examined in a mixture of alcohol and glycerine under a dissecting microscope. Generalized drawings or paintings of caterpillars usually have no scientific value, although they may serve esthetic purposes. Reared adults should be labeled with host-plant name and collection and emergence dates, in addition to the normal label information. Reared butterflies may also be used for courtship studies or mark-recapture projects (see below).

Mapping Butterflies. One good way to learn about butterflies in your area is to do a mapping survey. First decide which area you can cover, then prepare a base map and overlay it with a grid that produces a manageable number of squares. (More than a hundred is probably too many.) For example, in Northern Virginia, the senior author chose an area near his

home that was 5 km on each side and divided into a hundred 0.5 × 0.5 km squares (Figure 1). Once your area is established, try to spend a half-hour to an hour in each square during early spring, late spring, summer, and fall. Not only will you discover some species you did not realize were in your vicinity, but also you will learn how the butterflies are distributed by local habitats or land-use categories. State surveys are for the advanced or ambitious. They may use counties or 50-km squares as mapping and sampling units.

Butterfly Walks. A technique for studying the local abundance and seasonal changes in butterfly communities was developed in England (Pollard 1977). It involves laying out a route of 2 or 3 km through appropriate local habitats and then walking the route twice a month or more frequently and recording all butterflies seen within 5–10 m of the route. Difficult species may be netted to confirm identification. It will add to your study if you can also note sex (if identifiable), time, behavior (mating, visiting flowers, etc.), and location along route. Although this method will not produce *absolute* population estimates, it will provide an excellent *relative index* that can be used to compare seasons and years (Figures 2–4).

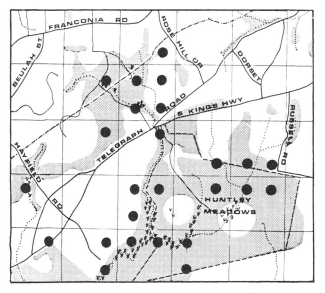

FIGURE 1. Local distribution of the Appalachian Eyed Brown (*Satyrodes appalachia*) in Fairfax County, Virginia, using Universal Transverse Mercator (UTM) grid. Each ruled square is 1 square kilometer. Shaded areas are woodland, and dotted lines are streams.

Mark-Release-Recapture. To mark adult butterflies with unique numbers or codes for mark-release studies, use either broad, square-tipped indelible marking pens or model paints. Specimens should be netted and then held with forceps while being marked. A standard marking system is to code different parts of the wings with different values. The 1-2-4-7 system (Ehrlich and Davidson 1960) is most widely used (Figure 5). Different positions on the wings have different values, and numbers from 1 to 99 may be used without duplicating any marks. As an example, one might place a mark—1—on the right forewing apex and another—70—on the posterior of the left hind wing. Thus, that butterfly would be "71." For numbers greater than 99, different colors may be used or one may add other marks near the basal part of the wings.

The usual purpose of mark-release-recapture programs is to obtain population estimates. For example, if one marks 30 butterflies in a field before lunch and then returns in the afternoon and captures 50, 10 of which bear previous marks, then the standard Lincoln Index for the population is:

$$\text{Population size} = \frac{30 \times 50}{10} = 150$$

Expressed in words, this equation is:

$$\text{Population Size} = \frac{\text{Total Marks} \times \text{Total Recaptured}}{\text{Recaptured Marks}}$$

More complex, multiple-day marking programs are used by advanced population-biology students. By using a mapped area and noting the locations of marked and recaptured butterflies, we may discover the magnitude and nature of their movements. We may find that some butterflies move very little during their lives, while others may move hundreds of meters daily. In the extreme, marking may be used to document the long-distance movement of migratory species, notably the Monarch. Urquhart (1960) and cooperators traced the Monarch's migration route by using recaptured butterflies and finally discovered their overwintering sites in central Mexico.

Marked butterflies have been used to document territoriality. The same butterfly, as determined by unique marks, may be found to occupy the same perch or nocturnal roost site day after day.

Butterfly longevity in nature is best studied by marking. The interval, in days, between initial marking of a freshly emerged adult and its last recapture is a minimum estimate of its adult lifespan. With sufficient recaptures over the flight period of a species, its maximum *potential* longevity may be closely estimated.

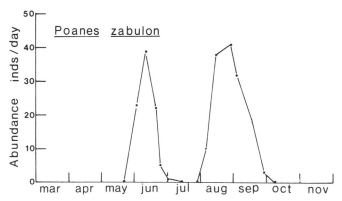

FIGURE 3. Seasonal abundance of a double-brooded butterfly, the Southern Golden Skipper (*Poanes zabulon*), in Fairfax County, Virginia.

Some butterfly species, particularly skippers, are not suitable subjects for marking, since their behavior may change with handling. Some species are easily injured. Choose a fairly common species and mark enough so that you will obtain adequate recaptures.

Flower Visitation. One may discover much about different butterfly flower preferences by carefully noting all of the visits seen in the field. Be sure to note flower identity, time of

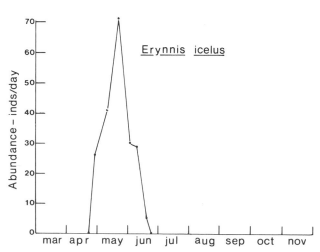

FIGURE 2. Seasonal abundance of a single-brooded butterfly, the Dreamy Dusky Wing (*Erynnis icelus*), in Fairfax County, Virginia.

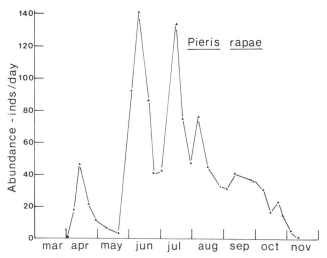

FIGURE 4. Seasonal abundance of a multiple-brooded butterfly, the European Cabbage Butterfly (*Pieris rapae*), in Fairfax County, Virginia.

day, height above ground, position of butterfly at flower, etc. With enough information, you may find some species have color preferences, visit certain sizes of flowers, or prefer flowers a certain height above the ground.

Photography. The advent of readily available 35-mm single lens reflex cameras and lightweight lenses has made it possible for many people to easily take excellent close-up pictures of butterflies or other insects. (The color photographs in this book were all taken in nature.) A micro lens, macro lens, or telephoto zoom macro lens will provide excellent results. Some people prefer to use electronic flash and low-speed film, such as Ektachrome 64, while others use high-speed film under available light conditions. Which combination is best depends upon one's experience and

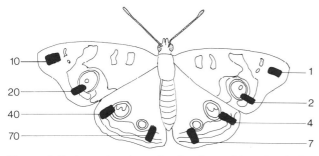

FIGURE 5. Buckeye (*Junonia coenia*), showing position of marks for capture-mark-recapture studies (see text for discussion).

preference. When working with a new camera–lens system, always shoot a trial roll at various lens apertures and speeds. Some photographers prefer to photograph a freshly emerged reared adult or a specimen that has been cooled in an ice chest, because the butterfly is relatively inactive and a tripod can be used. This may provide better pictures of the butterfly, but it will not be a record of natural behavior.

When attempting to approach wild butterflies, you must exercise a good deal of patience. Always approach slowly, and as low to the ground as possible. You may have to kneel or lie on your belly in mud or marshes if you want your picture badly enough. Butterflies visiting flowers or moisture or males perching on shrubbery are most easily approached, while patrolling males and ovipositing females usually stop so briefly that photography is nearly impossible. Some species are easy to approach, while others are extremely difficult.

Migration. When a mass movement of butterflies is discovered, it is most useful if observations can be quantified. Usually, one selects an easily viewed distance perpendicular to the direction of movement—for example, 100 m between two points along a road with a landmark at the far end. One then records the number of butterflies crossing this "front" during some unit of time, for example, 5 minutes. One might make a count every hour throughout the day, or make counts on several consecutive days. Always note the direction of movement, and if more than one species is involved, determine their proportional representation.

Classification and Names

The system of assigning scientific (Latin) names to species of animals and plants was first devised by Carolus Linnaeus of Sweden. For animals, including butterflies, the tenth edition of his *Systema Naturae*, published in 1758, is determined to be the beginning of scientific naming. All validly described species must have an original comparative description, a type specimen, and a type locality, and the species description must have been published in a book or recognized scientific periodical. Rules regarding scientific names and related matters may be found in the *International Code of Zoological Nomenclature*. Each Latin binomial consists of a genus name (always capitalized) followed by a species name. If subspecies names are used, they follow the species name, forming a trinomial. It is common usage to place the author's name after the species (in parentheses if the original genus to which the species was assigned was different from the present one) followed by the year of original publication. The classification of species follows a hierarchy. The standard group categories are kingdom, phylum, class, order, family, genus, species, and subspecies. All butterflies belong to the kingdom Animalia, phylum Arthropoda, class Insecta, and order Lepidoptera. Butterflies belong to the most advanced of three suborders and to two superfamilies within it. These two

superfamilies are the true butterflies (Papilionoidea) and skippers (Hesperioidea). Older classifications that divide the Lepidoptera into Heterocera (moths) and Rhopalocera (butterflies), or Jugatae and Frenatae, are archaic and have not been current since about 1940. The superfamilies of butterflies and skippers are based upon detailed study and analysis of external and internal anatomy by Ehrlich (1958).

True butterflies are composed of five families: Papilionidae (swallowtails and parnassians), Pieridae (whites and sulphurs), Lycaenidae (harvesters, coppers, hairstreaks, blues, and metalmarks), Libytheidae (snouts), and Nymphalidae (brushfoots, leaf wings, satyrs and wood nymphs, and milkweed butterflies). Skippers have only one family. Discussion of the characteristics of these families, as well as of several subfamilies, is found in the species accounts.

Scientific names enable the clear identification of particular groups and allow scientists from all countries to communicate and to use the same name for the same species, wherever it is found. For these reasons, scientific names are useful worldwide and are universally employed. However, some taxonomic specialists (especially in Europe) tend to "split" genera into overly fine divisions. Because name usage may vary depending upon interpretation and because currently there are several conflicting interpretations being debated, the usage of generic names for our butterflies is relatively unstable. Thus, the authors of a recently published North American catalogue (Miller and Brown 1981) adopted the "splitter" view, raising many subgenera to generic standing and dividing several genera of long standing into smaller genera. In many cases this may obscure relationships of large groups of related organisms. We have selected a middle ground, accepting some changes that seem appropriate. Generally, the nomenclature followed is that found in Howe's *Butterflies of North America* (1974), a standard reference with nomenclature accepted by most American lepidopterists.

Common names, which we use extensively in this book, are much less stable than scientific names. There is no code to give them stability, and they vary greatly from book to book and from country to country. In most instances we have chosen to use the names found in Klots (1951). For species not treated in that book, we use names from Riley (1975) or Pyle (1981). In a few instances we have coined new common names where none had existed.

We hope that by providing both scientific and common names, the usefulness of our book will be increased. Common names for some groups, such as birds, have gained a high degree of stability in English-speaking countries, thus eliminating a seeming barrier for neophytes who otherwise might shy away from the field. A new army of naturalists are taking up butterfly watching and photography, and this book is intended as much for them as it is for the serious butterfly collector.

Most authors of scientific names do not provide explanations or derivations for the genera and species they describe. We felt it would add to the butterfly student's interest and knowledge to provide derivations or *etymological* interpretations of the included scientific names. However, for some names we could find no cogent derivation and for others we may have missed the original author's intent.

Butterfly Structure

This section presents general descriptions of the external structures and functions of the various life stages. It is not intended to be a detailed morphological treatise, nor do we discuss internal anatomy of butterflies in more than a few instances, and there solely for the sake of understanding later life-history sections.

Familiarity with several terms describing direction or position helps biologists more readily explain the location of different structures: *Dorsal* refers to the upper side or top of a structure, *lateral* refers to the sides, *ventral* refers to the lower surface or bottom, *anterior* refers to the front or head end, *posterior* refers to the rear or tail end, *basal* refers to the portion of a structure near its point of attachment, and *apical* refers to the tip of a structure or that portion farthest from the point of attachment.

Adult. Butterflies, like other insects, have segmented bodies divided into three major regions: head, thorax, and abdomen. The head bears the compound eyes, segmented antennae, palpi, and the proboscis (Figure 6).

Butterfly eyes are composed of many separate visual receptors, termed *ommatidia*. A butterfly probably sees a mosaic image. Butterflies, like other insects, have been shown to be most sensitive to color at the ultraviolet end of the spectrum. Many species can detect ultraviolet wavelengths beyond those seen by humans. Some species have "hidden" ultraviolet reflectance patterns on their wings that constitute courtship signals: male whites and sulphurs are the most notable examples (Silberglied and Taylor 1978).

Butterfly antennae are usually clubbed. There may be a bare patch (nudum) under the club. Skipper antennae may have the club bent back at the tip or drawn out in a narrow extension (apiculus). Antennae serve the senses of smell and sound.

The paired palpi serve mainly to house the coiled proboscis. The proboscis is composed of two centrally grooved, appressed extensions that form a long, siphonlike tube through which liquids are taken into the digestive system. The proboscis varies greatly in length, depending upon the nature of the adult diet. Nectar-feeding butterflies generally have long proboscises, while those that feed on sap, carrion,

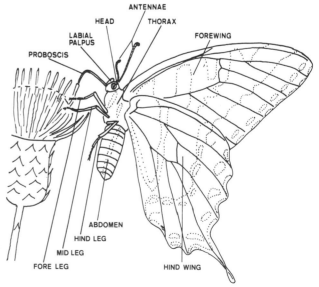

Figure 6. Tiger Swallowtail (*Papilio glaucus*), showing major features of an adult butterfly.

or rotting fruit have relatively short ones. Head muscles provide a "pumping" action to draw liquids up through the proboscis.

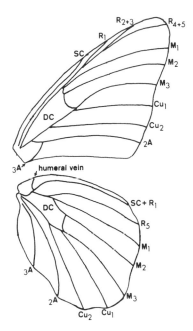

FIGURE 7. Wings of Red-spotted Purple (*Limenitis arthemis astyanax*), showing names of principal wing veins. SC = subcostal, R = radial, M = median, Cu = cubitus, A = anal or vannal, DC = Discal cell.

The thorax bears two pairs of wings, for powered flight, and three pairs of legs. The legs are jointed, with coxa, trochanter, femur, tibia, tarsus, and apical claw being the order of the segments from base to tip. The three pairs of legs are referred to as fore-, mid- and hind legs. Although the legs serve principally for walking, the forelegs of all brushfoot butterflies and many Lycaenidae are strongly reduced and have sensory cells that are used principally for "tasting" potential host plants or adult food sources.

The wings are the most notable structures of a butterfly. There are two pairs of wings. The anterior pair are termed *forewings* and the posterior pair are termed *hind wings*. The wings are thin, translucent, chitinous structures with a system of hardened veins (Figure 7) and a covering of modified setae or scales, which may be transparent or colored. The coloring may be due to either deposited pigments or diffraction caused by the scales' physical structure, usually a series of ridges and grooves (Downey and Allyn 1975). Further discussion of color adaptations is given in a later section, "Color and Pattern."

There are several systems of terms applying to different areas of the butterfly wings. The system we use mostly refers to various areas by position (Figure 8), while another names different wing regions (Figure 9).

The wing veins have their most obvious circulatory uses during development and shortly after adult emergence, when they are used to pump out the wings before they harden. Later, they function more as structural braces, but they still carry some blood (hemolymph). Note that the anterior or costal margin of the forewings is most heavily strengthened with veins (Figure 7).

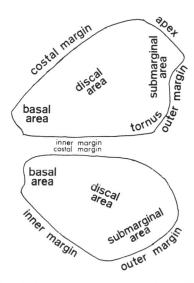

FIGURE 8. Terminology for wing areas, margins, and angles of an adult butterfly.

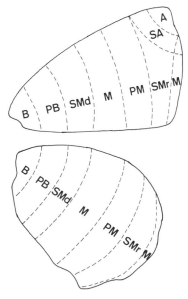

FIGURE 9. Terminology for wing regions of an adult butterfly. B = basal, PB = postbasal, SMd = submedian, M = median, PM = postmedian, SMr = submarginal, M = marginal, SA = subapical, A = apical.

The butterfly abdomen contains the principal portions of the food reserves and the entire reproductive system. The external chitinized portion of the genitalia is found on the posterior segments. The male's genitalia consist most obviously of paired lateral valvae (for clasping), dorsal tegumen and uncus, and the central aedeagus, or penis (Figure 10). On the female, the only visible genitalia are the pads (papillae anales), at either side of the ovipository opening on the tip of the abdomen, while the recessed copulatory (ostium bursae) occurs on the venter just anterior to the abdomen tip (Figure 11). Taxonomists make much use of the highly contorted genital structures in the separation of species.

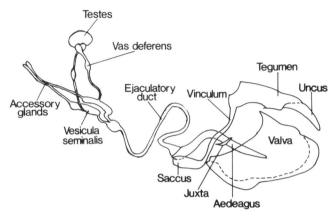

Figure 10. Internal and external reproductive system of a male Large Marble (*Euchloe ausonides*).

Egg. The eggs of butterflies are small and their shapes range from ovoid to hemispherical to columnar. They have a transparent chitinous *chorion*, which is incredibly sculptured. The detailed structures are best seen by using a scanning electron microscope, but some appreciation may be gained by viewing them with a hand lens or light microscope. The area with the pores through which the sperm pass is on the summit of the egg and is termed the *micropyle*.

Larva. The basic design of butterfly caterpillars (or larvae) includes a head and a relatively long body. The head has two semicircles of tiny eyes, or *ocelli*, and a pair of very short, simple antennae to either side of the mouth. The mouth consists of an upper lip (*labrum*), lateral chitinized and toothed mandibles, and a lower lip (*labium*). The silk glands open on small spinnerets located to either side of the labium. The first three segments represent the thorax, and each segment bears a pair of jointed legs. The remaining segments constitute the abdomen, and fleshy unjointed *prolegs* occur on

the third, fourth, fifth, sixth, and tenth segments. The distal portion of each proleg bears an oval series of hooks, or crochets. The exit of the digestive system occurs on the last segment. There is a pair of small lateral breathing holes (*spiracles*) on the first thoracic segment and on each of the first eight abdominal segments.

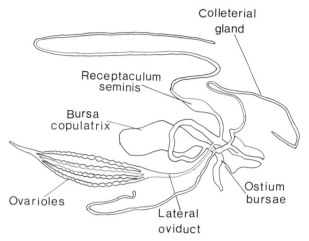

Figure 11. Internal and external reproductive system of a female Monarch (*Danaus plexippus*), after Ehrlich and Davidson (1961).

Larvae are usually cylindrical in cross-section, but some, notably the Lycaenidae, are dorsoventrally flattened.

Although some caterpillars, e.g., swallowtails, are naked, most bear simple hairs (*setae*), tubercles bearing many stiff hairs, or fleshy projections. Caterpillars of some families have structural specializations, which are discussed in the heading for each family's species accounts.

Larvae are preeminent consumers of food. Their primary functions are to eat and to grow, although they may also serve as a dormant stage for some groups.

Pupa. Often termed the *chrysalis*, the pupa is the transformation stage wherein the larval structures are broken down and those of the adult are formed. With the exception of Giant Yucca Skippers, pupae are immobile and appear to be elongated hardened structures. There is usually a series of small hooks, the *cremaster*, at the posterior terminus of the abdomen. In true butterflies, the wing pads and appendages are fastened down, but in skippers the proboscis case is usually free for part of its length.

As mentioned above, the pupa is primarily a transformation stage, but it may also be an overwintering or resistant dormant stage.

Color and Pattern

The wings of butterflies are wondrous objects, with their bold or subtle patterns of variously colored scales. Since these

are the most remarked-upon features of butterflies, they deserve full appreciation of their basis and adaptive value.

The colors may be due to either pigments or diffracted color from ridges and grooves in the scales. Most butterflies have their wings entirely clothed with pigmented scales, but some have combinations of both types. Some scales may be both pigmented and diffractive.

Common butterfly-scale pigments are melanins, which are usually black; flavonoids, which may be brown, yellow, orange, or green (rarely); and pteridines, which may be white, yellow, or red. Some pteridines produce ultraviolet colors, which may be seen by the human eye only in darkness under a strong ultraviolet light. A mixture of scales of several colors may give the appearance of a different color. Yellow and black scales intermixed give the appearance of green, as on the ventral hind wings of the Olympia Marble, while white and black scales give the impression of gray.

Structural colors usually are silver, blue, purple, green, or violet. If they change depending on the angle of light, they are termed *iridescent*. The dorsal wing surfaces of most male coppers and blues and of both sexes of some hairstreaks, as well as the silver spots on the ventral wing surfaces of some fritillaries, are good examples of structural colors.

The color patterns on the ventral wing surfaces of most butterflies do not differ significantly between the sexes, but they are often different on the dorsal surfaces. Some species have nearly identical color patterns for both males and females—for example, the Mourning Cloak, Red Admiral, and Silver-spotted Skipper.

Species whose color patterns differ between the sexes are termed *sexually dimorphic*. Most species differ at least to a small degree, while the sexes of others have very different colors or patterns. Examples of species with strikingly different colored or patterned sexes are most blues, the Falcate Orange-tip, many sulphurs, and a few skippers, most notably the Southern Golden Skipper. In a few species, one sex, usually the female, has two color forms. These include the Tiger Swallowtail, the Clouded and Orange Sulphurs, and the Northern Golden Skipper.

There is also a geographic element to color, as many species vary at least slightly in their color and pattern from one geographic area to another. These variations often form the basis for the naming of geographic subspecies. An almost limitless number of such forms could be named. Of greater interest are species with more or less discrete forms in different regions. One example that we illustrate is the Red-spotted Purple (*Limenitis arthemis astyanax*) and its second form, the White Admiral (*Limenitis arthemis arthemis*). Other examples are the Viceroy, which has a brown form or subspecies in peninsular Florida and an orange form elsewhere in the eastern states, the Barred Sulphur, and the Ox-eyed Wood Nymph.

Most species with more than one annual generation vary seasonally in their color or pattern. This phenomenon is termed *seasonal polyphenism* (Shapiro 1976). Most of these species vary slightly between broods, but in a few the differ-

ences are striking. In some temperate and northern species, spring and/or fall adults have increased black scaling and are usually smaller (see below). Some of these, such as the Pearl Crescent, Tiger Swallowtail, Clouded Sulphur, Veined White, and Spring Azure have discrete forms that have been named. These forms have been shown to be stimulated primarily by the day-length (photoperiod) the larvae experience. The spring forms are produced by larvae that are subjected to short day-lengths, usually less than 12 hours. The darker colors produced by the more extensive black scaling of the spring forms allow them an advantage over the larger, lighter summer forms. During the cooler periods of spring, the darker butterflies warm up and fly sooner in the day than they might be able to if they were light-colored.

Another group of seasonally variable species is adapted to seasonal tropical conditions. This group includes the Buckeye, Goatweed Emperor, Barred Sulphur, and Dwarf Sulphur. These species also have their seasonal forms triggered by different day-length regimes, but have their dry-season or winter form adapted to overwintering in a nonreproductive state. The dry-season form of these species is often larger and more cryptically marked beneath.

The two "punctuation" angle wings, the Question Mark and the Comma, both have an orange-hind-wing overwintering form and a black-hind-wing summer form. Both species migrate south in the fall but also overwinter in hollow trees within their northern breeding areas. Their polyphenism seems most like that of tropical butterflies but the adaptiveness of the two color forms is unknown.

The adaptiveness of color in butterflies seems to relate to three important functions—sexual recognition, predator avoidance, and temperature regulation (which is discussed above and under seasonal size differences later on). The first main function of butterfly color, the enabling of ready recognition by prospective mates, including the ability to discriminate between related species, facilitates searching and reduces the frequency of courtship with the wrong species. Males do occasionally solicit females of a different species, particularly among skippers, but females usually reject such advances. Rarely, matings between unrelated species occur, but the production of viable eggs is not likely, and even if offspring result, they are usually sterile. Visual cues are important in initial courtship phases, but chemical cues are important in the ultimate steps ending with mating.

Color may play several roles in predator avoidance. In *crypsis*, the butterfly is colored and patterned to blend into its background or to resemble an inedible object. The lower surfaces of the White Mountain Butterfly and Polixenes Arctic closely resemble the lichen-mottled rocks upon which they usually rest, while the underwings of angle wings, tortoise shells, purple wings, and pearly eyes closely resemble the bark of the trees upon which they rest. The pupae of many butterflies also resemble tree bark. Dead-leaf patterns are shown by the Eastern Snout Butterfly, Florida Leaf Wing, and Goatweed Emperor. The "snout" of the Snout (Plate 20) and tails of the leaf wings (Plate 33) give the added effect of a leafstem. The mottled-green ventral wing surfaces of the Olive and Hessel's Hairstreaks (Plate 14) allow them to blend into the foliage upon which they rest, while the larvae of many foliage-feeding butterflies are green.

Another form of crypsis is the diversionary effect of the false eyes and tails of hairstreaks and swallowtails, which lure birds to attack the nonvital end. Evidence of this is the large number of these insects found with the posterior part of the hind wings neatly clipped or torn off (Robbins 1981).

A second form of predator avoidance is the use of bright colors that warn predators that the species is distasteful. Many brightly colored butterflies are probably at least slightly distasteful, while others, such as the Monarch, Queen, Pipe Vine Swallowtail, and Baltimore, are both distasteful and emetic (that is, they cause vomiting if eaten). Such brightly colored distasteful insects are termed *aposematic*. If a bird once tries to eat one, it will avoid all similarly colored butterflies in the future.

Batesian mimicry is the term for the close resemblance to distasteful aposematic butterflies that is adopted by perfectly edible or only slightly distasteful species. The edible *mimics* benefit from this deception, since would-be bird and mammal predators avoid them to a greater degree than if they were not so patterned. The distasteful *models* lose a slight portion of their advantage by being mimicked. Normally, mimics must be less abundant than their models for the association to be successful in the long run.

Three recognized Batesian mimicry complexes occur among eastern butterflies. The most famous instance is that of the Monarch (Plate 37) and the Viceroy (Plate 32). A parallel situation is represented by the Queen (Plate 37) and the Floridian population of the Viceroy. A third large group of mimics resembles the distasteful Pipe Vine Swallowtail. Each of these groups is discussed under the respective model's species account.

Mimics often differ strikingly from their closest relatives in color and pattern. Witness the differences between the Viceroy, Red-spotted Purple, and White Admiral (Plate 31); between the Diana (Plate 21) and Regal Fritillary (Plate 22); and between the two female forms of the Tiger Swallowtail (Plate 3).

A single mutation that effects major changes in a species' color pattern so that it roughly approximates that of a distasteful species is the initial stage leading to mimicry. Further small evolutionary changes perfect the color pattern and other aspects of the resemblance. The gene controlling the major appearance shift is called a *switch gene*.

Another form of predator deception involving color is *flash coloration*. Butterflies that have somber colors ventrally and bright, often iridescent, color dorsally may confuse would-be avian predators. When the butterfly flies, bright flashes of color are seen, but when it alights, the insect suddenly seems to disappear, since the bird's search image, which centered on a bright flashing object, is lost to view. The Great Purple and White M Hairstreaks are good examples of users of this ploy.

Size

The size of butterflies is an important trait and has a number of important life-history implications. Each butterfly has a normal degree of size variation, and, in addition, there are often sexual, geographic, and seasonal differences in size. The limits of butterfly size are determined genetically, but very small individuals may result if they have barely enough food to escape starvation.

The best standard measurement of butterfly size is dry weight, which is obtained by weighing a dead butterfly after it has been thoroughly desiccated in a drying oven. A more easily obtained *relative* size indicator is forewing length, since wing size is usually a direct function of body weight due to aerodynamics. Forewing length, expressed in centimeters, is used throughout this book. Body length and wet weight are *not* good measures.

Sex Differences. One of the first observations made by most observers is that female butterflies are usually larger than their corresponding males. Although this is generally true for most families, in somewhat less than half of the Lycaenidae, the species show no significant difference in size.

This is also true for some Pieridae, and in some species, especially of tropical derivation, the males are actually larger. There is no widely accepted explanation for females' larger size, but a possible explanation may be their greater abdominal bulk, due largely to the precious load of eggs they carry. The size difference is usualy less than 10%, but a few, notably brushfoots, exhibit much greater disparities. The greatest size difference we found was shown by the Harris' Checkerspot (*Chlosyne harrisii*)—22%—while the Silvery Checkerspot (*Chlosyne nycteis*) and the *Speyeria* fritillaries (except *S. atlantis*) show differences varying between 15% and 21%. In other families, the Great Purple Hairstreak (*Atlides halesus*) has a 20% difference between the sexes, and male Orange-barred Sulphurs (*Phoebis philea*), with their forewings 10% longer than the females', show the greatest difference for butterflies with larger males.

Several of the butterflies with large intersexual size differences, notably the Diana, Regal Fritillary, Great Spangled Fritillary, and Common Wood Nymph, also have large differences in life expectancy. Females of these butterflies live

TABLE 2.   Seasonal differences in size as shown by mean forewing length (cm)

Species	Sex	Season		Seasonal difference[a]
		Summer	Spring–Winter	
Barred Sulphur	♂	1.59	1.70	7%
(*Eurema daira*)	♀	1.65	1.83	11%
Comma	♂	2.55	2.49	2%
(*Polygonia comma*)	♀	2.69	2.62	3%
Eastern Tailed Blue	♂	1.22	1.19	2%
(*Everes comyntas*)	♀	1.21	1.12	7%
European Cabbage Butterfly	♂	2.48	2.25	10%
(*Pieris rapae*)	♀	2.39	2.27	5%
Goatweed Butterfly	♂	3.05	3.08	1%
(*Anaea andria*)	♀	3.41	3.36	2%
Gray Hairstreak	♂	1.45	1.45	—
(*Strymon melinus*)	♀	1.47	1.46	<1%
Horace's Dusky Wing	♂	1.94	1.88	3%
(*Erynnis horatius*)	♀	1.96	1.91	3%
Orange Sulphur	♂	2.50	2.34	6%
(*Colias eurytheme*)	♀	2.83	2.45	15%
Pearl Crescent	♂	1.56	1.52	3%
(*Phyciodes tharos*)	♀	1.68	1.69	<1%
Question Mark	♂	2.90	2.78	4%
(*Polygonia interrogationis*)	♀	2.98	3.22	8%
Tiger Swallowtail	♂	5.54	4.56	18%
(*Papilio glaucus*)				

[a] Dash indicates no difference.

several months longer than males and wait until the end of the summer before laying their eggs.

Seasonal Differences.   Seasonal differences can also be marked (Table 2). Northern multiple-brooded butterflies usually have significantly smaller spring- and fall-brood individuals. The Pearl Crescent, Tiger Swallowtail, Orange Sulphur, Eastern Tailed Blue, and Horace's Dusky Wing can be cited as examples. Small size, as well as more extensive black scaling (see above), are adaptive during cooler periods; the butterflies warm up and fly sooner in the day than they might be able to do otherwise.

In contrast, the winter forms of more southern butterflies with tropical affinities are usually *larger* than their corresponding summer forms. *Eurema* butterflies, the large *Phoebis* sulphurs, and brushfoots, such as the *Anaea* leaf wings, are good examples. These winter forms are actually adapted to different factors. They are very long-lived, spending at least 3 or 4 months resting while in reproductive arrest (diapause), and their larger winter size is probably best explained by their longer lives and lethargic life style.

Geographic Differences.   Geographic variation in size is to be expected as different populations of each species adapt to different host plants, climates, and other environmental factors. In addition, a general rule for warm-blooded animal species is that size increases with increasing altitude or elevation (Bergmann's Rule), while the opposite is generally

true for butterflies, perhaps due to their differing thermoregulatory requirements. However, one butterfly does appear to comply with Bergmann's Rule, at least in part: the Mustard White (*Pieris napi oleracea*). In Michigan, individuals of more southern populations are significantly smaller (Wagner and Hansen 1980). In the eastern states, swallowtails and a number of brushfoot butterflies, such as the Great Spangled Fritillary (*Speyeria cybele*), Silver-bordered Fritillary (*Boloria selene*), and Green Comma (*Polygonia faunus*), are good examples of species whose average size is larger in more southern or lowland populations.

Despite the general trends described above, the average butterfly size in many populations often varies from site to site in no logical pattern, especially among butterflies that have more or less permanent local populations finely attuned to local conditions. An intriguing case is presented by a number of species that have populations in Florida and adjacent states. These species often change in size abruptly between two adjoining populations. In some cases, the Florida individuals are larger (Tiger Swallowtail, Banded Hairstreak, Hessel's Hairstreak, Southern Broken dash [J. M. Burns, unpublished]), while in others they may be smaller (Sleepy Dusky Wing). These differences probably represent a former discontinuity in the species' distribution dating back to the time when peninsular Florida was a series of low islands.

ern New York, and most of southern New Jersey south-westward in a belt along the upper Piedmont from the lower Delaware Valley of southeastern Pennsylvania to northern Georgia. West of the Appalachians it extends in a broad belt through the Midwest south of the Great Lakes. The latter area is much more extensive than that east of the Appalachians. Although the Upper Austral Life Zones is almost as rich in butterflies as the Lower Austral, few species are limited to it. East of the Great Plains, the Smoky Eyed Brown (*Satyrodes fumosa*), Mitchell's Satyr (*Neonympha mitchellii*), Regal Fritillary (*Speyeria idalia*), Northern Metalmark (*Calephelis borealis*), and Dusky Blue (*Celastrina ebenina*) are more or less limited to it, although none fits its limits very well.

Transition Life Zone. The Transition Life Zone (Figures 27–30) lies between the Upper Austral and Canadian Life Zones. Average summer temperatures vary from 18°C to 22°C, and there are about 4 or 5 months of freezing temperatures. The Transition Zone occurs in a broad swath from the New England coast west across New York and the Great Lakes states and in a southwestward band down the Appalachians to northern Georgia and western North Carolina. At least 14 species occur primarily within this zone (Table 3), and almost all belong to genera with a northern flavor. Yet the majority are limited to areas east of the Rocky Mountains. Four species (21%) range across the continent, and one species is Holarctic, occurring in the Old World as well.

Canadian Life Zone. The Canadian Life Zone (Figures 31 and 32) represents the major portion of boreal habitat in the eastern United States. Average summer temperatures rang-

FIGURE 17. Hardwood hammock, Tropical Life Zone (Upper Matecumbe Key, Florida). Habitat for many tropical butterflies, including the Florida White (*Appias drusilla neumoegenii*), Julia (*Dryas iulia largo*), Ruddy Dagger Wing (*Marpesia petreus*), and Hammock Skipper (*Polygonus leo*).

ing from 14°C to 18°C, and there are at least 6 months with freezing temperatures each year. In our area, this zone is limited to northern Maine, Vermont, and New Hampshire, the Adirondack Mountains of New York, and the northern portions of Michigan, Wisconsin, and Minnesota. Isolated

FIGURE 16. Second-growth scrub, Tropical Life Zone (Big Pine Key, Florida). Habitat for Martialis Hairstreak (*Strymon martialis*) and Columella Hairstreak (*Strymon columella modesta*).

FIGURE 18. Tropical scrub, Tropical Life Zone (Biscayne National Park, Florida). Habitat for Schaus' Swallowtail (*Papilio aristodemus ponceanus*).

FIGURE 19. Hardwood hammock, Tropical Life Zone (Key Largo, Florida). Habitat for the Zebra (*Heliconius charitonius tuckeri*), Orange-barred Sulphur (*Phoebis philea*), and the Large Orange Sulphur (*Phoebis agarithe maxima*).

FIGURE 21. Tupelo swamp, Lower Austral Life Zone (Great Dismal Swamp, Virginia). Habitat for Palamedes Swallowtail (*Papilio palamedes*) and Creole Pearly Eye (*Enodia creola*).

FIGURE 20. Live oak–yucca scrub, Tropical Life Zone (Florida). Habitat for the Giant Yucca Skipper (*Megathymus yuccae*) and White M Hairstreak (*Parhassius m-album*).

FIGURE 22. Southern broadleaf forest, Lower Austral Life Zone (Jericho Ditch, Great Dismal Swamp Wildlife Refuge, Virginia). Habitat for Hessel's Hairstreak (*Mitoura hesseli*), Pearly Eye (*Enodia portlandia*), and Carolina Roadside Skipper (*Amblyscirtes carolina*).

"islands" of this zone occur in the Appalachians from Pennsylvania south to Virginia.

About twenty butterflies are limited to this Life Zone in the eastern United States (Table 4), and the majority are either transcontinental (55%) or holarctic (35%). The remaining two, Macoun's Arctic (*Oeneis macounii*) and the Pink-edged Sulphur (*Colias interior*), occur primarily in Canada, and are close relatives of western species.

Hudsonian and Arctic-Alpine Life Zones. Two small areas, representing the Hudsonian and Arctic-Alpine Life

Zones (Figure 33), occur atop the Presidential Range in New Hampshire and Mt. Katahdin in Maine. There, freezing temperatures may occur during 10 or more months each year, and average midsummer temperatures are 14°C or less. The White Mountain Butterfly (*Oeneis melissa semidea*), found on the bare, windswept summits of the Presidential Range, and the Polixenes Arctic (*Oeneis polixenes katahdin*), found atop Mt. Katahdin, are the only true Arctic-Alpine species in the eastern U.S. Some of the butterflies found in black spruce muskegs of northern Minnesota are more normally inhabitants of Hudsonian Life Zone areas far to the north of our borders.

Variation with latitude.   Only a few butterflies are resident throughout most of the geographic area covered by this book. As one proceeds from north to south or east to west, the ranges of new species are encountered and old ones are left behind. It is well known that arctic areas are poor in butterfly species and that tropical areas are very rich. In the eastern United States, the portion between 35° and 40° north latitude is richest, with 147 resident species and 19 regular colonists (Table 5), while the area north of 45° north latitude is poorest, with only 107 resident species and 10 regular colonists. Family representation also changes latitudinally. Skippers and swallowtails increase in richness with decreasing latitude, both in species and relative proportion. Skippers account for 33% or less of all butterflies above 40° north latitude, and more than 40% southward. Although few species are involved, the proportion of resident swallowtails

FIGURE 23. Open freshwater marsh, Upper Austral Life Zone (Montgomery County, Maryland). Habitat for the Bronze Copper (*Lycaena hyllus*). River forest in background is habitat for Zebra Swallowtail (*Eurytides marcellus*).

FIGURE 25. Tall-grass prairie, Upper Austral Life Zone (Loess Hills, Iowa). Habitat for Regal Fritillary (*Speyeria idalia*), Ottoe Skipper (*Hesperia ottoe*), and Arogos Skipper (*Atrytone arogos*).

FIGURE 24. Coastal dunes and salt marsh, Upper Austral Life Zone (Cape Henlopen, Delaware). Habitat for Aaron's Skipper (*Poanes aaroni*) and Salt Marsh Skipper (*Panoquina panoquin*).

FIGURE 26. Deciduous forest, Upper Austral Life Zone (Difficult Run, Virginia). Habitat for Red-spotted Purple (*Limenitis arthemis astyanax*), Hackberry Butterfly (*Asterocampa celtis*), and Gold-banded Skipper (*Autochton cellus*).

gradually increases from 2% in the north to 7% (of all species) south of 30° north latitude (Table 5).

Resident satyrines and brushfoot butterflies decrease in their proportional representation southward, although absolute species number of the latter remains about the same. Lycaenids make up about the same proportion in all areas, while resident whites and yellows are poorest in the central latitudes.

TABLE 3. Eastern butterflies limited to the Transition Life Zone

Butterfly		Range[a]
Common name	Latin name	
Acadian Hairstreak	*Satyrium acadica*	T
Aphrodite Fritillary	*Speyeria aphrodite*	M
Black Dash	*Euphyes conspicua*	E
Bog Copper	*Lycaena epixanthe*	E
Dreamy Dusky Wing	*Erynnis icelus*	T
Edwards' Hairstreak	*Satyrium edwardsii*	E
Grizzled Skipper	*Pyrgus centaureae wyandot*	H
Harris' Checkerspot	*Chlosyne harrisii*	M
Indian Skipper	*Hesperia sassacus*	E
Long Dash	*Polites mystic*	T
Melissa Blue	*Lycaeides melissa*	T
Poweshiek Skipper	*Oarisma poweshiek*	E
Tawny Crescent	*Phyciodes batesii*	M
West Virginia White	*Pieris virginiensis*	E

[a] E = endemic to eastern North America; M = from Rocky Mountains eastward; T = transcontinental; H = holarctic.

Most butterflies in the northern latitudes are widespread transcontinental or holarctic species, while many of those in the south are tropical in their origins. The central Atlantic region has a mixture of northern and southern species, but is richest in butterflies endemic to the area covered by this book or limited to areas east of the Rockies.

Disturbed Habitats. Man's activities have wrought great changes in the composition and distribution of butterflies. The cutting of the primeval forests and introduction of weedy plants from Europe, the tropics, and the west have allowed the introduction or invasion of butterflies not originally

FIGURE 28. Mixed deciduous forest and open field, Transition Life Zone (Poverty Hollow, Virginia). Habitat for Diana (*Speyeria diana*) and Northern Metalmark (*Calephelis borealis*).

FIGURE 27. Second-growth mesic woodland, Transition Life Zone (Catoctin Mountains, Maryland). Habitat for West Virginia White (*Pieris virginiensis*) and Appalachian Blue (*Celastrina neglectamajor*).

FIGURE 29. Shale barrens, Transition Life Zone (Green Ridge Park, Maryland). Habitat for Olympia Marble (*Euchloe olympia*), Silvery Blue (*Glaucopsyche lygdamus*), and Grizzled Skipper (*Pyrgus centaureae wyandot*).

TABLE 4. Eastern butterflies limited to the Canadian Life Zone

Butterfly		Range[a]
Common name	Latin name	
Arctic Skipper	*Carterocephalus palaemon mandan*	H
Bog Fritillary	*Boloria eunomia dawsoni*	H
Chryxus Arctic	*Oeneis chryxus strigulosa*	T
Disa Alpine	*Erebia disa mancinus*	T
Freija Fritillary	*Boloria freija*	H
Frigga Fritillary	*Boloria frigga saga*	T
Green Comma	*Polygonia faunus*	T
Hoary Comma	*Polygonia gracilis*	T
Hoary Elfin	*Incisalia polios*	T
Inornate Ringlet	*Coenonympha inornata*	E
Jutta Arctic	*Oeneis jutta*	H
Laurentian Skipper	*Hesperia comma laurentina*	H
Macoun's Arctic	*Oeneis macounii*	E
Mountain Silver-spot	*Speyeria atlantis*	T
Nabokov's Blue	*Lycaeides argyrognomon nabokovi*	T
Pink-edged Sulphur	*Colias interior*	M
Purple Lesser Fritillary	*Boloria titania*	H
Red-disked Alpine	*Erebia discoidalis*	T
Saepiolus Blue	*Plebejus saepiolus amica*	T

[a] E = endemic to eastern North America; M = from Rocky Mountains eastward; T = transcontinental; H = holarctic.

native to our area. The European Cabbage Butterfly (*Pieris rapae*) and European Skipper (*Thymelicus lineola*) were accidentally introduced from Europe, and the American Copper (*Lycaena phlaeas*) probably also was introduced similarly. All three feed primarily on exotic weeds.

Other butterflies have colonized the eastern states on their own in recent decades, but were aided by the prior introduction of their weed hosts or by some other habitat alteration. The Inornate Ringlet (*Coenonympha inornata*), Large Marble (*Euchloe ausonides*), and Saepiolus Blue (*Plebejus saepiolus amica*) invaded and have successfully colonized the northern states from the west and north, while a number of tropical species have entered Florida as a result of the ever-increasing weediness and exotic nature of its flora. Documented colonizations that have occurred in Florida and that would not have been successful without man's indirect assistance include the Soldier (*Danaus eresimus tethys*), Malachite (*Siproeta stelenes biplagiata*), Orange-barred Sulphur (*Phoebis philea*), Dorantes Skipper (*Urbanus dorantes*), Brazilian Skipper (*Calpodes ethlius*), and Monk (*Asbolis capucinus*).

Most of the eastern United States is now occupied by a "weedy" butterfly fauna of native and introduced species that have become well adapted to abandoned fields, road edges, pastures, and similarly disturbed situations. The native species are ones pre-adapted to feed on a wide array of plants, including introduced exotics. Most of these were probably native to more southern and western open habitats, including prairie habitats, marshes, river flood plains, and barrens, and they were probably uncommon or rare species before man came along. Native butterflies that are probably now much more widespread and dominant include the Monarch (*Danaus plexippus*), Variegated Fritillary (*Euptoieta claudia*), Meadow Fritillary (*Boloria bellona*), Pearl Crescent (*Phyciodes tharos*), American Painted Lady (*Vanessa virginiensis*), Clouded Sulphur (*Colias philodice*), Orange Sulphur (*Colias eurytheme*), Eastern Tailed Blue (*Everes comyntas*), Silver-spotted Skipper (*Epargyreus clarus*), Common Sooty Wing (*Pholisora catullus*), Checkered Skipper (*Pyrgus communis*), Satchem (*Atalopedes campestris*), Fiery Skipper (*Hylephila phyleus*), several *Polites,* and the Delaware Skipper (*Atrytone logan*).

Several butterflies are at present expanding their ranges and may soon become more common elements of this weedy community. The Great Copper (*Lycaena xanthoides dione*) is spreading westward, the Southern Sooty Wing (*Staphylus hayhurstii*) is slowly invading northward after adapting to introduced lamb's quarters (*Chenopodium album*), and the Wild Indigo Dusky Wing (*Erynnis baptisiae*) can now include exotic crown vetch (*Coronilla varia*), commonly sown along highway verges, in its diet.

There have been notable declines in some eastern butterfly populations that have gone hand in hand with the expansions and introductions described above. Although the original

FIGURE 30. Deciduous woodland, Transition Life Zone (Appalachian Mountains, West Virginia). Habitat for Aphrodite Fritillary (*Speyeria aphrodite*), Edwards' Hairstreak (*Satyrium edwardsii*), and Tawny-edged Skipper (*Polites themistocles*).

mature forests were probably not rich butterfly habitats, species of other habitats are probably now more restricted, some enough to noticeably diminish their ranges. Many of these are noted in the individual species accounts. Prairie species, such as the Smoky Eyed Brown, Regal Fritillary, Dakota Skipper, and Powesheik Skipper have all suffered range contractions, as have native butterflies of peninsular Florida and the Keys, areas that are being catastrophically altered. Drainage of marshes has probably greatly reduced the occurrence of several species, including the Northern Eyed Brown (*Satyrodes eurydice*), Dion Skipper (*Euphyes dion*), and others. The cutting of southeastern bottomland forests, drainage of swamps, and conversion of pine flatwoods and savannahs to even-aged pine plantations and cropland have probably had extensive effects on the endemic butterflies of the Deep South.

It is now difficult, if not impossible, to reconstruct a picture of what many of our native butterfly communities were originally like. However, if accurate records and maps are kept in the future, it should be much easier to document the continuing distributional changes that are likely to occur.

FIGURE 31. Mixed evergreen forest and heath barrens, Canadian Life Zone (Dolly Sods, West Virginia). Habitat for Pink-edged Sulphur (*Colias interior*) and Mountain Silver-spot (*Speyeria atlantis*).

FIGURE 32. Black spruce and cranberry bog, Canadian Life Zone (Ossipee, New Hampshire). Habitat for Bog Copper (*Lycaena epixanthe*), Bog Elfin (*Incisalia lanoraieensis*), and Jutta Arctic (*Oeneis jutta*).

FIGURE 33. Alpine rock gardens, Arctic-Alpine Life Zone (Mt. Washington, New Hampshire). Habitat for the White Mountain Butterfly (*Oeneis melissa semidea*)

TABLE 5. Number of resident and colonist eastern butterfly species per family in relation to latitude

Family or Subfamily	Residency Status	Latitude				
		>45°	40°–45°	35°–40°	30°–35°	25°–30°
Satyrs	Resident	12	12	9	9	7
(Satyrinae)	Colonist	—	—	—	—	—
	Stray	—	—	—	—	—
Monarchs	Resident	—	—	—	—	3
(Danainae)	Colonist	1	1	1	2	—
	Stray	—	—	—	—	—
Brushfoots	Resident	25	29	27	23	22
(Nymphalinae)	Colonist	5	4	6	2	3
	Stray	4	7	10	4	10
Whites and Sulphurs	Resident	10	9	8	13	16
(Pieridae)	Colonist	—	5	6	2	1
	Stray	7	3	6	3	5
Swallowtails	Resident	2	6	7	7	10
(Papilionidae)	Colonist	—	—	—	—	1
	Stray	3	1	1	1	—
Gossamer Wings	Resident	26	37	32	25	26
(Lycaenidae)	Colonist	2	1	1	3	—
	Stray	—	3	5	7	2
Skippers	Resident	32	48	64	63	59
(Hesperiidae)	Colonist	2	7	5	5	—
	Stray	6	7	8	3	5
	Resident	107	141	147	140	143
TOTAL	Colonist	10	18	19	14	5
	Stray	20	21	30	18	22

Habitats

The local occurrence of most butterflies within the broad patterns discussed in the previous section depends on the occurrence of their host plants in particular plant communities or formations. The presence of any butterfly species depends not only on climatic dicta and the presence of suitable caterpillar foods, but also on appropriate adult nectar sources or other food, suitable arenas for flight and courtship, and, in some instances, the presence of certain symbiotic species, notably ants. In fact, these considerations may explain the occurrence not only of most butterflies found in more than a single Life Zone, but also of those that appear to be Life Zone–limited.

Habitats are divided into four basic groupings—forest, treeless areas, aquatic regions, and disturbed areas. Of course, some habitats, such as forested swamps, may fall into more than one category.

Forest. *Broadleaf deciduous forest* covered most of eastern North America prior to its settlement by man. Relatively few butterflies are adapted to life in mature forest, particularly in upland situations. A few species, such as the Little Wood Satyr (*Megisto cymela*), Northern Pearly Eye (*Enodia an-*

thedon), Red-spotted Purple (*Limenitis arthemis astyanax*), Tiger Swallowtail (*Papilio glaucus*), Spicebush Swallowtail (*Papilio troilus*), Banded Hairstreak (*Satyrium calanus*), Hickory Hairstreak (*Satyrium caryaevorum*), and Juvenal's Dusky Wing (*Erynnis juvenalis*), are widespread forest-adapted species (see Figure 26). Most forest species, however, require special situations. Some of these are enumerated below.

Northern mixed conifer forest, which covers much of the Canadian Life Zone, is dominated by pines, hemlock, spruce, birch, poplars, and a few oaks (see Figure 31). Characteristic butterflies include the Compton Tortoise Shell (*Nymphalis vau-album*), the White Admiral (*Limenitis arthemis arthemis*), Persius Dusky Wing (*Erynnis persius*), and the Arctic Skipper (*Carterocephalus palaemon mandan*).

Northern conifer forest contains few broadleaved trees and is usually found on acidic soils. The Green Comma (*Polygonia faunus*) is widespread in this region, but otherwise the habitat is very poor in species.

Jack pine forest is dominated by *Pinus banksiana* and is found on sandy soils in the Great Lakes area. Macoun's Arctic

(*Oeneis macounii*) is limited to this habitat, usually being found within forest glades.

Moist deciduous forest is often dominated by maples and beech and is usually found on lower north-facing slopes. Characteristic butterflies are the Gray Comma (*Polygonia progne*), West Virginia White (*Pieris virginiensis*), Early Hairstreak (*Erora laeta*), Dusky Blue (*Celastrina ebenina*), and Appalachian Blue (*Celastrina neglectamajor*).

Bottomland, or *riparian forest*, with rich, deep alluvial soils, is the most diverse in plant and animal life. While it may vary greatly in species composition from area to area, certain butterflies are frequent associates. Among these are the Comma (*Polygonia comma*), Question Mark (*Polygonia interrogationis*), Red Admiral (*Vanessa atalanta rubria*), Zebra Swallowtail (*Eurytides marcellus*), and Southern Golden Skipper (*Poanes zabulon*). The Diana (*Speyeria diana*) is found in coves or hollows in a subtype of this habitat (see Figure 28).

Southern deciduous forest is found along rivers or on the periphery of swamps, usually in the Lower Austral Life Zone (see Figure 22). Many of its butterfly associates feed on Giant Cane (*Arundinaria gigantea*) or maiden cane (*Arundinaria tecta*) in their larval stages. Typical butterflies include the Pearly Eye (*Enodia portlandia*), Creole Pearly Eye (*Enodia creola*), Gemmed Satyr (*Cyllopsis gemma*), Great Purple Hairstreak (*Atlides halesus*), King's Hairstreak (*Satyrium kingi*), Carolina Roadside Skipper (*Amblyscirtes carolina*), and Lacewinged Roadside Skipper (*Amblyscirtes aesculapius*).

Southern pine flatwoods usually occur on sandy or clay soils (see Figure 15). A relatively open understory is usually maintained by periodic fire. Hard pines, such as loblolly and long-leaf, are the usual dominants. Small groves of hardwoods, such as live oak or magnolia, are often scattered within this habitat. Some characteristic butterflies are the Georgia Satyr (*Neonympha areolatus*), Little Metalmark (*Calephelis virginiensis*), and Giant Yucca Skipper (*Megathymus yuccae*).

Tropical hardwood hammocks are limited to low limestone formations in southern Florida and the Upper Keys, particularly Key Largo and Matecumbe Key (see Figures 17–19). These forests contain a rich mixture of tropical hardwoods, including gumbo limbo, torchwood, figs, mahogany, and lignum vitae. This habitat will soon disappear from all areas except a few parks and preserves. Endemic to this habitat are the Florida Purple Wing (*Eunica tatila tatilista*), Ruddy Dagger Wing (*Marpesia petreus*), Atala (*Eumaeus atala florida*), Miami Blue (*Hemiargus thomasi bethunebakeri*), Schaus' Swallowtail (*Papilio aristodemus ponceanus*), and Hammock Skipper (*Polygonus leo*).

Mangrove forests in the intertidal zone of southern Florida and the Keys consist of red, black, and white mangroves as well as buttonwood (see Figure 14). The only characteristic butterfly is the Mangrove Skipper (*Phocides pigmalion okeechobee*).

Barrens. Somewhat intermediate between forested habitats and open habitats are the *barrens* (see Figure 29). These habitats are limited to certain geologic formations, and usually have vegetation consisting of scrubby oaks and pines, as well as various shrubs belonging to the heath family. These areas are often adapted to fire and thus have open areas with lupines, wild indigo, other more herbaceous plants, and grasses. Subtypes include sand barrens, shale barrens, serpentine barrens, dunes, and rocky river bluffs. A distinctive set of butterflies is found on many barrens, although all are not ever found in any single area. The Northern Metalmark (*Calephelis borealis*), Northern Hairstreak (*Fixsenia ontario*), Edward's Hairstreak (*Satyrium edwardsii*), Frosted Elfin (*Incisalia irus*), Hoary Elfin (*Incisalia polios*), Melissa Blue (*Lycaeides melissa samuelis*), Silvery Blue (*Glaucopsyche lygdamus*), Olympia Marble (*Euchloe olympia*), Grizzled Skipper (*Pyrgus centaureae wyandot*), Sleepy Dusky Wing (*Erynnis brizo*), and Wild Indigo Dusky Wing (*Erynnis baptisiae*) are characteristic.

Treeless Regions. Open, dry habitats are predominantly *grasslands.* The native prairie of the Great Plains was the largest original piece of this habitat prior to its cultivation. Much of Illinois and Iowa, as well as large areas of southwestern Minnesota and northwestern Missouri, were originally covered with tall-grass prairie (see Figure 25). An area of sand prairie is also found on Michigan's lower peninsula. Prairies must be maintained by periodic fires. Currently, native prairie is almost entirely limited to tiny preserves. The Powesheik Skipper (*Oarisma powesheik*), Dakota Skipper (*Hesperia dacotae*), Ottoe Skipper (*Hesperia ottoe*), and Byssus Skipper (*Problema byssus*) are typical prairie species. The Regal Fritillary (*Speyeria idalia*) is most abundant in native prairies, although it is found elsewhere. The Chryxus Arctic (*Oeneis chryxus strigulosa*) is found on Michigan's sand prairies, and several western butterflies are found sparingly on short-grass prairie in western Minnesota.

Beard-grass prairies dominated by beard grasses (*Andropogon*) often occur in close association with various barren situations and are found east of the extensive prairie province. These areas are often small, localized patches. Again, skippers predominate. Leonard's Skipper (*Hesperia leonardus*), Cobweb Skipper (*Hesperia metea*), and the Dusted Skipper (*Atrytonopsis hianna*) are frequently found at these small sites, while the Dotted Skipper (*Hesperia attalus*) and Arogos Skipper (*Atrytone arogos*) are limited to this habitat in portions of their ranges.

The Alpine rock gardens found on Mt. Katahdin, Maine, and in the Presidential Range of New Hampshire are dominated by scattered sedges and low-growing herbs (see Figure 33). The White Mountain Butterfly (*Oeneis melissa semidea*) and Polixenes Arctic (*Oeneis polixenes katahdin*) are limited to this habitat.

Aquatic Habitats. Aquatic habitats are characterized either by open water, perennially wet soil, or periodic tides. *Freshwater marshes* are probably the most widespread of the aquatic habitats (see Figure 23). Typical vegetation includes cattail, sedges, cord grass, small willows, various roses, and buttonbush. Typical butterflies include the Northern Eyed Brown (*Satyrodes eurydice*), Smoky Eyed Brown (*Satyrodes fumosa*), Acadian Hairstreak (*Satyrium acadica*), Bronze Copper (*Lycaena hyllus*), Mulberry Wing (*Poanes massasoit*), Black Dash (*Euphyes conspicua*), Two-spotted Skipper (*Euphyes bimacula*), and Dion Skipper (*Euphyes dion*).

Tamarack-sumac bogs, with central open areas of sedge, are typical of the southern terminus of the Great Lakes, especially Lake Michigan and Lake Huron. The Mitchell's Satyr (*Neonympha mitchellii*) and Swamp Metalmark (*Calephelis muticum*) are largely limited to this habitat.

Acid bogs are relictual areas of glacial lakes or depressions now filled with sphagnum moss, dwarf willows, blueberries, Labrador tea, cranberry, cotton grass, and sedges (see Figure 32). Various trees, often black spruce and tamarack, are found on their edges. Acid bogs are most widespread in northern Minnesota and Maine, but they may be found as far south as West Virginia. Typical butterflies are of the Jutta Arctic (*Oeneis jutta*), Red-disked Alpine (*Erebia discoidalis*), Bog Fritillary (*Boloria eunomia dawsoni*), Bog Elfin (*Incisalia lanoraieensis*), and Bog Copper (*Lycaena epixanthe*). Several other northern butterflies enter our area only in these acid bogs.

Wet meadows usually do not have standing water, but are wet during most of the year. They may have small thickets of alders and willows, as well as various herbaceous plants that thrive on wet soil. The Baltimore (*Euphydryas phaeton*) and Harris' Checkerspot (*Chlosyne harrisii*) are often limited to this habitat, while other more widespread species are also present.

White cedar swamps are highly acid and are often similar to acid bogs. They occur primarily on the coastal plain in sandy regions and are often near pine barrens. Hessel's Hairstreak (*Mitoura hesseli*) is entirely limited to this localized habitat.

Southern hardwood swamps are often dominated by bald cypress, water oak, magnolias, water tupelo (black gum), red bay, sweet pepperbush, and palmettos (see Figure 21). These areas are often adjacent to southern deciduous forest and are rich in butterflies. The Palamedes Swallowtail (*Papilio palamedes*), Yehl Skipper (*Poanes yehl*), and Duke's Skipper (*Euphyes dukesi*) are more or less limited to this habitat.

Brackish river and coastal marshes are located along the coast, often around coastal bays and near the mouths of large rivers. The Rare Skipper (*Problema bulenta*) and Broad-winged Skipper (*Poanes viator*) are found in such situations, the former only near the mouths of a few large rivers, and the latter throughout.

Sawgrass marshes are found mainly in peninsular Florida. They are inhabited by several skippers, most notably the Sawgrass Skipper (*Euphyes pilatka*).

Salt marshes, dominated by reed, pickerelweed, and cord grass, are extensive along protected portions of the Atlantic and Gulf coasts (see Figure 24). Aaron's Skipper (*Poanes aaroni*) and the Salt Marsh Skipper (*Panoquina panoquin*) are found through most of this formation.

Disturbed Habitats. The term *disturbed habitats* here refers to areas strongly influenced by man's activities. These range from city lots to pastures, old fields, roadsides, suburban plantings, lawns, power lines, and agricultural land. Many of these habitats actually support much higher densities of butterflies than "native" habitats. Disturbed areas often have fairly large numbers of imported plants ("weedy exotics") and a diverse array of herbaceous species.

Each kind of disturbance and setting favors a somewhat different array of plants and butterflies. A large proportion of native butterflies have become adapted to weedy situations, and the introduced butterflies are commonest in such habitats. Abandoned croplands, clover or alfalfa fields, lightly grazed pastures, hayfields, roadsides, drainage ditches, and power-line right of ways (mowed, not sprayed), are among the richest butterfly habitats. The preferences of individual species are given in the habitat sections of the individual species accounts. Butterflies now found in such situations include the Monarch (*Danaus plexippus*), Variegated Fritillary (*Euptoieta claudia*), Great Spangled Fritillary (*Speyeria cybele*), Meadow Fritillary (*Boloria bellona*), American Painted Lady (*Vanessa virginiensis*), Viceroy (*Limenitis archippus*), Gray Hairstreak (*Strymon melinus*), American Copper (*Lycaena phlaeas*), Eastern Tailed Blue (*Everes comyntas*), Clouded Sulphur (*Colias philodice*), Orange Sulphur (*Colias eurytheme*), Cloudless Sulphur (*Phoebis sennae eubule*), Little Sulphur (*Eurema lisa*), European Cabbage Butterfly (*Pieris rapae*), Black Swallowtail (*Papilio polyxenes asterius*), Silver-spotted Skipper (*Epargyreus clarus*), Checkered Skipper (*Pyrgus communis*), Common Sooty Wing (*Pholisora catullus*), Least Skipper (*Ancyloxypha numitor*), Fiery Skipper (*Hylephila phyleus*), Satchem (*Atalopedes campestris*), Peck's Skipper (*Polites coras*), Dun Skipper (*Euphyes ruricola metacomet*), and Ocola Skipper (*Panoquina ocola*).

Adult Behavior

Mate Location. The primary function of adult butterflies is reproduction. Other activities relate to maintenance and dispersal.

Eastern butterflies exhibit two stereotyped mate-locating behaviors (Scott 1975), termed *perching* and *patrolling*. The majority of species use one of these strategies.

Perching. Perching species include some satyrines, the majority of nymphalines, all hairstreaks and coppers, a few swallowtails, and most skippers. A perching male sits and awaits passing females, and he will fly out and investigate any passing object of about the correct size. When two males of a perching species encounter each other, they will often engage in a long spiral flight, or "chase," after which one male will usually return to the general vicinity of the original perch. These chases have been frequently interpreted as aggressive encounters, but are now usually felt to be short circuits of normal courtship. Each recognizes the other as the correct species, but since neither is female, the next releaser in the mating sequence does not take place.

Each species selects a particular kind of perch. The tips of twigs, branches, and low vegetation are frequent choices. Perch areas may be hilltops, wood edges, or sunlit forest glades. In general, species with low population density and wide dispersion tend to be perchers. Most perching takes place in the afternoon, although some species, notably skippers, will alternate between perching and feeding for most of the day. Some species, notably angle wings (*Polygonia*) and lady butterflies (*Vanessa*), do not take perches until shortly before sunset.

Patrolling. Patrolling males fly slowly around the environment where receptive females might be found. Patrollers include most satyrines, a few nymphalines, most blues, most swallowtails, all eastern whites and yellows, and a few skippers. Encounters between patrolling males often lead to brief chases, but these are neither as prolonged nor as intense as those of perchers. Females of patrolling species may occasionally initiate courtship (Rutowski 1980).

Most patrollers occur in fairly dense populations, usually in open habitats. Patrolling usually occurs from late morning to midafternoon.

Freshly emerged males of patrolling butterflies are often seen taking up moisture at wet spots along stream margins or by mud puddles (Downes 1973). Often, a number of males will be seen tightly clustered in a single small area (Figures 34–36). It has been shown that sodium ions (salt) are the cue for such "mud puddling," but the function of this behavior is not understood (Arms et al. 1974). Since perching species and females and older males of patrolling species only rarely engage in this activity, a reasonable thesis would be that it serves to activate some temperature-regulating system in the soon-to-be-active patrolling males.

Once a female is located, the male usually initiates the courtship by forcing the female to alight while he hovers above her. In most species, the male then alights behind or beside his intended mate and curves his abdomen laterally to join her. Often the female will fly a short distance, and the sequence will be repeated. Males of many species have odors (pheromones) that are used at close range to stimulate female receptivity. Pheromones often associated with the male's androconial scales, which occur in localized patches, often in the forewing cell.

Females are usually not receptive and have a variety of rejection postures, or "dances," to signal this fact to the male. A frequently used rejection posture is the female's spreading her wings and lifting her abdomen high to prevent coupling. Wing fluttering is another rejection sign, while frequent short flights are yet another method of discouragement.

If coupling occurs, the pair often takes a brief nuptial flight to the edge of the colony area. Either sex may be the carrier,

FIGURE 34. Mud-puddle club with Zebra, Tiger, and Spicebush Swallowtails. Great Falls, Maryland.

FIGURE 35. Mud-puddle club with Spicebush Swallowtails (*Papilio troilus*) and Tiger Swallowtail (*Papilio glaucus*). Potomac River, Maryland.

but it is usually the male—as in the whites and yellows (Miller and Clench 1968; Shull 1979). The "non-carrier"often appears torpid.

Copulation.   During copulation, the male passes a sperm packet, or *spermatophore*, into the female's *bursa copulatrix*, a pouch lying adjacent to her *oviduct* (egg canal). Mating may last from 20 minutes to several hours, even all night in some species (Burns 1970). Females are usually mated during their first day of flight. Some females mate only once, while others will mate several times. Frequency of mating may be determined by dissecting the bursa copulatrix and counting the empty spermatophores. Males can usually mate several times (Burns 1968).

Like mate-location efforts, mating usually takes place only during a few hours each day. Some species mate in the morning, some at mid-day, some during the afternoon, and others only at dusk. A few butterflies may mate at any time of day.

Oviposition.   A female in search of a suitable site to lay her eggs usually uses visual cues to first select an appropriate habitat of suitable shadiness and vegetation height. Then she may fly slowly and purposefully about, touching down periodically. When she touches down, she "tastes" the substrate with her front legs. In some species, tasting may be accomplished with nothing more than a brief "touch and go" landing, while in others the female spends a great deal of time walking around and probing the plant.

Figure 36. Mud-puddle club with Tiger Swallowtails (*Papilio glaucus*) and Palamedes Swallowtails (*Papilio palamedes*). Great Dismal Swamp, Virginia.

The female usually accomplishes oviposition by curving her abdomen ventrally and slowly depositing one or more eggs on the chosen substrate. Most species lay their eggs precisely on the particular part of the plant that the young caterpillar will eat. In other cases, the female will lay her eggs on the ground or on other plants near the correct host. Then it will be up to the young caterpillar to find the right plant. The species most noted for this are the fritillaries (*Speyeria*). Female Regal Fritillaries have been observed laying eggs on leaves and sticks in the general vicinity of dried violets, the host (Clark 1932). In recent years it has been discovered that female Red-banded Hairstreaks (*Calycopis cecrops*) lay their eggs on the underside of dead leaves on the ground. Female Commas (*Polygonia comma*) and Silver-spotted Skippers (*Epargyreus clarus*) flutter about, touching down briefly on various plants. Whenever they touch down on the "correct" plant, they then lay an egg on the next plant they alight on—no matter whether it is a fern, grass, or morning glory. Most species lay a single egg on each plant or branch, then move on to lay their next egg elsewhere, but some will lay several eggs on different parts of the same plant. More specialized are those butterflies that lay their eggs in groups or masses. These are primarily nymphalines, specifically checkerspots (*Euphydryas* and *Chlosyne*), crescents (*Phyciodes*), Question Mark (*Polygonia interrogationis*), Mourning Cloaks, tortoise shells (*Nymphalis*), and emperors (*Asterocampa*), but the Atala (*Eumaeus atala florida*) and Gold-banded Skipper (*Autochton cellus*) also have this habit. The caterpillars of all these, save for the Question Mark, feed colonially in the first several instars.

Like other activities, oviposition is limited to specific times of day, some very brief, by different butterflies. For example, whites and yellows lay their eggs about midday, while others, such as most skippers, lay theirs in the afternoon. Few species lay their eggs in the morning.

Thermoregulation.   Thermoregulation is another readily observed aspect of butterfly behavior. Since butterflies are poikilothermic (cold-blooded), they cannot regulate their body temperature internally. Instead, they regulate their temperature externally through a variety of behavioral actions. This is termed *behavioral thermoregulation*.

Generally speaking, butterflies cannot fly when air temperatures are much below 16°C. In addition, temperatures much above 38°C, in combination with high humidity, are unsuitable for most species.

Techniques used by butterflies to change their temperature include wing orientation, abdominal elevation or depression, perching and flight height, and movement into or out of shade (Rawlins 1980).

Wing orientation includes *basking*, to increase body temperature, or sun avoidance, accomplished with wings closed and oriented parallel to the sun's rays, so as to cast the least shadow. Butterflies usually use either dorsal or lateral basking for warming up. *Dorsal baskers* perch with their wings wide open and oriented perpendicular to the sun, while *lateral baskers* perch with wings closed and one side perpendicular to the sun. Some lateral baskers may tilt their bodies at an angle, to maximize the angle of exposure. Most species bask in the morning or all day during cool weather, while some bask during late afternoon to extend the daily period of activity. Most satyrines, all nymphalines, all swallowtails, a few lycaenids (especially coppers and metalmarks), and virtually all skippers are dorsal baskers, while some satyrines, all pierids, and some lycaenids (especially hairstreaks) are lateral baskers. As mentioned above, the spring or fall forms of many species have more extensive black scaling, which increases the efficiency of basking.

Dorsal baskers may warm up by elevating their abdomens so that they are exposed to the sun. At high air temperatures, dorsal baskers may shade their abdomens by depressing them below wing level, thus avoiding unnecessary heat load (Rawlins 1980).

Air temperatures are almost always warmer close to the ground and cooler farther from it. As a consequence, on cool days butterflies may perch and fly close to the ground, while on warmer days they can be cooler by perching and flying farther above the ground. Convective cooling may be increased by their spending a greater proportion of time in flight on hot days. Butterflies that live in a mosaic of sunny and shaded habitats may fly more in the sun during cool conditions and more in the shade during hot air temperatures. Finally, when air temperature and humidity are too high to permit flight, many butterflies may cease flying altogether and seek out shaded perches where they remain until conditions are tolerable again.

At night, during cloudy weather, and during very cool days, butterflies take up roosting positions, usually with closed wings, although metalmarks and some skippers roost with open wings. Dusky wings (*Erynnis*) are exceptional in that they roost with their wings deflexed so that they curl around a stem or twig (Burns 1969). Typical roosting sites are the undersides of large leaves, the stems of herbaceous plants, or atop flowers. Some species, such as the Zebra (*Heliconius charitonius tuckeri*) [Young and Thomason 1975] and some dagger wings (*Marpesia* [Benson and Emmel 1973]), are known to return to the same communal roosts night after night. Other species, such as the Eastern Tailed Blue, some sulphurs, and the European Cabbage Butterfly,

may be found roosting in small, loose aggregations (Clench 1970), but these may be more the result of roosting in the same small area receiving late-afternoon sun, rather than an intentional social function.

Another form of roosting is seen in species that hibernate or estivate. In these species, roosting in hollow tree trunks or on dead branches may last for weeks or months at a time. Mourning Cloaks, Compton Tortoise Shells, and angle wings may spend much of the year in hollow tree roosts, while species with alternating dry and wet season forms, such as the large sulphurs (*Phoebis*), small sulphurs (*Eurema*), leaf wings (*Anaea*), and others, may spend most fall and winter days in torpor, broken only by periodic feeding episodes. Some of these species may roost communally.

Overwintering Monarchs form huge festoons or clusters at their overwintering sites. Their tightly packed bodies assist the group in avoiding extreme winter temperatures.

Migration and Emigration. *Migration* and *emigration* are mass movements that enable butterfly populations to avoid unfavorable climatic conditions or crowding. Migration is a regular, two-way movement, such as that undergone by Monarchs, while emigrations are irregular, one-way movements.

The Monarch is the most obvious and spectacular migratory butterfly in our area, and its migration is discussed in its species account. Other butterflies that seem to have at least weak two-way movements are the Comma, Question Mark, Mourning Cloak, Red Admiral, American Painted Lady, Painted Lady, and Cloudless Sulphur. Other species in the far South may have regular movements, but these are poorly understood.

Many southern species have irregular northward movements, often detected in late summer. These may be no more than the result of locally high population pressure, but may also represent more innate behavior. Typical emigratory species include the Gulf Fritillary, Variegated Fritillary, Buckeye, Reakirt's Blue, Sleepy Orange, Little Sulphur, Checkered Skipper, Fiery Skipper, Satchem, and Ocola Skipper. These movements are usually so sparse that they are not noted, yet are so regular that almost every year most species successfully colonize areas far to the north of their permanent ranges. These species cannot survive most northern winters, and their progeny succumb when freezing weather arrives.

Seasonality

The times of year that butterflies are found, as well as the number of generations each species may have (*voltinism*), are mostly determined by the length of the growing season (number of days between last killing frost in spring and first

frost in fall), the availability of suitable larval food, and the peculiarities of each butterfly's life-cycle adaptions. In the far North, or in Alpine Zones, where the growing season may be less than 2 months, the majority have a single brood annually, and, in fact, some have flights every other summer, since it may require more than a single season for their caterpillars to complete development. On the other hand, in Tropical Zones, where there are virtually no freezing temperatures, adults fly at all times, and the majority of species may have more than one generation. As one moves between these extremes from north to south, two processes reflect the changeover. In one case, there is a gradual turnover of species (see "Distribution"), so that, on average, the new additions southward tend to have more annual generations than the ones left behind. Second, for other species of wide latitudinal range, the farther south one proceeds the more generations they may have. Still, there are some species with broad latitudinal ranges that have a fixed generation number wherever found.

The numbers of species of resident butterfly families falling into various voltinism (broodedness) categories are shown on Table 6. The greatest number, 80 (34%), fall into the single-brooded (univoltine) category, while at least 74 (32%)

show an increasing generation number as they range south. Particularly high proportions of satyrines (62%), nymphalines (42%), and hairstreaks (63%) have one or fewer generations each year, while swallowtails (64%) and skippers (40%) are most flexible in their generation number under different climatic conditions.

The number and percentage of butterflies in selected broodedness categories at a selected latitudinal sequence of localities are shown in Table 7. Note the gradual, yet steady, *decrease* in univoltine species southward, and the concomitant *increase* in multiple-brooded species. Single-brooded species are predominant in the northern states, species with two generations are most frequent in the middle states, and butterflies with three or more generations dominate the communities to the south.

As one proceeds southward, the first date of appearance of many butterflies occurs progressively earlier, since similar weather conditions and host-plant foliation take place at concomitantly earlier dates (Clench 1965). Many of these

TABLE 6. Voltinism of resident eastern butterflies (number of species per family per voltinism category)

Family or Subfamily	Voltinism Category							
	Biennial	Univoltine	Univoltine (north—more in south)	Bivoltine	Bivoltine (north—more in south)	Trivoltine	Four or more broods	TOTAL
Satyrines (Satyrinae)	4	9	4	0	2	2	0	21
Monarchs (Danainae)	0	0	1	0	0	0	2	3
Brushfoots (Nymphalinae)	0	18	5	9	2	0	9	43
Whites and Sulphurs (Pieridae)	0	5	0	1	0	6	8	20
Swallowtails (Papilionidae)	0	0	2	0	6	0	3	11
Metalmarks (Riodininae)	0	1	1	0	0	0	1	3
Coppers (Lycaenini)	0	3	0	1	2	0	0	6
Hairstreaks (Eumaeini)	0	16	1	2	2	3	3	27
Blues (Plebejini)	0	6	1	1	0	1	5	14
Pyrgines (Pyrginae)	0	6	2	5	6	0	4	23
Branded Skippers (Hesperiinae)	0	15	15	7	12	6	6	61
Giant Yucca Skippers (Megathyminae)	0	1	1	0	0	0	0	2
TOTAL	4	80	33	26	32	18	41	234

TABLE 7. Change in proportion of broodedness in a North–South sequence of localities

Locality	Latitude	Broodedness				
		Biennial	1 Brood	2 Broods	3 Broods	4+ Broods
Clearwater Co., Minnesota	48°	3(4.5%)	40(61%)	18(27%)	5(7.5%)	—
Marathon Co., Wisconsin	45°	—	40(64%)	14(22%)	9(14%)	—
Tompkins Co., New York	42°	—	35(66%)	11(21%)	7(13%)	—
Staten Island, New York	41°	—	39(48%)	25(31%)	12(15%)	5(6%)
Fairfax Co., Virginia	39°	—	25(33%)	28(36%)	16(21%)	8(10%)
Great Dismal Swamp, Virginia	37°	—	16(19%)	32(37%)	29(34%)	9(10%)
Hinds Co., Mississippi	32°	—	8(10%)	11(14%)	35(44%)	25(32%)
Dade Co., Florida	25°	—	2(2.5%)	2(2.5%)	20(24%)	59(71%)

trends are described in the individual species accounts, so the reader will know the approximate date that a particular species should appear at any given latitude. In contrast, the few single-brooded fall-flying butterflies, such as Leonard's Skipper (*Hesperia leonardus*) emerge progressively later at more southern latitudes or at lower elevations. For multiple-brooded species not only does the first date of appearance occur earlier in the year, but the last day of flight becomes progressively later as one proceeds to the south. Each species of butterfly overwinters in a particular stage. Day-length (photoperiod) is the usual cue that triggers the onset of hibernation (diapause) long before cold weather actually arrives.

An intriguing phenomenon not previously noted is that, for many multiple-brooded species, especially skippers, the period between generations seems to *increase* as one goes southward in their range. The result is that generation number may not increase as rapidly in the South as one might expect. It is possible that the caterpillars of such species feed less under very hot, humid summer temperatures, thereby delaying development.

Synchrony of emergence is important, so that the two sexes are present at the same time and successful mating can occur. The flights of biennial species, such as Macoun's Arctic (*Oeneis macounii*) and the Red-disked Alpine (*Erebia discoidalis*), only on even-numbered years or only on odd-numbered years at particular localities, with no emergence of adults at all during "off" years (Masters 1974), is an extreme example of such timing. The virtually simultaneous emer-

gences and brief flights of other butterflies, particularly those of early spring or the far North, are other appropriate examples.

The males of most species begin their emergence at least a few days before the females, probably due to their smaller average size and, hence, more rapid development. It is important that males establish their perch arenas or patrolling areas by the time the first females appear.

The emergence schedule of a few butterflies, namely the Common Wood Nymph (*Cercyonis pegala*), Diana (*Speyeria diana*), Regal Fritillary (*Speyeria idalia*), and Great Spangled Fritillary (*Speyeria cybele*), is so offset that the males emerge at least 2 or 3 weeks before the females. In these butterflies, the females are much longer-lived than the males, and wait until late summer to lay their eggs.

Another aspect of seasonality is the differential longevity and behavior of butterflies that overwinter as adults. Some multiple-brooded northern butterflies, namely the Question Mark, Commas, Milbert's Tortoise Shell, Red Admiral, and lady butterflies, hibernate as adults, and often migrate as well. In these species, the overwintering adults live much longer than the summer adults, and are in reproductive diapause much of the time. Many southeastern species survive the relatively dry winter months by having separate forms, which spend much of the time resting, but also fly and feed at flowers on warm days. These dry-season forms often live as long as 4 or 5 months, in contrast to the wet-season (or summer) forms, which live no more than a few weeks.

Longevity

The life span (longevity) of adult butterflies may differ by several orders of magnitude. The actual or *expected life span* of most individuals of a species is usually much shorter than the *maximum life span*. This is due largely to intense predation by birds, other insects, and spiders, as well as to losses caused by rain and cold weather. In order to discover the expected and maximum life spans of butterflies, one must carry out lengthy mark-release-recapture programs and have a good understanding of each species' life cycle and movements. Expected life span for butterflies ranges from about 2 days to 14 days (Scott 1973), while maximum life span ranges from about 4 days (Spring Azure) to 10 or 11 months (Mourning Cloak). Reported life spans for eastern butterflies are shown in Table 8.

Females generally live longer than males (despite the apparent trend shown by Figure 8), since the latter may perform their function rapidly, whereas females require more time to search our host plants and mature their eggs. In general, large butterflies live longer than smaller ones, but there are many exceptions. Butterflies that undergo hibernation or reproductive diapause as adults live longer than those that do not. Furthermore, adults of species that have both normal and diapausing or hibernating generations have different intraspecies life spans. For example, Monarchs (*Danaus plexippus*) that migrate and overwinter may live as long as 5 or 6 months, whereas nonmigratory individuals in midsummer probably live no more than 1 month. The Barred Sulphur (*Eurema daira*) lives 4 months or more when in reproductive diapause during unfavorable seasons, but lives no more than 10 days when continuously active and reproductive. Population studies of many more species must be conducted before we fully understand butterfly longevity and its implications.

TABLE 8.   Maximum longevity of selected eastern butterflies[a]

Species	Sex	Maximum Longevity (days)	Location	Species	Sex	Maximum Longevity (days)	Location
Satyrodes appalachia (Appalachian Eyed Brown)	♂	7	Virginia	*Eurema daira* [WET] (Barred Sulphur)	♂	9	Costa Rica
					♀	10	
Megisto cymela (Little Wood Satyr)	♂	4	Virginia	*Eurema daira* [DRY]	♂	110	Costa Rica
					♀	74	
Speyeria cybele (Great Spangled Fritillary)	♂	22	Virginia	*Papilio glaucus* (Tiger Swallowtail)	♂	12	Maryland
					♀	3	
Phyciodes tharos (Pearl Crescent)	♂	4	Virginia	*Mitoura gryneus* (Olive Hairstreak)	♂	6	Virginia
	♀	10					
Limenitis archippus (Viceroy)	♂	17	Virginia	*Everes comyntas* (Eastern Tailed Blue)	♂	10	Virginia
					♀	9	
Anthocharis midea (Falcate Orange-tip)	♂	14	Virginia	*Celastrina ladon* (Spring Azure)	♂	2	Virginia
	♀	4			♀	4	
Colias eurytheme (Orange Sulphur)	♂	25	Virginia	*Thorybes bathyllus* (Southern Cloudy Wing)	♂	8	Virginia
	♀	14					
Colias philodice (Clouded Sulphur)	♂	17	Virginia	*Poanes zabulon* (Southern Golden Skipper)	♂	7	Virginia
	♀	11					
Pieris rapae (European Cabbage Butterfly)	♂	17	Virginia				
	♀	3					

[a] All butterflies collected by the mark-recapture technique.

Adult Resources

Adult butterflies feed on various liquids or dissolved solids, primarily for energy needs, since butterflies are fully grown upon emergence, and tissue repair does not take place. Some long-lived butterflies, such as the Zebra, may convert some food toward the development of eggs. The availability of appropriate adult foods is at least equal in importance to the presence of suitable caterpillar hosts in determining whether a specific area will support populations of particular butterflies.

Although butterflies are thought of by many only as flower visitors, many species in fact rely primarily on other foods, including dung, bird droppings, carrion, tree sap, rotting fruit or fungus, aphid "honeydew" secretions, and, rarely, pollen. All of these foods contain available sugars or proteins for conversion to glucose for immediate energy or to fat storage for later use.

Species that feed on substances other than flower nectar tend to be forest-dwelling butterflies, particularly from the satyrines (except the Inornate Ringlet and possibly the Arctics [*Oeneis*] and Alpines [*Erebia*]). Commas (*Polygonia*), Mourning Cloak, Compton Tortoise Shell, Red Admiral, Viceroy, Red-spotted Purple, White Admiral, Emperors (*Asterocampa*), and Leaf Wings (*Anaea*) also feed primarily on nonfloral foods. Some, such as the Commas, Mourning Cloak, Compton Tortoise Shell and Emperors, feed primarily on tree sap or fungus fluxes, while others, such as the Viceroy, Red-spotted Purple, and White Admiral, favor dung, carrion, and rotting fruit. Some of these species will visit flowers when their preferred foods are unavailable, especially during very dry conditions. Adult harvester butterflies probably feed almost entirely on honeydew produced by aphids, which themselves serve as food for the butterfly's caterpillars. The Zebra is unique among eastern butterflies in that it collects clumps of pollen with its proboscis. The pollen is dissolved by a secretion and the protein-rich liquid is then taken up by the proboscis.

Although Saunders (1932) studied flower visitation relationships in New York, this facet of butterfly natural history has been little studied. In our studies, we have found a

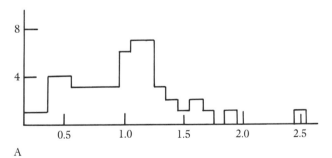

A

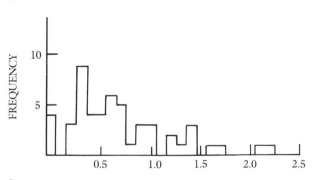

B

Figure 37. Distribution of proboscis (tongue) lengths (A) and flower-tube (corolla) lengths (B) in a northern Virginia butterfly community. Measurement in centimeters.

number of factors that help explain the flower-visitation patterns shown by butterflies.

First, one must realize that the nectar flowers available to butterflies vary with time of year and from site to site. Observations of the flowers that are available at a particular time and place and which are visited or not visited, and what the patterns are for simultaneously occurring butterflies, allow one to draw conclusions.

Multiple-brooded species display different propensities for flower visitation during different seasons. The proportion of individuals observed at flowers is lowest in spring and greatest during the summer, when energy demands are greatest. For example, the Southern Golden Skipper is seldom observed at flowers during its spring generation, but is an avid flower visitor during its summer flight.

Some butterflies are very broad in the spectrum of flowers they visit; examples are the Pearl Crescent, European Cabbage Butterfly, Spring Azure, and Silver-spotted Skipper. Examples of others that predominantly visit only a narrow range of flowers are: the Northern Hairstreak (New Jersey tea, dogbane), Coral Hairstreak (butterflyweed, dogbane, common milkweed), Southern Golden Skipper (selfheal, everlasting pea), and Peck's Skipper (red clover, New York ironweed).

Flower size, particularly the length of the flower tube (corolla), is one of the main restrictions in flower choice. Obviously, a butterfly cannot remove nectar from a flower tube longer than its proboscis unless it crawls inside, a rare event for most species. Second, the amount of nectar in flowers is proportional to their size, and it is most efficient for butterflies to visit flowers with the highest nectar content. The modal tube length of flowers visited by butterflies is about 0.5 cm, while the modal proboscis length of flower-visiting butterflies is 1 cm, about twice the length of flowers visited (Figure 37).

For each butterfly, the size of flowers visited is in direct proportion to the length of its proboscis (Figure 38, Table 9). Thus, on the average, butterflies with short proboscises visit short flowers, and butterflies with long proboscises visit long-tubed flowers. This is not to say that butterflies with long proboscises never visit plants with small flowers (but the reverse is certainly true). A "long-tongued" butterfly may visit short flowers, particularly if they are tightly clumped so that the nectar from many flowers may be withdrawn by merely withdrawing the proboscis from one flower and inserting it into another. The visitation of goldenrods by Monarchs on migration is an excellent example of this.

It is important to stress that the size of flowers visited relates to proboscis length, not butterfly size, since large butterflies may have short proboscises and some small butterflies have long proboscises. The importance of the relationship between proboscis length and forewing length in flower visitation by different butterflies is shown in Table 10. A proboscis:forewing ratio of *less* than 0.3:1 is typical for the

TABLE 9. Selected butterflies, showing relationship of proboscis length to size of flowers visited

Butterfly	Proboscis Length (cm)	Average Corolla Length of Flowers Visited (cm)
Monarch	1.55	0.50
Great Spangled Fritillary	1.45	0.66
Pearl Crescent	0.61	0.33
American Painted Lady	1.04	0.60
Orange Sulphur	1.00	0.66
Cabbage Butterfly	0.95	0.61
Tiger Swallowtail	1.71	0.79
Spicebush Swallowtail	2.31	1.00
Gray Hairstreak	0.62	0.43
Eastern Tailed Blue	0.49	0.40
Silver-spotted Skipper	1.37	0.76
Juvenal's Dusky Wing	0.97	0.69
Least Skipper	0.81	0.52
Southern Golden Skipper	1.36	1.02

satyrines and nymphalines, which specialize in nonfloral resources. True butterflies, which regularly visit flowers, have ratios between 0.3 and 0.5:1, while most skippers have ratios greater than 0.45:1. (A few of the hesperiine skippers have proboscis:forewing ratios greater than 1:1.)

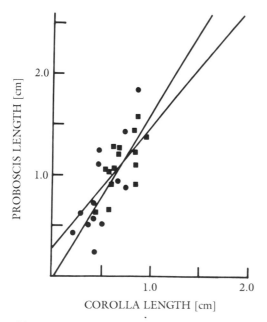

FIGURE 38. Graph of the relationship between proboscis lengths and mean flower-tube lengths utilized by butterflies in a northern Virginia study area. ● = True butterfly species (Papilionoidea), ■ = Skippers (Hesperioidea). Lines are regressions showing 'best fit' of the relationship.

The relative abundance of nectar sources at a particular time and place is another extremely important factor in determining which flowers are visited. Among flowers for which a given butterfly has approximately equal preference, the one that is most abundant is usually selected. It is often observed that flowers favored at other sites and times may be spurned entirely in favor of another, more abundant flower. As a species' flight period goes by, individuals often switch from one flower to another as different flowers come into bloom and others succeed them. Naturally, individuals of different broods are faced with completely different arrays of flowers to choose from.

Flower color is also an important determinant of flower visitation. Most butterflies favor white, blue, pink, or purple flowers, although oranges, reds, and yellows may be favored by a few (Table 11). Most flower-visiting nymphalines and pierids favor white or yellow flowers, although some also have a predilection for the pink milkweeds when they are available. Lycaenids generally favor small white flowers, although many hairstreaks will swarm to common milkweed,

TABLE 10. Proboscis: forewing ratios and adult feeding habits of selected butterflies

Butterfly	Sex	Proboscis: Forewing Ratio	Feeding Habit
Common Wood Nymph (*Cercyonis pegala*)	♂	0.34 : 1	Rotting fruit
	♀	0.31 : 1	
European Cabbage Butterfly (*Pieris rapae*)	♂	0.41 : 1	Flowers
	♀	0.43 : 1	
Gray Hairstreak (*Strymon melinus*)	♂	0.38 : 1	Flowers
	♀	0.43 : 1	
Great Spangled Fritillary (*Speyeria cybele*)	♂	0.37 : 1	Flowers
	♀	0.31 : 1	
Harvester (*Feniseca tarquinius*)	♂	0.11 : 1	Honeydew
	♀	0.09 : 1	
Juvenal's Dusky Wing (*Erynnis juvenalis*)	♂	0.49 : 1	Flowers
	♀	0.49 : 1	
Little Wood Satyr (*Megisto cymela*)	♂	0.24 : 1	Honeydew, sap flows
	♀	0.26 : 1	
Pearl Crescent (*Phyciodes tharos*)	♂	0.39 : 1	Flowers
	♀	0.42 : 1	
Question Mark (*Polygonia interrogationis*)	♂	0.40 : 1	Sap flows, rotting fruit
	♀	0.38 : 1	
Red Admiral (*Vanessa atalanta*)	♀	0.40 : 1	Sap flows, rotting fruit, flowers
Southern Golden Skipper (*Poanes zabulon*)	♂	0.95 : 1	Flowers
	♀	0.79 : 1	
Tiger Swallowtail (*Papilio glaucus*)	♂	0.32 : 1	Flowers
	♀	0.36 : 1	

TABLE 11. Flower color preferences of selected butterflies in Fairfax County, Virginia, expressed as number of visits per number of flower species (in parentheses)[a]

Species	Flower Color					
	Red/Orange	Yellow	White	Pink	Purple	Blue
Danaus plexippus (Monarch)	—	38(3)	4(2)	10(1)	16(3)	—
Speyeria cybele (Great Spangled Fritillary)	—	—	6(2)	8(2)	3(2)	—
Phyciodes tharos (Pearl Crescent)	1(1)	116(7)	159(11)	11(2)	—	2(1)
Anthocharis midea (Falcate Orange-tip)	—	29(2)	28(6)	—	—	—
Colias eurytheme (Orange Sulphur)	—	67(6)	281(7)	7(3)	2(2)	8(3)
Pieris rapae (European Cabbage Butterfly)	5(2)	33(10)	253(13)	32(8)	14(4)	6(3)
Strymon melinus (Gray Hairstreak)	—	3(3)	12(8)	8(2)	2(2)	—
Calycopis cecrops (Red-banded Hairstreak)	—	5(3)	72(11)	6(2)	3(2)	—
Everes comyntas (Eastern Tailed Blue)	—	15(3)	40(9)	8(1)	—	1(1)
Epargyreus clarus (Silver-spotted Skipper)	1(1)	2(1)	69(9)	45(4)	121(6)	1(1)
Erynnis icelus (Dreamy Dusky Wing)	—	16(1)	32(5)	3(1)	29(1)	4(1)
Hesperia leonardus (Leonard's Skipper)	—	1(1)	—	3(1)	26(4)	1(1)
Atalopedes campestris (Satchem)	5(1)	40(2)	3(2)	10(3)	15(5)	1(1)
Polites origenes (Cross Line Skipper)	—	7(2)	14(4)	2(2)	17(5)	—
Polites coras (Peck's Skipper)	—	8(1)	9(5)	23(4)	46(6)	—
Poanes zabulon (Southern Golden Skipper)	—	7(1)	6(4)	2(2)	58(6)	—

[a] Unpublished data from Opler.

and the Coral Hairstreak especially favors the orange-flowered butterflyweed. Most swallowtails and skippers favor white, pink, and purple flowers.

The height of flowers above the ground is also of importance, since some species, such as the Little Sulphur, Common Sooty Wing, and Least Skipper seldom rise more than a half meter above the ground for any purpose, while others, such as the Red-spotted Purple, Great Purple Hairstreak, Spring Azure, Tiger Swallowtail, and Silver-spotted Skipper, will readily visit flowers more than 3 m above ground. Most visitation takes place between 1 and 3 m, but in early spring, when air temperatures are cool, there is proportionately more visitation close to the ground.

The positioning of flowers and the shape of inflorescences may dictate the butterflies that visit particular plants. Some plants have flowers that are directed upward, in flat or slightly domed inflorescences; in others, the flowers project laterally, while some have spherical inflorescences or random arrangements, and a few are directed downward. In each case the visiting butterfly may have to assume a different position. Some butterflies may adapt to all of these circumstances, while the behavior of others is so fixed that they will perch only in a particular position, usually horizontal or at a 45° angle. Only a few species will perch upside down on flowers. Among common eastern species, the Great Spangled Fritillary, Tiger Swallowtail, Banded Hairstreak, and Silver-spotted Skipper are a few that seem able to assume a variety of

positions at flowers. Others, such as the Monarch, European Cabbage Butterfly, and other swallowtails, usually perch at 45°–90° angles, while some, including the Pearl Crescent, American Copper, and several hesperiine skippers, prefer to perch atop an inflorescence in a horizontal posture. Butterflies may also assume basking positions relative to the sun while taking nectar.

Some butterflies may visit the flowers of their caterpillar hosts and thereby contribute to the pollination of plants they eat at another life stage. Notable examples are the Monarch, Queen, Spring Azure, Falcate Orange-tip, European Cabbage Butterfly, Orange and Clouded Sulphurs, and the Barred Sulphur.

A number of eastern plants have flowers that attract a variety of butterflies, and a student of butterfly natural history or a "butterfly gardener" would benefit by learning their names and locating local colonies of: milkweeds (*Asclepias*), dogbane and Indian hemp (*Apocynum*), New Jersey tea (*Ceanothus americanus*), sweet pepperbush (*Clethra alnifolia*), pickerelweed (*Pontederia cordata*), red clover (*Trifolium pratense*), selfheal (*Prunella vulgaris*), lantana (*Lantana*), mountain mint (*Pycnanthemum*), buttonbush (*Cephalanthus occidentalis*), buckwheat (*Fagopyrum sagittatum*), viper's bugloss (*Echium vulgare*), shepherd's needle (*Bidens pilosa*), tickseed sunflower (*Bidens comosa*), thistles (*Cirsium*), joe-pye weed (*Eupatorium*), ironweed (*Vernonia*), and selected goldenrods (*Solidago*).

Caterpillar Foods

With the exception of the Harvester (*Feniseca tarquinius*), which feeds on wooly aphids (*Schizoneura* and *Pemphigus*), all eastern butterfly species feed on plants in their caterpillar stage. A few species are very broad in their host choice, but most are specific, feeding on only a few plants in one or two families, and most caterpillars eat only one part of their host (e.g., leaves or flower buds). The Painted Lady has a broader range of caterpillar hosts during its emigration cycle than at other times. After overwintering, Baltimore caterpillars may eat plants other than their normal hosts. Host choice also varies in different geographic areas and at different times of year.

Species whose caterpillars feed on only a few plants in one genus are *monophagous,* those that feed on several plants in different genera of the same or a few closely related families are *oligophagous,* and, finally, those that select many plants in a number of unrelated plant families are *polyphagous.*

In the eastern United States, most species are either monophagous (43%) or oligophagous (48%), and feed on plants in only a few closely related families. A relatively small proportion are polyphagous (9%). Most polyphagous butterflies are nymphalines, including the Variegated Fritillary (*Euptoieta claudia*), Comma (*Polygonia comma*), Question Mark (*P. interrogationis*), Mourning Cloak (*Nymphalis antiopa*), Painted Lady (*Vanessa cardui*), and Buckeye (*Junonia coenia*). Most other broad-spectrum feeders are lycaenids. The Gray Hairstreak (*Strymon melinus*) has by far the broadest host range: it will eat a variety of monocotyledonous and dicotyledonous plants.

Polyphagy allows species to be opportunistic exploiters of a variety of resources in time and space, but often these opportunities are not closely attuned to any single host. They usually cannot achieve a long-term stable existence in any habitat and must continually seek new situations with temporarily rich food supplies. All eastern polyphagous species are multiple-brooded, broad-ranging, and excellent colonists. Almost all show little geographic variation.

In contrast, monophagous butterflies are often limited to particular habitats, and form local colonies where their hosts occur. Furthermore, the plant part or stage upon which they feed may be available only for a short time each year; as a result, most monophagous species are single-brooded. Monophagous species show the greatest amount of geographic variation.

Oligophagous butterflies can be more flexible than monophagous species. They can afford to be less dependent on, and are often less restricted by, habitat or times of year. Some oligophagous species are almost as widespread as polyphagous butterflies. Notable examples are the European Cabbage Butterfly, Eastern Tailed Blue, and Fiery Skipper. On the other hand, other oligophages, such as the Ap-

palachian Eyed Brown, Striped Hairstreak, Silvery Blue, and Leonard's Skipper, display the predominant traits of monophagous butterflies.

The disruption of native environments in the eastern United States has clearly tipped the scales in favor of polyphagous and adaptive oligophagous butterflies. Only a few monophagous species, whose hosts have thrived with disturbance, are more widespread. Some monophagous species that can accept introduced weeds have also benefited. The American Copper is an outstanding example, but of course it is most likely an introduction as well.

The study of host-restricted species has allowed biologists to speculate on butterfly evolution and speciation (Ehrlich and Raven 1965). Examination of the overall preferences of butterfly families and the preferences of closely related species or genera makes possible a few conclusions. Flowering plants were probably highly diversified when butterflies first appeared nearly 100 million years ago. Thus, it is not likely that butterflies evolved completely in tandem with the dominant plant groups we now see. Nevertheless, some butterfly groups demonstrate that they have been associated with specific plant groups for long periods, probably many millions of years. Examples include the satyrines (with grasses and sedges), "yellow pierids" (with legumes), "white pierids" (with mustards and capers), and hesperiine or "branded" skippers (with monocotyledons, primarily grasses and sedges in the eastern U.S.).

The evolution of host-plant specificity and choice indicates that host switches or additions are fairly frequent as measured in evolutionary time. As plant species and populations wax and wane in their geographic range, abundance, and the variety of habitats in which they occur, the number of specialist plant-eating insects, including butterflies, rises and falls accordingly.

Widespread butterflies with different hosts in different portions of their range are most likely to produce new species, with new host associations following isolation of their component populations through some stretch of geologic time. Among present-day eastern butterfly species, the Baltimore and Henry's Elfin are most likely to spawn new host species after some future isolation. Species pairs, such as the Spring Azure and Dusky Blue, Dorcas Copper and Bog Coppers, and Olive Hairstreak and Hessel's Hairstreak, probably have had relatively recent (1–2 million years) common ancestors and relatively recent host changes. The introduced European weeds have allowed small changes in the diets of some native butterflies in the last 200 years, as well as significant changes in their distributions. For example, the Southern Sooty Wing (*Staphylus hayhursti*) and Common Sooty Wing (*Pholisora catullus*) now both feed on lamb's quarters (*Chenopodium album*) and have much larger ranges in the eastern states as a result. The Orange Sulphur was able to spread throughout the eastern seaboard states largely due to the presence of introduced clovers, vetches, and alfalfa.

Butterflies not only feed on particular plants, but also usually eat only certain parts or stages of their hosts. Most caterpillars feed on young leaves, while a few will eat older leaves as well. Others, such as the blues, many hairstreaks, and marbles, feed on the flowers or young fruits of their hosts. Finally, the yucca skippers (*Megathymus*) usually bore into the stem or roots of their yucca host plants.

Several butterflies are polyphagous, if one considers all the plants they have been recorded to eat, but are effectively monophagous or oligophagous if one considers regional or seasonal specificity. An example of a butterfly with regional specificity is the Henry's Elfin, which feeds on redbud in most of its range, but feeds on hollies (*Ilex*) in the Atlantic coastal plain and blueberry (*Vaccinium*) in the northern states and Florida. Similarly, the Baltimore feeds on turtlehead (*Chelone glabra*) in more northern areas but eats beardtongue (*Aureolaria*) in the south.

The Spring Azure is a good example of a butterfly with seasonal specificity, since there are usually only two or three suitable hosts for each brood of a given population. For example, in northern Virginia, the spring females select common dogwood and wild cherry, whereas second-generation females select New Jersey tea, Osier dogwood, and viburnum.

In most species, host choice is determined by the egg-laying females. However, in a few species that do not lay their eggs directly on a host, it is up to the young caterpillars to choose among an array of nearby plants. Mistakes are also made by the females of some species, who lay their eggs on plants that are toxic to their caterpillars. Examples of butterflies and toxic host plants are: some whites and winter cress (Chew 1977), Tiger Swallowtails and peach, and, finally, the Spring Azure and several different plants (C.G. Oliver, unpublished). In these cases, the signals that indicate a plant is a suitable oviposition substrate do not forewarn of the plant's toxicity to larvae.

Life History

The basic life cycle of a butterfly from egg to adult varies from 3 weeks (Harvester) to 2 years (Macoun's Arctic and others). Some butterflies have the ability to remain in the pupal stage for several years in order to bypass unfavorable seasons. In the eastern United States, some Schaus' Swallowtails and Olympia Marbles might take 3 or 4 years from egg to emergence. Between the deposition of the egg and the adult's emergence, many variations of the basic life cycle may occur. Many of these variations involve periods of obligatory growth cessation (*diapause*) whose onset is usually triggered by specific day-length (photoperiod) regimes. Diapause allows butterflies to pass unfavorable periods, usually winter, in a relatively resistant stage.

Each butterfly passes winter or other unfavorable periods in a specific stage, but each of the four life-cycle stages, including the adult, is used. These stages, in combination with the variation in generation number and flight time, allow various kinds of yearly cycle, even within species. For example, several single-brooded early summer hairstreaks have egg diapause lasting about 10 months, while the other three stages undergo growth, metamorphosis, and reproduction in the remaining 2 months. In other species, the chrysalis is the resistant stage, and the adults emerge the year following larval growth. Other species have continuous broods in the warm part of the year, and overwinter as adults or larvae. The species accounts reveal many other variations.
A Generalized Life Cycle. In the following generalized life-cycle description, many of the alternative adaptations are discussed. After oviposition, eggs complete embryonic development and hatch in 4 to 10 days. As noted above, the eggs of some species enter a long diapause after oviposition, and in others, notably the coppers, the young larva completes embryogenesis but remains in the eggshell until the next spring. Upon hatching, young caterpillars of many species dine on the eggshell for their first meal, while in other species the larva exits and begins host-plant feeding. The caterpillars of most eastern butterflies construct a shelter within which they rest by day and out of which they venture at night to feed. The shelter may be a folded leaf or several leaves folded together. The young caterpillars of a few species spin webs or bore into host-plant material, wherein they are protected. A few species rest openly on leaves or stems; the Monarch, the Black Swallowtail, and the European Cabbage Butterfly caterpillars are examples. Several nymphalines lay egg clusters, and their young feed together in tight groups, especially during their early stadia. The caterpillars of many lycaenids, especially blues and hairstreaks, have secretory glands and associated structures on the dorsum of abdominal segments 7 and 8. These glands produce a sugary liquid secretion that is sought by ants, who in turn may protect the caterpillars from parasites and predators. (Some lycaenid caterpillars are taken into ant nests and fed by the ants.) As larvae grow, they become too large for their skins, and they molt. This process, termed *ecdysis*, begins with the splitting of the head capsule and then the rearward splitting of the body's skin or integument. Repeated contractions by the caterpillar move the old skin backward over the body. Often the caterpillar rests for a day or two after ecdysis before it resumes feeding. Each stage of the caterpillar is termed an *instar* and the concomitant time periods are called *stadia*. Most caterpillars pass

through five stadia, but some undergo only four, and others pass through six or seven. Normally, 3 or 4 weeks are required to complete feeding, but the carnivorous Harvester may require only 10 days, and others may spend several months feeding. Of course, some butterflies pass the winter as partially grown caterpillars, usually among dead leaves or litter near the base of the host. When feeding is complete, the mature caterpillar seeks a place to pupate. In many instances this may be in the larval shelter or elsewhere on the host plant, but in others the caterpillar may form its chrysalis in litter near the host or wander considerable distances to find a suitable site. Once a site is found, caterpillars may spin a loose silk cocoon (e.g., as do skippers) or spin a silk button upon which to hang (e.g., as do most true butterflies). The swallowtails, whites, sulphurs, and some lycaenids also spin a silk girdle that encircles the body for additional support. After the caterpillar rests for some time its skin splits for the last time, revealing the chrysalis. In species with pendant chrysalids ecdysis is a critical time; as the posterior end is cleared, the new pupa must quickly, in one deft movement, latch its cremaster hooks into the silk button before it falls to the ground. The pupa may wriggle freely at first, but it usually hardens completely and remains motionless. Pupae of the Yucca Skippers are exceptional in that they are able to move freely up and down their burrows, but normally they remain in their tough silk 'tents.'

The pupal period lasts from 1 to 2 weeks, unless, of course, diapause or hibernation intervenes. Inside the pupa, larval structures are broken down and the adult butterfly is formed. Upon emergence the adult's wings are small and moist, the two halves of the proboscis are unattached, and the intestinal passage is blocked by a structure called the *meconium*. The newly emerged butterfly hangs with its wings downward and slowly pumps its blood, more properly called *hemolymph*, through its veins to expand its wings. Simultaneously, the new adult works the two halves of the proboscis together until they are connected, and it passes the meconium and the final contents of the larval gut. When its wings are expanded, it slowly opens and closes them until they are dry. Normally, the butterfly rests about an hour or more before taking its first flight.

SPECIES ACCOUNTS

Introduction

Format

Species accounts are given for all butterflies known to reside or occasionally stray into the area of the United States east of the Plains. (A list of butterflies whose occurrence in our area is dubious is presented after the species accounts.) The format of the species accounts allows presentation of the maximum biological information available for each butterfly. This format follows the order of topics in the introductory section.

Etymology and Synopsis. Both the scientific name, including author and date of original description, and the preferred common names are given. We present a subspecies epithet when only one subspecies is known in our area. For most resident species, we have given our etymological derivation of the scientific name, although in some instances our translation may well differ from the original author's unstated intent. The synopsis states some noteworthy trait of the butterfly.

Butterfly. Next, a lengthier discussion of the adult butterfly presents its distinguishing traits, its geographic variation, and the mean and range of the forewing length of both sexes (in most cases at least ten individuals of each sex were measured).

Range. The next section gives the butterfly's overall range, in general terms, beginning with the northeastern extreme and then proceeding westward and southward. Notable historical declines or gains in range are also pointed out. An accompanying map presents the eastern United States portion of the historical range in greater detail. Many species expand their range temporarily each year beyond the area where they are normally resident. For such species the areas of temporary occupation are shown by lighter shading. Small isolated colonies or strays found out of range are shown as black circles. The maps are based upon an extensive literature review and county records contributed by more than one hundred lepidopterists (see Acknowledgements).

Habitat. The habitat description mentions general as well as specific vegetation types, plant formations, or soil types where the butterfly may be found. In addition, limitation to certain Life Zones is cited in many cases.

Life History. The life history portion of each account includes topics relating to each butterfly's biological attributes. Under behavior, we include mate-location behavior, courtship and mating, oviposition, flight behavior, and whether the species engages in mud-puddling behavior. Where they are available, we give the times that the species has been observed partaking in each activity.

The paragraph on broods includes a statement on the number and timing of adult flights each year. If the butterfly occurs over a wide latitudinal range, flight dates and any variation in brood number are discussed for several sites so that the reader may have a good idea of when to look for the insect in his area.

In the section concerning early stages, we mention the number of eggs and their placement by wild females, the seasonal timing of the life cycle, the overwintering stage, and a general description of the egg, caterpillar, and chrysalis.

Food Sources. Adult food habits, including the kinds of flowers used, are presented in a separate section. Since several plants are used by many butterflies, only common names are given in the text, and scientific names and plant families are given in the index.

The caterpillar host plants are also enumerated, together with indications of when previously reported hosts may be in error or require further documentation. Host plants that are fed upon only in laboratory conditions are specifically pointed out. Although caterpillars may feed and develop on such plants, females may never lay eggs on them in nature.

References. Finally, for the advanced reader we make reference to scholarly papers that present more detailed information on aspects of the butterfly's distribution or biology. The full reference citations are given in the Bibliography.

Plates

The illustrations are all photographs of free-living butterflies filmed in the wild. Almost all of the species in our area of interest are included, with the few exceptions being rarer or difficult to locate species not photographed by the time of book publication. For most species we were fortunate enough to "capture" both males and females, and for many species we include more than one illustration, to show representative behaviors (feeding, dorsal basking, etc.). Plate 1 includes photographs of some fine examples of caterpillars.

SUPERFAMILY PAPILIONOIDEA: TRUE BUTTERFLIES

Family Papilionidae: Swallowtails and Parnassians

Of the true butterflies, only the swallowtail subfamily (Papilioninae) occurs in the eastern United States. Adult swallowtails and other true butterflies are characterized by a tibial spur or *epiphysis* on the front leg and only a single vannal vein near the inner margin of the hind wing. The swallowtails are large and brightly colored and usually have a long tail projecting from the hind wing (first cubital vein).

The males of most swallowtails patrol long routes in search of females, although the Black and Missouri Woodland Swallowtails are both perching species. Courtship and mating take place in late afternoon and egg-laying usually occurs near midday. Dorsal basking is the norm, and flight is often strong and rapid. The mouthparts are long, and adult feeding is often restricted to flowers with long tubes. Adult males often gather in groups at wet mud or sand.

The larval hosts belong to several families with aromatic leaves. The citrus, umbel, laurel, magnolia, annona, and pipevine families are the most frequent hosts. The hemispherical eggs are usually laid singly on host leaves. The caterpillars are usually naked but may have several rows of fleshy tubercles (see Figure 1, Plate 1). There is a forked eversible organ, the *osmaterium,* on the prothoracic dorsum. When disturbed, the caterpillar everts this bright yellow or orange organ, which emits a pungent odor. It is presumably an antipredator device. The caterpillars feed on leaves, usually at night, and rest under leaves or in folded leaf shelters by day. Most eastern species are at least bivoltine. The brown or green roughened chrysalises are slung from a girdle as well as the usual silk pad. The chrysalis is the overwintering stage in all our species.

Pipe Vine Swallowtail
Battus philenor (Linnaeus), 1771
PLATE 2 · FIGURES 7 & 8 · MAP 1

Etymology.   The genus is named after Battus, a regal Greek family from Cyrene (North Africa). The species name is derived from the Greek *philenor,* "fond of husband, conjugal."

Synopsis.   This species is the distasteful model for six mimetic, palatable butterflies in the eastern United States. It and the Polydamas Swallowtail are the only regular United States representatives of a diverse tropical American genus, all of which feed on various pipe vines *(Aristolochia).*

Butterfly.   Male forewing: $\bar{X}$ = 4.53 cm, range 3.2–5.6 cm; female forewing: $\bar{X}$ = 4.75 cm, range 3.8–6.2 cm. The Pipe Vine Swallowtail may be distinguished by the following combination of characters: black dorsal surface with iridescent blue-green reflectance on hind wings; submarginal row of white spots on both dorsal wings; a strongly curved submarginal row of seven black-edged, circular, orange spots in a field of iridescent blue on the ventral hind wing; and no red-orange mark or patch at anal angle of the dorsal hind wing. The male differs from the female above in its more reflective green hind-wing iridescence, and its smaller, somewhat obscured, submarginal white spots. Spring individuals tend to be smaller, have longer body scales, and more distinct submarginal white spotting above (Clark 1932).

The species that clearly mimic the Pipe Vine Swallowtail are the Spicebush Swallowtail (*Papilio troilus*), the dark female form of Tiger Swallowtail (*Papilio glaucus*), the female Black Swallowtail (*Papilio polyxenes asterius*), the female Missouri Woodland Swallowtail (*Papilio joanae*), the female Diana (*Speyeria diana*), and the Red-spotted Purple (*Limenitis arthemis astyanax*). All of these species are mostly black dorsally, with iridescent blue or blue-green on the outer portions of the hind wings. In addition, all except the female Diana have submarginal rows of orange markings on the ventral hind wing. All except the female Tiger Swallowtail

41

have fairly direct flight, with rapid, low-amplitude wing beats. When visiting flowers or otherwise at rest, all except the female Diana flutter their wings or at least slowly lower and raise their wings, mimicking the behavior of their model. The effectiveness of this mimicry system has been demonstrated in the laboratory by J.V.Z. Brower (1958) and in the field by Jeffords et al. (1979). Brower showed that birds avoid the palatable mimics after first experiencing distasteful Pipe Vine Swallowtails. Jeffords and co-workers showed that palatable Prometheus moths that were painted to resemble Monarchs or Pipe Vine Swallowtails and then were released in the field with unpainted controls suffered less predation by birds. The highest density of mimics occurs in the southern Appalachians and Ozarks, where the Pipe Vine Swallowtails are abundant.

Range.   The Pipe Vine Swallowtail's range extends from central New England (rarely coastal Maine) west through southern Ontario and the Great Lakes states to Colorado and the southwest. The butterfly ranges south through the remainder of the east (although it is rare in the prairie biome and in peninsular Florida), and farther south through Mexico. It has been reported to occur in Costa Rica, although it has not been seen there recently. An isolated population occurs in central California.

Habitat.   The Pipe Vine Swallowtail is most abundant in deciduous forests of the Appalachians, but it is found in a variety of other situations, ranging from open fields to roadsides, gardens, desert scrub, and tropical forests. It occurs from the Tropical Life Zone through the Transition Life Zone.

Life History.   *Behavior.* Males patrol during most warm daylight hours. The flight is low and rapid, with low-amplitude wing beats. One mated pair was observed in late morning (1050 hr) with the female as the active member of the pair, but courtship and mating may occur over a much broader period. Females have been observed to oviposit from 1050 to 1700 hr, an unusually long period. On small host species, 2–4 eggs are laid in a group on the lower side of a host leaf, but larger groups may be laid on larger vine hosts. Huge groups of freshly emerged males may be seen in wet sand or mud along country roads or in fields adjacent to woods. They flutter their wings weakly while mud-puddling, but they beat their wings rapidly when taking nectar.

Broods. In the northern part of its range it is bivoltine, but there are three or more farther south. In Illinois, flight dates range from April 24 to May 31 and from July 2 to September 23. In Virginia and the Delaware Valley there are three broods. In the Deep South (Florida, Georgia, Louisiana, Mississippi), flight dates from early February to October or November suggest four generations, but this requires confirmation.

Early Stages. Young caterpillars feed in tightly clustered groups on the lower surfaces of host leaves, but they feed alone when older. The larvae are similarly colored in all instars, being black or dark, brownish red with long, fleshy, red tubercles in rows along the body. The chrysalis is green or brown with yellow splotches. There are strongly fluted lateral keels along the abdomen and paired spatulate protuberances projecting from the dorsum of the last three or four abdominal segments.

Adult Nectar Sources.   Adults favor pink, purple, or orange flowers, although not exclusively. Thistles are favorite flowers, especially in mid- to late summer, but others, such as bergamot, lilac, viper's bugloss, common milkweed, azaleas, phlox, teasel, dame's rocket, and petunias, are also visited with some frequency.

Caterpillar Host Plants.   All reliable reports of larval hosts are of various pipe vine species (*Aristolochia*), including ornamental Dutchman's pipe (*A. durior*) as well as Virginia Snakeroot (*A. serpentaria*), *A. longiflora,* and *A. reticulata* (Scriber and Feeny 1976). Rausher (1980) has shown that first-brood females in Texas oviposit preferentially on *A. reticulata,* while at the same locality second-brood females lay more eggs on *A. serpentaria.* There have been several reports of larvae on wild ginger (*Asarum*), but these are very doubtful, as are suggestions of knotweed (*Polygonum*), bindweed (*Ipomoea*), and morning glory (*Convolvulus*).

1. *Battus philenor*

Polydamas Swallowtail
Battus polydamas lucayus (Rothschild & Jordan), 1906

PLATE 2 · FIGURE 9 · MAP 2

Etymology. The species is named after Polydamas, son of Nicias, a Thessalian athlete (Greek myth).

Synopsis. This tropical species is the only eastern swallowtail that completely lacks tails or similar projections on the hind wings. It is widespread in the American Tropics.

Butterfly. Male forewing: $\overline{X} = 4.60$ cm, range 4.2–4.9 cm; female forewing: $\overline{X} = 5.09$ cm, range 4.8–5.4 cm. The species is distinguished by the black dorsal wing surfaces, which are relieved by a submarginal band of yellow chevrons on the forewing and a postmedian band of large yellow spots on the hind wing. There is a submarginal row of sinuous red lines on the ventral hind wing. The forewing apex is produced, and the hind wing outer margin is crenulate.

Range. The Polydamas Swallowtail occurs from peninsular Florida and southern Texas south through the Caribbean Islands, Mexico, Central America, and most of South America to Argentina. The species wanders widely, and strays or temporary populations have been found as far north as Missouri and Kentucky.

Habitat. The Polydamas Swallowtail has a predilection for open areas, particularly abandoned fields, open pine woods, and suburban areas. Normally, the species is restricted to Tropical and Subtropical Life Zones.

Life History. *Broods.* The number of generations is indefinite, but two or three are possible in Florida, since adults have been found there from early April until mid-November. Adults live as long as 27 days.

Early Stages. Early stages similar to those of the Pipe Vine Swallowtail. The larvae are gregarious when young. The last-instar caterpillar has lateral yellow thoracic bands; otherwise, it is similar to that of the Pipe Vine Swallowtail.

Adult Nectar Sources. Lantana is favored in Florida and other tropical areas. In Costa Rica, *Stachytarpheta frantzii* and *Cordia* are also favored. This swallowtail was observed visting honeysuckle in Arkansas and soapberry in Kentucky.

Caterpillar Host Plants. The Polydamas Swallowtail feeds only on various pipe vines (*Aristolochia*). In Arkansas, larvae were found on Virginia snakeroot (*A. serpentaria*) together with those of the Pipe Vine Swallowtail. In Texas, Kendall (1964) found larvae feeding on calico-flower (*A. elegans*), a plant that seems to be toxic to Pipe Vine Swallowtail caterpillars.

2. *Battus polydamas lucayus*

Zebra Swallowtail
Eurytides marcellus (Cramer), 1777

PLATE 2 · FIGURES 10–12 · MAP 3

Etymology. The genus is named after Eurytus, King of Oechalia. Marcellus is a Roman surname.

Synopsis. This "kite-tailed" swallowtail is the only regular United States representative of a diverse pantropical genus. The butterfly has two strikingly different seasonal forms. The caterpillars feed only on various species of pawpaw (*Asimina*).

Butterfly. Male forewing: $\overline{X} = 3.8$ cm, range 3.0–4.6 cm;

female forewing: $\overline{X} = 4.0$ cm, range 2.9–4.9 cm. The Zebra Swallowtail can be confused with no other eastern United States species. The early-spring form is smaller and lighter than the summer form ("lecontei") and has tails only half as long as summer individuals. Late-spring individuals ("telamonides") are somewhat intermediate. Mather (1970) gives an extensive analysis of this species' seasonal variation.

Range. The Zebra Swallowtail ranges westward from

southern New England across the southern portion of the Great Lakes to southern Minnesota. It then ranges south to the Gulf Coast and southern Florida.

Habitat. The breeding habitat of this species is rich broad-leaved woodland along rivers and in swamps where its host occurs. In late spring and summer this swallowtail may wander out of its woodland habitat into open brushy areas and abandoned fields in search of nectar plants. The species ranges from the Lower Austral to the Transition Life Zone (rarely).

Life History. *Behavior.* The males patrol with a rapid, direct flight about 1–2 m above the ground through the understory in the vicinity of their hosts. Courtship takes place from midday until the evening (1240–2100 hr). Mating females have been seen ovipositing near midday (1130–1145 hr), during which time they flutter from spot to spot near a woods' edge and lay single eggs on low leaves of young pawpaw. An observer in Ohio saw a female lay eggs on grass blades at the base of a pawpaw. Large aggregations of freshly emerged males may be observed imbibing moisture from wet sand and along stream and river banks, particularly in early spring.

Broods. In the northern part of the Zebra Swallowtail's range, it is bivoltine, although the first is always more numerous. The Clarks (1951) report that, in northern Virginia, there is only one generation in some years. In New York the dates for the two broods are April 20 to June 5 and June 20 to August 18. In the South, the species flies earlier and continues to fly later in the season; in these areas there may be three or four generations. In Georgia, the Zebra Swallowtail flies from February 13 to October 21, in Mississippi it flies from February 9 until August 27, and in Florida it flies from March until December.

Early Stages. The caterpillars feed on the leaves of pawpaw. The last instar is pea green, with a yellow-edged, black, broad dorsal band on the third thoracic segment. The remaining segments have five transverse alternating bands of yellow and black. The chrysalis is either bright green or brown, and is smooth, short, and stout.

Adult Nectar Sources. The proboscis of the Zebra Swallowtail is much shorter than those of other swallowtails, and it is unable to utilize long-tubed flowers. First-generation adults rely upon blueberry, redbud, and blackberry, as well as lilac. Second-generation adults favor common milkweed, dogbane, verbena, and viper's bugloss.

Caterpillar Host Plants. All known hosts belong to the genus *Asimina* (Annona family). Through most of its range, pawpaw (*Asimina triloba*) is the only host, but in the Southeast (North Carolina, Florida), other pawpaw species are fed upon, including *Asimina longifolia*, *A. parviflora*, *A. pygmaea*, *A. reticulata*, and *A. speciosa*.

3. *Eurytides marcellus*

Black Swallowtail
Papilio polyxenes asterius (Stoll), 1782
PLATE 3 · FIGURES 13–15 · MAP 4

Etymology. The genus is derived from the latin *papilio* (butterfly), while the species is named after Polyxena, daughter of Priamos, King of Troy (Homerian myth, *Iliad*).

Synopsis. This is the common parsleyworm found in most surburban and country gardens. It belongs to the confusing *Papilio machaon* Complex, which is most diverse in the western states. The female is a member of a mimicry ring patterned after the distasteful Pipe Vine Swallowtail.

Butterfly. Male forewing: $\bar{X}$ = 4.3 cm, range 3.6–4.8 cm; female forewing: $\bar{X}$ = 4.7 cm, range 4.5–5.3 cm. A Florida female had a 6.4-cm forewing. The Black Swallowtail has a small red patch with a centered black spot at the anal angle of both hind wings, clearly differentiating it from all other eastern swallowtails except the Missouri Woodland Swallowtail. The male has a narrow, postmedian, yellow band on both wings, which may be represented on the female by a series of small yellow spots. As mentioned above, the female is a member of the Pipe Vine Swallowtail mimicry ring, and it cannot be readily distinguished in flight from other dark eastern swallowtails.

Range. The subspecies *asterius* ranges from the Maritime Provinces west across southern Canada to the central Rocky Mountains and south through the entire eastern United States, as well as the western mountains of New Mexico and Arizona. It also occurs in northern Mexico. Other subspecies range south through the mountains of Mexico, Central America, and northern South America.

Habitat. The Black Swallowtail occurs in a wide range of open habitats in almost every conceivable situation, including gardens, old fields, pastures, roadsides, and marshes. It occurs in all Life Zones from the Lower Austral through the Canadian.

Life History. *Behavior.* Male Black Swallowtails perch and periodically patrol open areas as their mate-locating strategy. They normally perch atop low vegetation (0.5–1.5 m) from which they have an unobstructed view of surrounding terrain. In the West, males often select hilltops as perching arenas to which receptive females may orient. A male may occupy a given territory for most of a day (0900–1530 hr). Rawlins (1980) has described the thermoregulatory behavior of Black Swallowtails. At low temperatures, the butterflies perch close to the ground with wings spread and abdomens raised (dorsal basking). At higher temperatures, they perch higher, with abdomens lowered in shade, and fly more frequently. Courtship and mating take place from midday to late afternoon (1230–1730 hr). The mated pair rests in shade with wings closed, and, if they are disturbed, the female is the carrying sex. Oviposition takes place most often at midday (1030–1420 hr), but is occasionally observed in late afternoon (1630–1710 hr). The female flies low, over low vegetation, periodically dipping down; if she locates a suitable host, she lays a single egg near the tip of a young host leaf. More than one egg may be laid on each plant.

Roosting takes place in late afternoon on grasses or other herbaceous vegetation, usually on a western slope with late-afternoon sun exposure (Rawlins and Lederhouse 1978). The butterflies roost with wings closed and may suffer fairly high mortality overnight. Freshly emerged males may occasionally visit damp spots along roads or streams.

Broods. The Black Swallowtail is bivoltine in the northern portion of its range. For example, in New York's Finger Lakes region, flights occur from April 17 to June 30 and July 22 to October 13. Farther south, three generations seem to be the rule. For example, in Virginia the broods extend from early April to late June, early July to mid-August, and mid-August through September (rarely mid-October). The timing of the broods is imprecise, since pupal diapause is not broken synchronously, and individual pupae of the second generation may also overwinter.

Early Stages. Development to the chrysalid is always direct, with no period of arrest, and winter is passed in this stage. The pale-green egg is spherical. Of the four larval instars, the first two consist of bird-dropping mimics, while the last two are cryptic. The first instar is predominantly black, with a median, dorsal, white saddlemark. The second has the same basic pattern, to which is added a few lateral orange dots and white markings. In the third stadia, the caterpillar is black, with the anterior and posterior of each segment narrowly edged with pale mint green. There are also five longitudinal rows of orange dots. The head is patterned black and white. The last-instar larva is predominantly green, with black greatly restricted to the center of each segment. The orange dot rows are also present. The caterpillars are variable, and some may be predominantly black. The pupa, like those of all swallowtails, is attached at the posterior end with silk and is slung at about a 30° angle from a silk girdle. The pupa is either green or is light brown splotchily marked with black and blends well when attached to a dead plant stem or bark. Experiments have shown a combination of genetic and environmental factors to be involved with pupal color determination. During the summer, either green or brown pupae may result from the genetic threshold and the nature of the pupation site selected. In the fall, the short photoperiod leads only to brown diapausing pupae (West et al. 1972).

Adult Nectar Sources. Adults visit a variety of plants for nectar. Among those that seem to be visited most often are milkweeds, thistles, and red clover. The adults continue to flutter their wings while taking nectar.

Caterpillar Host Plants. A variety of umbelliferous plants (family Apiaceae) are utilized. This butterfly and the Missouri Woodland Swallowtail are the only two eastern butterflies that regularly feed on these plants. Cultivated or wild carrot (*Daucus carota*) is most often selected, but dill, parsley, and celery are also used. Occasionally, species in the citrus family (*Rutaceae*) are chosen, e.g., common rue (*Ruta graveolens*) and Texas turpentine broom (*Thamnosma texana*).

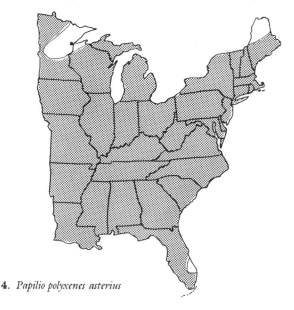

4. *Papilio polyxenes asterius*

Missouri Woodland Swallowtail
Papilio joanae Heitzman, 1974

MAP 5

Etymology. The butterfly is named in honor of Joan Heitzman, wife of Richard, the species' describer.

Synopsis. Most members of the *machaon* complex interbreed where their ranges overlap, but the Missouri Woodland Swallowtail is apparently a local woodland endemic that is reproductively or behaviorally isolated from the widespread Black Swallowtail, which occurs nearby in open habitats.

Butterfly. Similar in size and other features to the Black Swallowtail, the Missouri Woodland Swallowtail may be distinguished by the following features. 1) In males, the most apical yellow spot of the forewing median band is short and single. 2) The front is usually yellow (not black). 3) On the female's hind wing, the costal spot of the postmedian band is small and obscured. 4) On both sexes, the ventral postmedian bands usually have more orange than those of the Black Swallowtail. 5) The black spot in the hind wing anal angle orange patch is not centered, but often touches the wing margin.

Range. The Missouri Woodland Swallowtail is known only in Missouri and northern Arkansas.

Habitat. This species, in striking contrast to the ubiquitous open-field Black Swallowtail, is limited to cedar glades and woodland areas.

Life History. *Behavior.* Males select perching sites in forest openings on ridge tops, where they perch on small trees or shrubs from 0830–1730 hr. Females fly to the perching sites and mating occurs after a short, fluttering courtship flight. Mated pairs were seen from 1230 to 1600 hr. Males do not interact with male Black Swallowtails encountered along road edges through their habitat. Females refuse to lay eggs on carrot, dill, or parsnip, plants readily accepted as oviposition sites by Black Swallowtails (Heitzman 1973).

Broods. This swallowtail is bivoltine. The first flight extends from April 7 to June 11 and there are one or two smaller emergences until early September.

Early Stages. Early stages are similar to those of the Black Swallowtail. The larvae have a bright, light-green or an occasional bright, light-blue color form. The segmental spots are yellow orange, yellow, or white (not bright orange), and the segmental bands are broken on most individuals.

Adult Nectar Sources. Adult Missouri Woodland Swallowtails nectar at rose verbana, wood betony, pucoon, and false garlic.

Caterpillar Host Plants. To date, yellow pimpernel (*Taenidia integerrima*), meadow parsnip (*Thaspium barbinode*), and golden Alexander (*Zizia aurea*), all members of the family Apiaceae, are the only known hosts.

5. *Papilio joanae*

Giant Swallowtail
Papilio cresphontes Cramer, 1777

PLATE 4 · FIGURE 19 · MAP 6

Etymology. The species is named after Cresphonte, one of the Heraclids, persons who claimed descent from Hercules (Greek myth).

Synopsis. The Giant Swallowtail averages larger than other eastern swallowtails, although some female Tiger Swallowtails from Florida are larger than the largest Giant. The caterpillars are a minor pest of citrus and are called the Orange Dog in such situations.

Butterfly. Male forewing: $\overline{X}$ = 5.5 cm, range 4.6–6.9 cm;

female forewing: $\overline{X}$ = 5.8 cm, range 5.3–7.4. With its diagonal yellow bar crossing the dorsal forewing, and its predominantly yellow ventral wing surfaces, the Giant Swallowtail may be easily confused only with the rare Schaus' Swallowtail (*Papilio aristodemus ponceanus*). It may be distinguished from the latter by its tails, which are filled with yellow (they are completely black in the Schaus'), and the small, brick-red patch just interior to the blue median band on the ventral hind wing (the Schaus' has median, chestnut-brown band interior to postmedian band of blue lunules). First-brood individuals are smaller than those of the second.

Range. This species' range extends from southern New England west across the southern portion of the Great Lakes states and the central plains to the Rocky Mountain front. It ranges south throughout the remainder of the East, the Southwest, and on through the Caribbean, Mexico, and Central America to South America. In the North, it occurs in very localized colonies and is a fairly rare butterfly. It is more abundant in the South.

Habitat. The Giant Swallowtail inhabits a variety of semi-open or brushy situations, ranging from rocky or sandy exposed hillsides above streams in the North, to pine flat-woods, towns, and citrus groves in the Deep South. It occurs from Tropical through Transition Life Zones.

Life History. *Behavior.* Males patrol flyways along river bluffs, forest edges, or through pine woods or citrus groves. The flight is strong, yet leisurely, and they may glide long distances between wing beats. Courtship and mating take place in the afternoon. A courting pair rose 10 m in the air. Females have been observed ovipositing in the morning (1000 hr) about 2 m above ground. Small flights of emi-grants have been observed on occasion (Saverner 1908). Giant Swallowtails are often seen taking moisture at moist mud or sand.

Broods. The Giant Swallowtail is definitely bivoltine, having two clearly separated generations wherever it occurs in the eastern United States, although there may be three in Florida. As one proceeds south, the butterfly emerges earlier in the year, and there may be a greater hiatus between flights. In Illinois, the flights are from mid-May through June and from late July through August. In Georgia, the first brood occurs in April and the more numerous second brood occurs from late July through mid-September. In Florida, the Giant Swallowtail may be found throughout the year.

Early Stages. The young caterpillars rest on leaves, while the later instars rest on small branches. Most feeding occurs at night. Each instar is differently patterned, but all have dis-ruptive patterns reminiscent of bird droppings. The first

instar is predominantly black with a white saddle, while the mature caterpillar is mottled dark brown, with the posterior white or cream colored. There is a median white or cream patch composed of two adjoining lateral triangles, and a white or cream band almost encircling the lateral portion of the thorax. The osmateria vary from pale orange to wine red and release a pungent odor.

Adult Nectar Sources. Few nectar sources have been reported in the eastern United States. Those that have include swamp milkweed, Japanese honeysuckle, goldenrod, dame's rocket, bouncing Bet, azalea, and bougainvilla.

Caterpillar Host Plants. In the northern part of its range, prickly ash (*Zanthoxylum americanum*) is the Giant Swallowtail's principal host plant, although hop tree (*Ptelea trifoliata*) has been used in Illinois. Hercules club (*Zanthoxylum clavaherculis*) is also eaten with some frequency, especially farther south, and citrus species and torchwood (*Amyris elemifera*) are fed upon in Florida. Other trees in the citrus family are known to be hosts.

6. *Papilio cresphontes*

Schaus' Swallowtail
Papilio aristodemus ponceanus Schaus, 1911
PLATE 4 · FIGURE 20 · MAP 7

Etymology. The butterfly is named after Aristodemus, a king and hero in the war at Messenia (716 B.C.). The subspecies is named after Ponce de Léon, a Spanish statesman and explorer who sought the Fountain of Youth in Florida.

Synopsis. The Schaus' Swallowtail may be close to extinction in its native haunts on Florida's Key Largo and Elliot Key (Covell 1977). It is one of the few species endemic to the Caribbean area that has a distinct subspecies limited to the Florida Keys.

Butterfly. Male forewing: $\overline{X}$ = 4.7 cm, range 4.1–5.4 cm; female forewing: $\overline{X}$ = 5.1 cm, range 5.0–5.2 cm. The characteristics that distinguish this very rare butterfly are mentioned in the Giant Swallowtail account. The male's antennal clubs are yellow, while the female's antennae are entirely black.

Range. Formerly, the Schaus' Swallowtail ranged from Miami to Lower Matecumbe Key. Due to the developments in the Keys, the butterfly now may be found only on North Key Largo and the larger Keys within Biscayne National

Monument. Other subspecies occur on several of the Greater Antilles.

Habitat. The Schaus' Swallowtail occurs only within tropical hardwood hammocks and associated scrub areas.

Life History. *Behavior.* Adult males patrol by circling slowly in hardwood hammock clearings at 2–6 m above ground level. One courtship was reported in the morning (1015 hr), during which the female perched on the ground and the male hovered over her. Oviposition was also observed in late morning (1000–1200 hr). Females usually lay a single egg on top of young, expanding host leaves up to 2 m above ground; shaded plants are usually selected (Rutkowski 1971).

Broods. Schaus' Swallowtail is univoltine, with a single spring flight each year (April 30–June 21), although occasional adults have been seen in late July and early September. Emergence is triggered by rainfall, and pupae may pass one or two years before producing adults.

Early Stages. Development from egg to chrysalid requires about 5 to 6 weeks. The egg is pale green and spherical. Young larvae are glistening black, with anterior and posterior white saddlemarks, together with a white patch on each side near the middle of the body. The mature caterpillar has a broad, arcuate, pure white hood at the posterior end of the body and a series of about six lateral white and yellow patches. Its venter and prolegs are white as well. The remainder of the body is a rich brown. The stout chrysalid is dark brown and mottled.

Adult Nectar Sources. Schaus' Swallowtails have been seen nectaring at flowers of guava, cheese-shrub, and wild coffee, all of which are white-flowered woody plants.

Caterpillar Host Plants. The primary larval host is torchwood (*Amyris elemifera*), and wild lime (*Zanthoxylum fagara*) is also used on occasion. Both plants are members of the citrus family.

Bahaman Swallowtail
Papilio andraemon bonhotei (Sharpe), 1900
PLATE 5 · FIGURE 25 · MAP 8

Etymology. The species is named after Andraemon, father of Amphissus and husband of Dryope, who was changed into a lotus (Ovidius). The subspecies is named in honor of John Lewis James Bonhote (1875–1922), an English citizen who wrote a book on the birds of Britain.

Synopsis. Although the Bahaman Swallowtail is protected by the U.S. Endangered Species Act as a threatened resident species, it now seems that it is no more than a periodic colonist. Its permanent home is in the Bahamas.

Butterfly. The Bahaman Swallowtail may be distinguished

from the Giant and Schaus' Swallowtails, with which it may occur, by the uniform postmedian yellow bands on both wings and its relatively long tails.

Range. The full species occurs on Cuba, Jamaica, and rarely in the Florida Keys as a temporary colonist.

Life History. *Broods.* This swallowtail flies from at least April to October on the Bahama Islands and probably has at least three generations during the year.

Early Stages. The mature caterpillar is dark olive green to

7. *Papilio aristodemus ponceanus*

8. *Papilio andraemon bonhotei*

black with white saddlemarks on the anterior and posterior portions. There are two rows of blue spots along the back, and the venter and prolegs are lilac.

Caterpillar Host Plants. The Bahaman Swallowtail selects *Citrus, Ruta,* and *Zanthoxylum,* all from the citrus family, as caterpillar hosts.

Androgeus Swallowtail
Papilio androgeus epidaurus (Godman & Salvin), 1890
MAP 9

Etymology. Androgeus was the son of the Cretan King Minoas and his wife Pasifae. The subspecies *epidaurus* is named after a seaport in ancient Greece. This settlement is now called Pidarro.

Synopsis. This tropical swallowtail became established in the citrus groves of south Florida in 1976. The two sexes are dissimilar, the male largely yellow and the female largely black, with a large green hind-wing patch.

Butterfly. The male has elongated forewings, as in the Giant Swallowtail, but dorsally has only a single broad yellow band on each wing (those of the Bahama Swallowtail are narrower). The tails are relatively short and narrow. Above, the female is black, with a central small yellow patch on the forewing, and a large iridescent green patch occupying most of the hind wing. The normal tail is lacking, but three pointed projections occur along the hind-wing outer margin.

Range. The Androgeus Swallowtail occurs from Mexico south to Argentina and on the larger Caribbean Islands. The small south-Florida population was probably established by a female stray from Cuba.

Habitat. In south Florida, the species is found in overgrown citrus groves, but in its native haunts the species occurs in tropical forest.

Life History. *Behavior.* Males have a rapid, erratic patrolling flight in the canopy of their forest habitat and occasionally swoop down into sunlit clearings. Freshly emerged males may visit moist patches along roads or streams.
Early Stages. Early stages have been partially described by Ross (1964). The second-instar larva resembles a fresh bird dropping, being glossy tan brown with median and posterior

dorsal white patches. The final instar is more disruptively patterned, being dark gray to black with numerous irregular white markings and slightly raised tubercles on each segment that are marked with blue at their bases.

Caterpillar Host Plants. Larval hosts are orange (*Citrus sinensis*) and *Zanthoxylum elephantiasis.*

9. *Papilio androgeus epidaurus*

Tiger Swallowtail
Papilio glaucus Linnaeus, 1758
PLATE 3 · FIGURES 16–18 · MAP 10

Etymology. The species name is probably derived from the Latin *glaucus* "blue gray." Glaucus is also the name of several mythological characters.

Synopsis. This large, yellow, black-striped butterfly is one of the most familiar and widespread in North America. Notable features include its dimorphic female and a diminutive northern or spring form. It is believed to be the first

North American insect to have been portrayed: A recognizable drawing was sent back to England after Sir Walter Raleigh's third expedition to Virginia (see Introduction).

Butterfly. Male forewing: $\bar{X} = 5.1$ cm, range 4.0–6.5 cm; female forewing: $\bar{X} = 5.4$ cm, range 4.0–7.6 cm. Normal males and females are yellow with three vertical black stripes on each forewing and a broad black border on the outer margin of all wings. This black border contains a row of yellow spots. Normal females tend to be somewhat larger, to have broader black stripes, and to have extensive patches of iridescent blue scales in the black border of the hind wing. The dark, mimetic form of the female is largely black dorsally and ventrally save for the marginal row of spots, which are yellow above and orange below.

Both sexes of the spring form, which are close to individuals of the northern subspecies *P.g. canadensis* in size and appearance, are diminutive, have the relatively broad black stripes, and have more extensive orange scaling on the ventral hind wing. On the Georgia coastal plain and in peninsular Florida, individuals are larger and more orange than most northern *glaucus* and the subspecies name *australis* has been applied. Clark and Clark (1951), in their treatment of Virginia butterflies, provide a lengthy discussion of this species' variation, and Mather (1954) describes size variation in Mississippi.

The black female form is a presumed Batesian mimic of the Pipe Vine Swallowtail (*Battus philenor*) and the reader should refer to the discussion under that species. The black female is absent or infrequent in New England and peninsular Florida, areas where the Pipe Vine Swallowtail is also rare. The black forms are most frequent (virtually 100%) in areas where the Pipe Vine Swallowtail reaches its greatest abundance, i.e., the Carolinas. It has been shown elegantly by Burns (1966b) that normal yellow females are preferred as mates by males, but that black females are attacked by birds at a significantly lower rate. Thus, both forms are maintained in the population by different selective advantages. This is known as a *balanced polymorphism*.

Range. The Tiger Swallowtail ranges from the Maritime Provinces west and north through Canada to central Alaska. It extends south, keeping east of the continental divide, throughout our entire area to southern Florida and the Gulf Coast.

Habitat. This species occurs almost everywhere that there are deciduous woods, including towns and cities. It is most numerous, however, in woods along streams and rivers, as well as wooded swamps. It occurs in all Life Zones from the Tropical to the Canadian.

Life History. *Behavior*. The Tiger Swallowtail, being a true treetop species, is on occasion a "high flier." Individuals have been seen in numbers flying around the tops of tulip poplars (*Liriodendron*) more than 50 m above ground in a virgin deciduous wood in Maryland. The males patrol particular routes, often along streams, rivers, or woodland roads in search of receptive females. Mating and associated courtships may be seen in late afternoon, while oviposition takes place near midday. Newly emerged males of each generation seek out mud puddles or stream edges, where they congregate to imbibe water and mineral ions from moist sandy spots (Arms et al. 1974). Adults live for at least 12 days (Fales 1959).

Broods. In the northern part of its range the Tiger Swallowtail is strictly univoltine (*canadensis*). For example, in Maine, extreme flight dates range from May 2 to July 16. Farther south it is bivoltine, as in southern New York (May 17 to early July and mid-July to September 6). In the Deep South, there are three broods as flight dates range from mid-February or early March to mid-October or November. The fact that chrysalids of any generation may undergo diapause confuses matters (Clark and Clark 1951).

Early Stages. The first-instar larva is black with a white saddle, while the mature larva is predominantly dark green with two large eyespots on the swollen thoracic area. The chrysalis may be either brown and tan or green. Most feeding occurs at night, and the mature caterpillar rests on a silken pad within a curled leaf shelter. The chrysalis is attached to a limb or tree trunk and winter is passed in this stage.

Adult Nectar Plants. Although Tiger Swallowtails will occasionally feed on flowers near ground level, they prefer those 1.5 m high or higher. Especially favored are common and swamp milkweeds, Japanese honeysuckle, buttonbush, joe-pye weed, ironweed, and thistles. Among cultivated plants, lilac, butterflybush, and Abelia are sought avidly. Many other plants are also visited.

Caterpillar Host Plants. The Tiger Swallowtail female selects a variety of trees and occasionally shrubs as hosts (L.P. Brower 1958). These include wild cherry (*Prunus serotina* and *P. virginiana*), sweet bay (*Magnolia virginiana*), tulip poplar (*Liriodendron tulipifera*), ash (*Fraxinus nigra* and *F. americana*), hop tree (*Ptelea trifoliata*), *Carpinus caroliniana*, spicebush (*Lindera benzoin*), and lilac (*Syringa vulgaris*).

10. *Papilio glaucus*

Birch (*Betula*) and aspen (*Populus*) are utilized by more northern populations, whose caterpillars cannot survive on tulip poplar (Scriber et al. 1982). Several reported hosts, such as camphor tree and catalpa, require confirmation. Females will lay eggs on peach even though the leaves are toxic to the larvae.

Spicebush Swallowtail
Papilio troilus Linnaeus, 1758
PLATE 4 · FIGURES 21–24 · MAP 11

Etymology. Troilus was the son of Priam, King of Troy (Homerian epic, *Iliad*).

Synopsis. Within the swallowtails, this species is most closely related to the Tiger and Palamedes Swallowtails, as is indicated by their similar larvae, despite the fact that their adults are dissimilar. The Spicebush is a palatable mimic of the Pipe Vine Swallowtail.

Butterfly. Male forewing: $\bar{X}$ = 4.9 cm, range 4.1–5.5 cm; female forewing: $\bar{X}$ = 4.9 cm, range 4.4–5.6 cm. Both sexes of the Spicebush may be distinguished from the Pipe Vine, female Black, and dark female Tiger Swallowtails, which they superficially resemble, by several characters. There is a round, red-orange spot on the costal margin of the dorsal hind wing that is lacking in both the Pipe Vine and female Black Swallowtails, and a postmedial row of discrete orange spots on the ventral hind wing that is not present on any of the others. The males may be distinguished in flight by the median blue-green band on the dorsal hind wing, but the female must be observed closely. In some individuals the male's hind-wing band may be more blue than green. A form that has the submarginal spots greatly enlarged is most prevalent in the southern states, but it is not a geographically discrete unit and is not properly referred to as a subspecies.

Range. The Spicebush Swallowtail occurs from central New England west across southern Canada to southern Manitoba, thence south through the eastern states to Florida, the Gulf Coast, and Texas. The species becomes progressively rare west of the Mississippi River. Occasional strays have been found in Cuba.

Habitat. This butterfly is found in second-growth woodlands, open fields near woods, bottomland woods (including swamp forest), pine barrens, and other semi-open situations in which suitable hosts occur.

Life History. *Behavior.* The males patrol woodland edges, open woods, along roads, and in glades in search of receptive females. Courtship has been observed in the afternoon (1230–1530 hr), and often involves a slow, hovering, upward flight with the female rising and the male hovering around her. Mated pairs have also been observed in the afternoon (1240–1520 hr); the female is the carrying partner when the pair takes flight. Females oviposit in late morning (910–1230 hr), usually laying a single egg on each of several young host leaves. Young trees are usually selected, and eggs are placed 2–5 m above ground.

The normal flight is usually low (1–3 m), direct, and fairly rapid. The species seems to disperse widely, without any special home territory. Freshly emerged males sometimes may be seen in groups at moist spots near streams or along roads.

Broods. In the northern portion of its range, two generations are usual; for example, in New York's Finger Lakes region, flight extremes are May 10 to September 22 (Shapiro 1974). Farther south, although the yearly range of flight dates becomes broader, there are believed to be no more than two and a partial third generation. Virginia is a fairly typical example, where there are flights in mid-April to mid-June, late June to late August, and mid-September to mid-October. Some chrysalids from each generation overwinter, and the adult emergences are usually largely staggered.

Early Stages. The first-instar larva folds the edge of a host leaf over to form a shelter. Subsequently, the caterpillar spins a silk mat on the center of a host leaf that is folded up on both sides to form a shelter during the day as well as during molts, and from which the caterpillar ventures to feed at night.

11. *Papilio troilus*

The first-instar caterpillar is a bird-dropping mimic, being predominantly black with a white saddlemark. The second instar becomes yellowish brown, with more extensive white splotches, tiny dorsal thoracic eyespots, and two dorsal rows of tiny metallic blue spots. The third instar is similar, but the eyespots are more developed. The fourth instar is greenish above and whitish dorsolaterally. There are two black, orange-edged eyespots on the thoracic dorsum, with two oval orange patches just behind them. There are four lines of bluish black spots on the dorsal portion of the abdominal segments. The mature larva is similar, being dark green above, with the eyespots and patches enlarged, and the abdominal spots becoming blue spots encircled with black. The chrysalis is fairly smooth, and has a slight but distinct ridge along either side. Either green or mottled gray and brown forms occur.

Adult Nectar Sources. The Spicebush Swallowtail has a very long proboscis relative to its size, and it can therefore take advantage of some long-tubed flowers not available to most true butterflies. Japanese honeysuckle and jewelweed are two such plants that are visited frequently. In addition, milkweeds are often visited in the midsummer, while thistles are relied upon in the fall. Other species visited with some frequency are dogbane, mimosa, azalea, lantana, and sweet pepperbush.

Caterpillar Host Plants. Young sassafras trees (*Sassafras albidum*) are utilized most frequently, but spicebush (*Lindera benzoin*) is also selected with some frequency. Other reported hosts are tulip poplar (*Liriodendron tulipifera*), camphor, sweet bay (*Mangolia virginiana*), and prickly ash (*Zanthoxylum americanum*); most of these require confirmation.

Palamedes Swallowtail
Papilio palamedes (Drury), 1773
PLATE 5 · FIGURES 26 & 27 · MAP 12

Etymology. Palamedes was the son of Nauplius, King of Euboia (myth).

Synopsis. The Palamedes Swallowtail is a species of the lowland coastal plains and lower Mississippi Valley of the southeastern states. The sexes are similar. The butterflies fade from black to brown through time in museum collections. This is the only eastern butterfly that feeds on plants of the Laurel family.

Butterfly. Male forewing: $\bar{X}$ = 5.4 cm, range 5.0–5.6 cm; female forewing: $\bar{X}$ = 5.5 cm, range 5.0–6.0 cm. The Palamedes should be confused with no other eastern swallowtail. Dorsally, the basal halves of both wings are solid black and there is a broad yellow postmedian band on each hind wing. Ventrally, there is a yellow postbasal stripe and a postmedial band of yellow crescents on the hind wing. This large butterfly seems to show little geographic variation. First-brood individuals are smaller, and Virginia specimens are smaller than those from farther south. The butterfly occasionally has local migrations.

Range. This swallowtail ranges south along the fairly immediate Atlantic coastal plain and coastal marshes from southern New Jersey (rarely) to Florida, thence westward along the Gulf Coast to northern Veracruz, Mexico. The species also ranges north into the lower Mississippi River drainage.

Habitat. The Palamedes Swallowtail is found most abundantly in broadleaf evergreen swamp forests and wet woods near rivers. Its two hosts, red bay (*Persea borbonica*) and sweet bay (*Magnolia virginiana*), are common in both situations.

Life History. *Behavior.* Males patrol through woods, along streams, and at forest edges for much of the day. Oviposition occurs near midday, and other reproductive activities have not been reported. Females seldom appear in open areas. When freshly emerged, the males may congregate by the hundreds on wet sand.

Broods. In the northern part of its range, the Palamedes Swallowtail is bivoltine. For example, in Virginia the butterfly appears from late May through June and from mid-July to early October. Farther south, in Georgia, Louisiana, Mississippi, and Florida, the species first appears in early March and may fly as late as November or December. In these states there are probably three generations, although the last may be only partial.

12. *Papilio palamedes*

Early Stages. The mature caterpillar is pale green, with the first abdominal and the last two thoracic segments enlarged to give it a hump-backed look. There is a pair of black, orange-margined eyespots on the third thoracic segment.

Adult Nectar Sources. Sweet Pepperbush and thistles are major adult food sources, but azalea and blue flag are also favored with periodic visits.

Caterpillar Host Plants. By far the major larval host for the Palamedes Swallowtail is red bay. Sweet bay is utilized less frequently (Brooks 1962).

Family Pieridae

The Pieridae is a worldwide family that is most diverse at tropical latitudes. The butterflies are usually small to medium-sized, and are predominately white, yellow, or orange, often with some black or occasionally pink or red scaling. The white, yellow, red, and orange colors are due to pteridine pigments, which fluoresce in ultraviolet light. Males of many species have androconial scales, often in the forewing discal area.

All species in our area of interest use the patrolling mate-location strategy, although males of some tropical groups are perchers. Males patrol and court all day, but mating usually occurs in the afternoon. Females of most species oviposit near midday. All our species are lateral baskers, and some whites occasionally perch with partially opened wings.

Adults feed entirely on floral nectar, and freshly emerged males of many species often take moisture at wet sand or mud.

Larval hosts in our area are usually crucifers and legumes, but capers, composites, heaths, or willows are eaten by some. Eggs are usually laid singly on hosts and are columnar, being much taller than they are wide. The caterpillars do not build shelters and feed on host leaves, flowers, or fruits. The larvae are often green, with lateral and dorsal white or yellow lines. Their cylindrical bodies often have many small tubercles that bear many short hairs. The chrysalid is slung from a silken girdle, as in swallowtails, and in most species it has an anterior conical projection. Most species are multiple-brooded, but a few have only a single annual flight. The chrysalid or partially grown larva overwinters in more temperate species, while pierids of tropical affinities overwinter as adults in reproductive diapause.

Subfamily Pierinae: Whites

The whites are medium-sized butterflies that are predominately white with some black marks. Yellow scaling appears on the ventral hind wings of some and may appear green when intermixed with black scales. The male Falcate Orange-tip has an orange forewing apex, although sexual dimorphism is usually less pronounced in whites than in sulphurs.

Caterpillar hosts are usually crucifers, although capers are selected by some. The Pine White (*Neophasia menapia*), an occasional stray from the west, feeds on various conifers.

Pine White
Neophasia menapia (Felder & Felder), 1859
MAP 13

Synopsis. The Pine White is found in western conifer forests. Occasionally it has outbreaks that inflict serious defoliation on stands of yellow pine (*Pinus ponderosa*). The Pine White can be confused with no other eastern butterfly. It has a black band along the forewing costa and black apical markings similar to those of the marbles (*Euchloe*), but, of course, lacks the green marbling below. The Pine White was found once in Minnesota, and ranges west to southern Alberta and southern British Columbia, whence it ranges south to central New Mexico, northern Arizona, and central California.

54

Florida White
Appias drusilla neumoegenii (Skinner), 1894

PLATE 5 · FIGURE 28 · MAP 14

Etymology. The genus name is derived from an epithet for the nymph at the fountain of Aqua Appia, Italy. The subspecies was named in honor of Berthold Neumoegen (1845–1895), a professional banker and broker, who emigrated from Germany to New York. He was an amateur who collected in the West Indies and Arizona, while being very active in the Brooklyn Entomological Society.

Synopsis. The sexually dimorphic Florida White is shade-loving and shuns the sun except when engaged in migratory movements. It is the only eastern U.S. butterfly whose caterpillars feed exclusively on plants of the caper family.

Butterfly. Male forewing: $\overline{X}$ = 3.1 cm, range 2.4–3.4 cm; female forewing: $\overline{X}$ = 3.1 cm, range 2.9–3.1 cm. The male is solid white above and below save for a thin black edging of the forewing costal margin. The ventral hind wing has a satiny sheen. No other eastern U.S. butterfly is pure white dorsally, except for the male Mustard White (*Pieris napi oleracea*), which does not occur in the South. The females may be all-white (dry-season form) or have a black outer margin on the forewing and a yellow-orange hind wing (wet-season form). The male's forewing apex is more pointed than the female's, which is rounded.

Range. The Florida White and its related subspecies range from southern Florida and southern Texas south through the Antilles, Mexico, and Central America to Brazil. Rarely, the species wanders northward, since there are also records for Massachusetts, New York, Maryland, and Arkansas in our area. These may have been hurricane-blown waifs.

Habitat. The Florida White is restricted to the shaded understory of evergreen hardwood forests, except during the early dry season (October–November) when it may be emigratory. Farther south, in Latin America, the species is found in subtropical and tropical lowland forests, which have a distinct dry season; there, the butterfly may be found in riparian evergreen woodland.

Life History. *Behavior.* The males patrol actively, with a rapid, erratic flight, darting from one small forest clearing to another in search of receptive females. The adults may fly high in the canopy in search of nectar. Unlike most other pierids, adult males do not visit moist soil.

Broods. The Florida White probably has two or three generations: the reproductive wet-season form flies in the Florida summer, and the long-lived, migratory, dry-season form emerges in late fall and survives the Florida winter, reproducing in the spring. In Florida, the butterflies have been seen during every month except November (Kimball 1965).

Early Stages. Single eggs are laid at the tips of young, expanding host leaves. Development is direct. The caterpillar is dark green dorsally, and paler gray-green laterally, the two tones being divided by a thin white lateral line. The head is yellow-green and there are many tubercles (chalazae) bearing short yellow or black setae (Chermock and Chermock 1947).

13. *Neophasia menapia*

14. *Appias drusilla neumoegenii*

Adult Nectar Sources. Nectar sources have not been reported in Florida, but in Costa Rica the adults favor flowering trees and lianas, especially *Cordia* species and *Cissus* (grape family).

Caterpillar Host Plants. The larval host in Florida is Guinea plum (*Capparis lateriflora*). Elsewhere in its range other *Capparis* species are selected.

Checkered White
Pontia protodice (Boisduval & LeConte), 1829
PLATE 5 · FIGURES 29 & 30 · MAP 15

Etymology. The genus name refers to an area along the western edge of the Black Sea.

Synopsis. This is the common *Pontia* in the eastern United States, and it is reputed to have declined in abundance subsequent to the introduction of the European Cabbage Butterfly (*Pieris rapae*).

Butterfly. Male forewing: $\overline{X} = 2.3$ cm, range 2.2–2.6 cm; female forewing: $\overline{X} = 2.4$ cm, range 2.2–2.6 cm. Both sexes have a quadrate black mark in the dorsal forewing cell. The summer-form male has one to three black marks in the postmedian forewing area, and its ventral hind wing is weakly veined with gray-green. The summer female is boldly marked dorsally, with much black on the forewing apex and outer margins of both wings. Ventrally, the gray-green veining is much more distinct. The spring form ("vernalis") has strongly reduced black marks above but has very heavy green veining ventrally. The determination of these forms has been elucidated by Shapiro (1968), who showed that larvae subjected to long nights (greater than 14 hr) developed into spring-form adults. Shapiro also demonstrated the differential seasonal fitness of laboratory-reared adults of both forms released in the field simultaneously.

Range. The Checkered White occurs from central New England west to Montana and southern British Columbia thence south to southern Florida, the Gulf Coast, and Central Mexico. Occasional strays occur in Maine and southern Canada.

Habitat. Open, relatively dry sites are the preferred habitat of the Checkered White. In the East these include dry upland pastures, fallow corn fields, railroad tracks, sandy lots, and disturbed beach dunes. The species is at home in the Lower and Upper Austral Life Zones.

Life History. *Behavior.* Males patrol over flats or ridges in search of receptive females. Rutowski (1979) has studied courtship behavior in great detail. Successful courtship leading to mating is rapid, as in most pierids taking place in about 3 seconds. The male often carries the female on a post-nuptial flight, so that mating and spermatophore transfer may take place without interference from other males. Mated pairs have been observed from 0800–1400 hr, with most seen between 1000 and 1200 hr. Females may mate more than once. Virgin females or previously mated females with reduced spermatophores may actually flutter around males as a solicitation to mating (Rutowski 1980).

Spring individuals fly more weakly and visit flowers less often than summer-brood individuals (Clark 1932).

Broods. The Checkered White has three and sometimes a partial fourth generation. In the lower Delaware Valley, the broods occur from May to mid-June, in late July to early August, and September to mid-October. Farther south, in Georgia, the first generation flies in March and April, the second from the end of May to late July, and the third from mid-August to the end of September. Occasional individuals emerge during warm spells in the late fall, e.g., November 30 in Michigan.

Early Stages. Development is direct in all stages, and the pupae of the last brood overwinter. The larva is alternately striped yellow and purple-green. There are a number of setiferous tubercles over most of the body.

Adult Nectar Sources. Adults often visit host-plant flowers for nectar, especially those of winter cress and hedge mustard. They are also fond of centaury and milkweed flowers.

15. *Pontia protodice*

Caterpillar Host Plants. In the eastern states, wild peppergrasses (*Lepidium densiflorum* and *L. virginicum*) are preferred, but the adults also oviposit on shepherd's purse (*Capsella bursa-pastoris*), winter cress (*Barbarea vulgaris*), and a variety of other crucifers. An old record of fleabane (*Erigeron*) being utilized is highly suspect.

Western White
Pontia occidentalis (Reakirt), 1866

Synopsis. This butterfly of high western mountains has been found twice in Minnesota (R. L. Huber, personal communication). It is closely related to the Checkered White (*Pontia protodice*), but may be distinguished by its more regular markings, its more continuous submarginal black band on the dorsal forewing, and its gray-green veining on the ventral hind wing. Like the Checkered White, it has two seasonal forms. It is probably bivoltine in areas just to the north of Minnesota. Its caterpillars, which are similar to those of the Checkered White, feed on the flowers and seed pods of various plants in the mustard family.

Mustard White
Pieris napi oleracea Harris, 1829
PLATE 6 · FIGURE 31 · MAP 16

Etymology. The species name is derived from *napus*, which is Latin for "yellow turnip."

Synopsis. The Mustard White belongs to a diverse group of butterflies, the *Pieris napi* complex. These butterflies are extremely variable, have different biologies, and occupy diverse habitats throughout much of North America and Eurasia.

Butterfly. Male forewing: $\overline{X}$ = 2.2 cm, range 2.0–2.6 cm; female forewing: $\overline{X}$ = 2.3 cm. The Mustard White has two seasonal forms. The spring-brood butterflies have the ventral surface of the hind wings veined with gray-brown and gray-green, while the summer-generation individuals are immaculate. As in other seasonally dimorphic butterflies, the spring form results from caterpillars that develop during short photoperiods (Oliver 1970). In Michigan, the Mustard White becomes progressively smaller as it approaches its southern range limit; this is in opposition to the general rule that a given butterfly species becomes larger as one goes south (Wagner 1980).

This species may be distinguished from the European Cabbage Butterfly (*Pieris rapae*) by its more fragile-appearing wings, its lack of forewing spots, and its veining.

Range. Members of the *Pieris napi* complex are widespread in both Eurasia and North America, where the species ranges from New England across the Great Lakes area to the Rocky Mountains and central California. It ranges north through much of Canada to Alaska and is the only butterfly found on the Aleutian Islands.

Habitat. In the eastern states, the Mustard White is found in deciduous woods, bogs, and open fields. It is usually confined to wooded habitats in the spring. It is found in Transition and Canadian Life Zones in the East, but is found up to Arctic Alpine situations in Alaska. Before the introduction of the European Cabbage Butterfly it has been reported that the Mustard White occurred more widely in open areas.

Life History. *Behavior.* Male Mustard Whites patrol during most warm daylight hours in search of receptive females. Mating occurs from 1055 to 1730 hr. The flight is more rapid than the West Virginia White's, but is closer to the ground than that of the European Cabbage Butterfly. The spring-

16. *Pieris napi oleracea*

brood individuals restrict their activities more to woodland, and the later broods are found more in open areas.

Broods. The Mustard White has two full generations and sometimes a partial third; most pupae from the second generation overwinter. In Maine, the broods extend from May 29 to June 19 and from July 14 to September 16. In southern Michigan they fly as early as April 23.

Early Stages. Females lay single, pale-green eggs on the underside of host leaves, upon which the caterpillars feed. Development to the chrysalis is direct, and most chrysalids resulting from the second and all from the third brood enter diapause and overwinter. The mature caterpillar is green and covered with short, fine setae. There are a dorsal stripe and a lateral yellow-green stripe on each side. The chrysalis may be either green or tan flecked with black. The frontal projection is longer than that of the European Cabbage Butterfly, about one-seventh of the larva's total length. In the green form there may be a yellow dorsal line, and the frontal and lateral projections may be yellow-tipped as well.

Adult Nectar Sources. Mustard White adults favor the flowers of their cruciferous hosts, but also seek nectar from other plants on occasion.

Caterpillar Host Plants. Female Mustard Whites select and their caterpillars feed upon toothwort (*Dentaria diphylla*), rock cress (*Arabis*), water cress (*Nasturtium officinale*), mustard (*Brassica rapa*), and winter cresses (*Barbarea orthoceras* and *B. vulgaris*). Chew (1977) demonstrated that Mustard White caterpillars could not complete development on *Barbarea vulgaris*. In an intensive study in northern Vermont, Chew et al. (1977) demonstrated that Mustard Whites utilize different hosts in different subsites and at different times at the same site.

West Virginia White
Pieris virginiensis Edwards, 1870
PLATE 6 · FIGURE 32 · MAP 17

Etymology. The species name is derived from Virginia. The butterfly was originally described from what was then Virginia but is now West Virginia.

Synopsis. The West Virginia White is a close relative of the Mustard White (*Pieris napi oleracea*) but is restricted to moist deciduous woods, especially maple, and has only a single flight each spring.

Butterfly. Male forewing: $\overline{X}$ = 2.1 cm, range 2.0–2.3 cm; female forewing: $\overline{X}$ = 2.2 cm, range 2.0–2.4 cm. Although some have considered the West Virginia White to be a race of the Mustard White (*Pieris napi oleracea*), the species do not intergrade in their overlap zone in New England, although they occasionally mate with each other (Chew 1980). The West Virginia White may be distinguished by its semi-translucent wings, blurred veining on ventral hind wings, and lack of yellow tinting beneath.

The more northern populations (Great Lakes and New England) have individuals with darker veining ventrally than do those of the southern Appalachians, whose pale-gray veining barely contrasts with the general white wing color (Wagner 1978).

Distribution. The West Virginia White ranges from north central Wisconsin east through northern Michigan, southern Ontario, and New York to Vermont, Massachusetts, and Connecticut. It then extends southwest through the southern Appalachians to northern Georgia and Alabama.

Habitat. The species is restricted to moist deciduous or mixed woodland, usually in the Transition Life Zone. Most often maple and beech are elements of these forests, but on rare occasions it may be found in riparian woodland dominated by box elder.

Life History. *Behavior.* Males patrol through woodlands with a weak, low flight, although they may take off with a erratic, rapid flight if they sense danger. Mating occurs from midday to early afternoon (1320–1600 hr), with males acting as carriers. Females will accept Mustard White males as mates on rare occasions (Chew 1980). Single eggs are laid beneath young host leaves. Newly emerged males seek

17. *Pieris virginiensis*

moisture near streams or the damp margins of woodland roads.

Broods. The West Virginia White is strictly univoltine, with a flight in the spring. At the southern terminus of its range the butterfly flies from March 26 to May 15; at the northern extreme it flies from April 26 to June 19.

Early Stages. The caterpillar completes feeding by early summer, and the chrysalis undergoes diapause until the following spring. The mature caterpillar is yellow-green with a narrow dorsal stripe and two green lateral stripes. The frontal projection of the chrysalis is long and slender.

Adult Nectar Sources. The adults most often take nectar from flowers of their caterpillar host, toothwort, but will also sip from others such as white wake-robin, Canada violet, and garlic mustard.

Caterpillar Host Plants. Toothworts (*Dentaria diphylla* and *D. laciniata*) are the only known hosts.

European Cabbage Butterfly
Pieris rapae (Linnaeus), 1758
PLATE 6 · FIGURES 33 & 34 · MAP 18

Etymology. The genus name is derived from Pieris, a Greek muse. The species name is derived from *rapa*, Latin for turnip.

Synopsis. The European Cabbage butterfly was accidently introduced into Quebec, Canada, about 1860 (Scudder 1887). Thereafter, it spread rapidly and now occupies most of North America. It flies from early spring and is certainly the most ubiquitous species in our area. Next to the Orange Sulphur (*Colias eurytheme*), this butterfly causes the most serious economic damage to crops and home gardens, particularly to cabbage, Brussels sprouts, and cauliflower.

Butterfly. Male forewing: $\overline{X}$ = 2.4 cm, range 2.1–2.6 cm; female forewing: $\overline{X}$ = 2.3 cm, range 2.0–2.5 cm. The European Cabbage Butterfly may be distinguished from the Mustard White and West Virginia White by its pale yellow-green ventral hind wings, the black forewing tips, and the one (male) or two (female) black spots in the center of the forewing. The butterfly is seasonally dimorphic. The spring and late fall short-day form is smaller, less yellowish, and has strongly reduced black areas. Some spring males are immaculate, much like the summer form of the Mustard White. Males live as long as 13 days and females as long as 10 days.

Range. Various races of the European Cabbage Butterfly are native to Eurasia and North Africa. It has been introduced to Australia, North America, Bermuda, Hawaii, and other Pacific Islands. In North America, the species occurs from coast to coast and from central Canada to northern Mexico. It is rare in southern Florida and the southern portions of the Gulf states.

Habitat. The European Cabbage Butterfly is found in virtually every type of open country from Lower Austral to Hudsonian Life Zones. The species is found in dense woodland only in early spring before the canopy leafs out and is rare in certain native open habitats were suitable crucifers are rare or absent, e.g., habitats with acidic soils.

Life History. *Behavior.* The European Cabbage Butterfly is a patrolling species with an open population structure (Emmel 1972). Although the butterflies readily move distances of 250 m or more, males may remain in very small areas day after day and roost in the same bushes on consecutive nights. Males patrol with a deceptively meandering lazy flight, and will usually fly about a meter above the ground. The male courts females during most warm daylight hours (850–1500 hr), either by forcing a female down from above in flight or hovering over her, then landing lateral to her and bending his abdomen to mate. Most often the female's rejection pose is seen: she flutters her spread wings and raises her abdomen, making coupling nearly impossible. Most mated pairs are seen about midday and early afternoon, but mated pairs have been observed from 840 to 1830 hr. Usually both participants will be perched, facing in opposite directions,

18. *Pieris rapae*

and the male usually will take the female for a brief nuptial flight. The female flutters around the hosts, laying single, pale-green eggs on the underside of leaves. Most egg-laying occurs near midday (930–1340 hr). Freshly emerged males may be seen taking moisture at wet spots near streams or on dirt roads.

Broods. The European Cabbage Butterfly will continue to produce broods as long as temperature conditions and availability of suitable host leaves permit. In the northern part of its range it may have only two or three generations, but in the south, seven or eight broods are the rule. The European Cabbage Butterfly flies from May 20 to September 5 on Michigan's upper peninsula, from late March to early November in northern Virginia, and in all months in Mississippi. It is usually the first butterfly to emerge from chrysalids each spring at any given locality.

Early Stages. Development is direct given favorable temperatures and host leaves. Hibernating pupae overwinter. The egg is pale green and columnar. Larvae are green and covered with many short, fine setae. There are a weak, thin dorsal yellow line, and a thin, broken yellow spiracular line on each side. The pupa, which may be either green or tan, has a short pointed frontal projection, an interrupted dorsal ridge and lateral ridge, and a pointed medial projection. These ridges and projections may be tinged with black, red, or yellow.

Adult Nectar Sources. European Cabbage Butterflies visit an extremely wide variety of flowering plants, including some that produce no nectar. More than fifty kinds of plants are visited in just one area of northern Virginia. Interestingly, some plants that attract an array of other butterflies are virtually shunned by this species. Mustards, winter cress, and dandelion are favorites in spring; dogbane and red clover are preferred in midsummer and asters in later summer. At any season, a variety of mints, including hedge nettle (*Stachys*), mint (*Mentha*), wild bergamot (*Monarda*), and self-heal (*Prunella*), are visited avidly. Milkweeds and large-head composites are rarely visited, considering their wide use by other butterflies and the overall abundance of this butterfly. This species seems to prefer flowers with a lateral orientation.

Caterpillar Host Plants. The European Cabbage Butterfly oviposits on a wide variety of crucifers, and occasionally selects plants in the caper family (Capparidaceae). The chemical cues in these plants are mustard oils (glucosinolates); these are used by both adults and larvae to select suitable hosts. In the eastern states the leaves of winter cresses (*Barbarea orthoceras* and *vulgaris*), mustards (*Brassica*), peppergrass (*Lepidium*), and cultivated crucifer crops, such as cabbage, collards, and broccoli are favored. Some crucifers, such as kale, purple cabbage, and shepherd's purse (*Capsella bursa-pastoris*) are avoided.

Great Southern White
Ascia monuste phileta (Fabricius), 1775
PLATE 7 · FIGURE 37 · MAP 19

Etymology. The species name is derived from the Latin *ascius*, "without a shadow." Monuste, one of the Danaides, killed her husband Eurysthenes (Hyg. Fab. 170).

Synopsis. The Great Southern White is a well known, sporadic migrant in our southeastern states, but its movements are not well understood. The female is seasonally dimorphic, having a white winter form and a dark-gray or black summer form.

Butterfly. Male forewing: $\overline{X}$ = 3.3 cm, range 2.7–3.9 cm; female forewing: $\overline{X}$ = 3.1 cm, range 2.7–3.4 cm. The males are pure white dorsally, with black along the coastal margin and deeply indented black marks at the veins on the outer margin. Ventrally, the hind-wing and forewing apices are a pale tan. The female is seasonally dimorphic, with the long-lived winter (dry-season) form white, as in the male, but with a bit more marginal black and a black cell spot on the forewing. The summer (wet-season) female has wings almost completely clouded with black scaling, giving it a dark-gray or smoky black appearance. Pease (1962), in an early experiment of its kind, demonstrated that caterpillars raised under short day-length conditions produce the gray, smoky form. It

19. *Ascia monuste phileta*

should be noted that midsummer samples may include a few white females. The antennal tips of both sexes are vivid blue-green in live individuals.

Range. The Great Southern White is resident in Florida, the southern portions of the Gulf states, and southern Texas south through the Carribbean, Mexico, and Central America to Argentina. It emigrates sporadically and may establish temporary summer populations in coastal Georgia and South Carolina. Rare vagrants are found as far north as Virginia and Missouri.

Habitat. The butterfly prefers coastal dunes and salt marshes, but also breeds in other open sites, such as open fields and in gardens during the summer.

Life History. *Behavior.* Males patrol in search of receptive females. Mating pairs have been seen from 1230 to 1315 hr, the females carrying. Migrants usually follow the coastline, flying 3 to 5 m above the ground in a direct manner and seldom stopping to visit flowers.

Broods. Broods are poorly documented, but Klots (1951) refers to at least three generations. The winter form (November–February) is most likely long-lived and in reproductive diapause through most of the winter. The number of summer generations is unknown, but it is usually the first or second summer generation that undergoes the periodic emigrations (Nielsen and Nielsen 1950).

Early Stages. The female lays the eggs, in clusters of about 20, on the upper surface of host leaves. Eggs are pale yellow and flask-shaped, with eleven longitudinal ridges. The mature caterpillar has a tan-yellow head with an orange front and many black papillae bearing white setae. The body is covered with numerous setae and transverse rows of black papillae. The dorsal setae are predominantly black, while those below the spiracular line are white. The caterpillar has five longitudinal orange bands or lines that are separated by areas of mottled gray. The pupa is ivory white with many olive-black markings and black dots. Its wing cases are lustrous white with wide dark margins. There is a discontinuous yellow-orange dorsal stripe. The general shape is similar to that of the European Cabbage Butterfly, but there is a rounded dorsal thoracic bulge, the frontal projection is blunt, and there are no ridges. Notably, there is a pair of lateral, recurved, black horns on the third abdominal segment (Comstock 1954).

Adult Nectar Sources. Shepherd's needle, lantana, saltwort, and verbena are several plants whose flowers are visited avidly.

Caterpillar Host Plants. Beach cabbage (*Cakile maritima*) and saltwort (*Batis maritima*), both succulent crucifers, are used along the coast, while other crucifers, such as peppergrass (*Lepidium virginicum*) and cultivated cabbage, collards, and kale are selected and fed upon elsewhere. Several capers, such as spider flower (*Cleome spinosa*), *Cleome rufidosperma*, clammy weed (*Polanisia*), and nasturtium are also used with some frequency.

Large Marble
Euchloe ausonides (Lucas), 1852

MAP 20

Etymology. The species name is derived from Ausonius, a Roman poet of the fourth century A.D.

Synopsis. The Large Marble is primarily a species of more western and northern areas. It appears to have colonized the Great Lakes region around 1960. Its close Old World relative is *Euchloe ausonia*.

Butterfly. Male forewing: $\overline{X}$ = 2.0 cm, range 1.7–2.2 cm; female forewing: $\overline{X}$ = 2.0 cm, range 1.8–2.3 cm. The Large Marble may be distinguished from the Olympia Marble by its larger size, more extensively marked forewing apex, and denser green marbling on the ventral hind wing.

Range. The Large Marble occurs from southern Ontario, northern Michigan (Isle Royale), and Minnesota west to California and north to Alaska. It is found in most western mountain ranges south to northern New Mexico and central Utah.

Habitat. The Large Marble is found in a variety of open, sunny habitats. In Michigan, it is found in flat, sandy areas. Throughout its range it may be found from the Upper Sonoran (California) through the Hudsonian Life Zone.

20. *Euchloe ausonides*

Life History. *Behavior.* Males patrol open areas during most warm, daylight hours in search of receptive females. This flight is a bit higher (1 m) than that of the Olympia Marble. Courtship is described by J. A. Scott (1975a). The male hovers behind the female until she lands, whereupon the male lands behind her and bends his abdomen laterally to copulate. Mated pairs were found by Scott from 0930 to 1550 hr. Dissection of females showed that some had mated three times, while most had mated only once or twice. Egg-laying takes place from 0915 to 1500 hr. Females lay a single egg on unopened central flower buds, usually on a terminal flower cluster. Scott (1975a) described population parameters at a study site in California, and he found the butterflies lived an average of 4 days.

Broods. The Large Marble is univoltine everywhere in its range except in central California, where it is double-brooded. Flight dates for the Great Lakes population range from June 1 to 17.

Early Stages. Early stages are nearly identical to those of the Olympia Marble.

Adult Nectar Sources. In California, adults use a variety of small to medium flowered plants, but usually host crucifers are preferred.

Caterpillar Host Plants. In Michigan and Ontario, rock cress (*Arabis drummondi*) is the only recorded host plant. Elsewhere in its range, the Large Marble utilizes mustards (*Brassica*), wild radish (*Raphanus*), *Descurainea*, rock cresses (*Arabis*), and dyer's woad (*Isatis tinctoria*).

Olympia Marble
Euchloe olympia (Edwards), 1871
PLATE 6 · FIGURES 35 & 36 · MAP 21

Etymology. The generic name is derived from the Greek *eu* + *chloe*, "well colored in light green." The species is named after the site of the ancient Olympic games.

Synopsis. The Olympia Marble is the only primarily eastern United States member of its genus. Other relatives are found in the West and are mountain- or desert-adapted. This butterfly has several curiously isolated segregates or populations.

Butterfly. Male forewing: $\overline{X}$ = 1.8 cm, range 1.5–2.1 cm; female forewing: $\overline{X}$ = 1.9 cm, range 1.6–2.3 cm. The Olympia Marble may be distinguished from the Large Marble (*Euchloe ausonides*) where they co-occur in the northern Great Lakes area by its less strongly marked forewing apex, smaller size, and less dense green marbling on the ventral hind wing. Several geographic isolates occur (see Map 21), and there is a general trend for the dorsal black markings to become less estensive in western populations (Clench and Opler 1983). Wagner (1977) has shown differences between dark, diminutive individuals on lake-edge dunes and the paler, larger individuals of Michigan's interior sand prairies. In living butterflies there is a broad pink band along the costal margin of the ventral hind wing; this marking slowly fades after death and is absent from most museum specimens.

Range. The Olympia Marble occurs from the mountains of northern Virginia, Maryland, and southern Pennsylvania west through southern Ontario and Minnesota to central Montana and Colorado; it ranges south to central Kentucky, Arkansas, and central Texas. The species is not continuous within its range, but occurs as several geographic isolates, each consisting of several to many localized colonies.

Habitats. Habitats usually have well-drained soils and a semi-arid appearance. Examples range from shale barrens (Virginia, West Virginia, Maryland), lake-shore dunes (Indi-ana, Michigan), sand plains (Michigan), river bluffs (Wisconsin, Missouri), and open river forests (Arkansas, Oklahoma) to rocky foothills (Colorado). The butterfly occurs from the Upper Austral through the Transition Life Zone.

Life History. *Behavior.* The butterflies fly very close to the ground (0.5 m) with a steady, direct flight. Males patrol in search of receptive females during most daylight hours (J. A. Scott 1975a). Females oviposit in early afternoon (1430 hr), usually depositing a single egg on unopened host flower buds. In windy situations, like along the edge of Lake Michi-

21. *Euchloe olympia*

gan, Olympia Marbles keep to the leeward side of the dunes (Shull 1907).

Broods. The Olympia Marble is univoltine. Individuals of northern populations emerge later, as is shown by the following sample flight periods: Michigan (upper peninsula)—May 24 to June 25, Illinois—May 10 to 28, Virginia—mid-April to late May. In Texas flights may occur as early as April 4, and in Colorado as late as July 18.

Early Stages. Shull (1907) has described the early stages of the Olympia Marble in some detail, while Opler (1974) has presented drawings of the larvae and pupa. The chrysalis is pale brown and tan with short, dark-brown, lateral lines behind the head. The caterpillar is gray with subdorsal yellow stripes and subspiracular white stripes subtended by yellow. The caterpillars feed exclusively on flower parts and seed pods.

Adult Nectar Sources. Adults visit the flowers of various rock cresses, their hosts, as well as chickweed, houstonia, and phlox.

Caterpillar Host Plants. With the exception of one report of an egg laid on hedge mustard (*Sisymbrium*), all known hosts are rock cresses (genus *Arabis*), including *A. laevigata* (West Virginia), *A. glabra* (tower mustard Ontario), *A. drummondi* (Michigan), *A. lyrata* (Indiana, Michigan), and *A. serotina* (West Virginia).

Falcate Orange-tip
Anthocharis midea (Hübner), 1809
PLATE 7 · FIGURES 38 & 39 · MAP 22

Etymology. The genus name is derived from the Greek *antho*, "flower," and *charis*, "favor," i.e., "flower-loving."

Synopsis. The Falcate Orange-tip is a delicate pierid butterfly of early spring. The first appearance of the orange-tipped males is sure to delight the butterfly aficionado after a long cold winter.

Butterfly. Male forewing: $\overline{X}$ = 1.8 cm, range 1.6–2.0 cm; female forewing: $\overline{X}$ = 1.9 cm, range 1.7–2.1 cm. The butterfly is sexually dimorphic. Both sexes have a falcate forewing apex, a black dot in the middle of the dorsal forewing, and a solidly marbled ventral hind wing. The apical fourth of the male's dorsal forewing is orange bordered by black. The equivalent area of the female's forewing is usually white, although a few have a faint yellow-orange apex. Some individuals from coastal South Carolina and Georgia are pale yellow instead of the usual white. Dos Passos and Klots (1969) have discussed the geographic variation, primarily of the extent of orange on the male forewing. Several weak clinal subspecies have been named. A close relative, the Mexican Orange-tip (*Anthocharis limonea*), occurs several thousand kilometers to the south in the central Mexican highlands.

Range. The Falcate Orange-tip ranges from Massachusetts and Connecticut west to southern Wisconsin and Missouri and south to coastal Georgia, the northern portions of the Gulf states, and central Texas.

Habitat. This species always requires the presence of trees nearby but does not occur in densely shaded situations. Typical situations include open deciduous woodland, pine barrens, and low-lying young woods near streams or swamps.

Life History. *Behavior.* Adults have a narrow daily activity range. Individuals normally do not appear before 930 or 1000 hrs and usually cease flight by 1430. During this entire period males may be seen patrolling routes along woodland edges or clearings, usually a bit less than a meter above the ground. Their routes usually follow ridges or valley drainages. Individual males apparently repeat the same routes through their lifetimes, since marked males may be caught in the same general area more than a week after initial capture. In some circumstances males seek open ridge tops, where they patrol back and forth (Merritt 1952). Most courtships, and presumably mating as well, take place from 950 to 1200 hr. The perched female is rapidly circled by the male. The female's rejection posture (seen most often) is widespread quivering wings and raised abdomen. Oviposition takes place

22. *Anthocharis midea*

from the 955 to 1205 hr. The area selected is usually shaded and somewhat moist. The female flies slowly, dipping down to touch vegetation at frequent intervals. When she finds a suitably situated host she alights and deposits a single egg, usually at the base of a young bud, but sometimes on leaves, stem, or pods. Females must wander through their habitat throughout their lives, since marked ones are almost never recaptured. Maximum longevity for marked males, as measured between date of initial capture and last recapture, is 11 days.

Broods. The Falcate Orange-tip is usually univoltine. Emergence occurs earlier at southern and lowland locations. In New York, flight dates range from late April to June. In northern Virginia, flights normally run from mid-April to mid-May, and in coastal Georgia flights occur from March 17 to April 7. Emergence occurs as early as March 3 in Mississippi. In the Virginia mountains, two closely juxtaposed flights have been reported that extend to mid-June (Clark and Clark 1951). Whether these represent two broods is uncertain. Throughout most of the flight there is a continuous emergence of fresh individuals. A certain proportion of chrysalids do not produce adults the following spring, but wait a second year to emerge (dos Passos and Klots 1969). This adaptation helps the species survive unusually dry springs.

Early Stages. Eggs are laid in spring. The egg is orange and spindle-shaped with vertical ribs. Buds, flowers, pods, and, rarely, leaves are eaten. Development to the pupa is direct, with most feeding occurring at night. Pupation occurs on dead stems or branches near the ground, and adults usually emerge the following spring. Occasionally, they wait two years. The first-instar larva is dirty yellowish-green. The last-instar caterpillar is blue-green with lateral white stripes and a dorsal greenish-orange stripe. The chrysalis is extremely slender, with an elongated conelike extension at the anterior end.

Adult Nectar Sources. Both males and females favor crucifers and low, small, white-flowered plants as nectar sources, and generally they favor the most abundant crucifer. For example, in northern Virginia the host hairy bitter cress was favored at first in early April, but winter cress was favored in late April as it came into full flower. At the end of the flight in early May, peppergrass was favored. Wild strawberry, violet, and chickweed each were visited on occasion. Elsewhere, toothwort and wild plum are visited.

Caterpillar Host Plants. Plants of the crucifer family, Brassicaceae, are the sole hosts. In any one locality only a single species is favored. In northern Virginia, ovipositing females selected only hairy bitter cress (*Cardamine hirsuta*), although three or four other crucifers were present. In other localities the known hosts are: tower mustard (*Arabis glabra*), smooth rock cress (*Arabis laevigata*), lyre-leaved rock cress (*Arabis lyrata*), sicklepod (*Arabis canadensis*), mouse-ear cress (*Arabidopsis thaliana*), winter cress (*Barbarea vulgaris*—oviposits but possibly no larval survival), small-flowered bitter cress (*Cardamine arenicola*), shepherd's purse (*Capsella bursa-pastoris*), and cut-leaved toothwort (*Dentaria laciniata*).

Subfamily Coliadinae: Sulphurs

Sulphurs are predominantly yellow or orange butterflies, often with black borders on the wings. They range from small to large and usually are sexually dimorphic. Tropical species are often seasonally dimorphic as well.

Caterpillar hosts are predominantly legumes, although a scattering of species feed on other plants. The Dwarf Yellow (*Nathalis iole*) is unique in many ways and could be treated as a separate subfamily.

Clouded Sulphur
Colias philodice Godart, 1819

PLATE 7 · FIGURES 40–42 · MAP 23

Etymology. The genus is named after Kolias, an epithet of Venus (Greek myth), while the species is possibly named after Phyllodoce, a sea nymph, who was daughter of Nereus and Doris (Vergilius).

Synopsis. The Clouded Sulphur is closely allied to and often flies in the same fields with the Orange Sulphur (*Colias eurytheme*). This species occurs farther northward and was originally the dominant *Colias* over most of the eastern United States.

Butterfly. Male forewing: $\overline{X}$ = 2.4 cm, range 2.2–3.2 cm; female forewing: $\overline{X}$ = 2.6 cm, range 2.2–3.1 cm. This butterfly may be distinguished from the Orange Sulphur by its clear lemon-yellow dorsal wing color in both males and yellow females. The white ("alba") female form generally has narrower black borders than those of the Orange Sulphur, but cannot be reliably distinguished. Small short-day individuals occur in spring and rarely in the fall, as is true for the Orange Sulphur. White females are more frequent in the North, accounting for up to 95% of the population (Alaska). In the eastern United States they usually account for less than 50%. In a northern Virginia survey, about one-third of the females were white. A mark-recapture study (also in northern Virginia) showed a maximum longevity of 17 days for males, whereas a much more intensive study of Colorado populations indicated a maximum longevity of 24 days for males and 17 days for females (Watt et al. 1979). Average residence time ranged from 2 to 7 days; in this study this finding was assumed to be a fair measure of longevity. Most Clouded Sulphurs remain in an area of 40–100 acres during their lifetime.

Range. The Clouded Sulphur ranges from Quebec and the Maritime Provinces northwest across Canada to Alaska. It also ranges to the arid eastern portions of the Pacific states and south to the Gulf states and Florida, where it is very rare. An isolated subspecies occurs in the mountains of Guatemala.

Habitat. The Clouded Sulphur is found in a variety of open situations, but it is best adapted to clover fields, lawns, and moist meadows or fields.

Life History. *Behavior.* The Clouded Sulphur is very similar to the Orange Sulphur in its activities. Courtships have been observed from 1040 to 1320 hr, while mated pairs (male carrying) were usually seen from 1000 to 1100 hr, although one pair was seen at 1640. Females were observed laying eggs from 0945 to 1400 hr. Newly emerged males visit wet spots for moisture, and lateral basking is employed for warming up.

Broods. Two broods are produced in central Alaska each year, but three to five are usual in the eastern United States, the actual number depending upon latitude and elevation. Spring emergence ranges from early May (Michigan) to early March (Georgia) and the flight season usually ends with freezing temperatures in early to mid-November. Occasion-

ally, individuals are seen during unusual warm spells in midwinter.

Early Stages. The larvae are very similar to those of the Orange Sulphur, except that there may be either yellow-green or dark blue-green forms and the lateral stripe may contain broken red strips. Hibernation is usually as a chrysalis.

Adult Nectar Sources. The flowers visited are essentially the same as for the Orange Sulphur, with a similar seasonal sequence in northern Virginia.

Caterpillar Host Plants. The Clouded Sulphur's preferred host throughout most of the eastern states is white clover (*Trifolium repens*). In northern Virginia, it also selected trefoils (*Lotus*). It also uses other clovers, although females were never seen laying eggs on red clover, (*Trifolium pratense*) in northern Virginia. Alfalfa (*Medicago sativa*) and white sweet clover (*Melilotus alba*) are selected rarely.

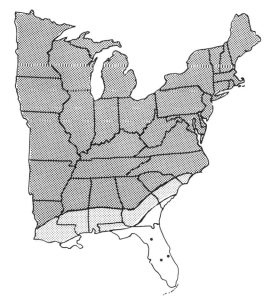

23. *Colias philodice*

Orange Sulphur, Alfalfa Sulphur
Colias eurytheme Boisduval, 1852
PLATE 7 · FIGURES 41 & 42 · MAP 24

Synopsis. The Orange Sulphur is one of the most common, most widespread butterflies in North America. At times its caterpillars may be serious pests in alfalfa fields. It was a very rare butterfly east of the Appalachians prior to 1890 but has subsequently colonized virtually the entire eastern United States.

Butterfly. Male forewing: $\overline{X}$ = 2.4 cm, range 2.1–2.8 cm; female forewing: $\overline{X}$ = 2.6 cm, range 2.3–3.1 cm. The Orange (or Alfalfa) Sulphur is very closely related to the Clouded Sulphur (*Colias philodice*) and frequently hybridizes with it. Any males and yellow females with *any* amount of orange on the dorsal wing surfaces may be referred to this species. The

Dog Face
Colias cesonia (Stoll), 1790

PLATE 8 · FIGURE 44 · MAP 27

Etymology. The species name is possibly derived from Cesson, the name of three or more cities in France.

Synopsis. The Dog Face belongs to a tropical group of sulphurs (subgenus *Zerene*). Its colloquial name is derived from the yellow dog's head on each forewing.

Butterfly. Male forewing: $\bar{X}$ = 2.9 cm, range 2.6–3.3 cm; female forewing $\bar{X}$ = 3.0 cm, range 2.7–3.4 cm. The abrupt, almost falcate forewing apex and the "dog's head" on the forewing will readily separate this butterfly from other sulphurs. The female has somewhat blurred marginal areas on the dorsal forewing. There are two seasonal forms. The summer reproductive form has largely clear yellow ventral wing surfaces, and the dorsal hind-wing surfaces are also yellow except for a black border on the outer margin. The winter form ("rosa") has ventral hind-wing and forewing apical areas suffused with reddish pink, and the dorsal hind wing clouded with black scaling. As is true for most butterflies that overwinter as adults, there is probably a striking difference in longevity between the seasonal forms. Winter adults may survive as long as 7 or 8 months.

Range. The Dog Face is resident from the Gulf states, Texas, and the arid Southwest south through the tropical seasonally dry lowlands of Mexico and Central America to Argentina. It also occurs in Cuba and Hispaniola. It cannot survive freezing winter temperatures, but it regularly emigrates northward through the plains as far north as Manitoba. It breeds most summers in the central Midwest. East of the Mississippi it is a rare colonist but has been found as far north as Ontario, and it bred in New York during 1896. In the southeast states it is sporadic and may not survive the winters in most years except in south Georgia and Florida.

Habitat. The species is at home in open, dry situations, such as hot, dry, scrub oak groves, open woodland, and short-grass prairie hills.

Life History. *Behavior.* Males patrol during most daylight hours, flying rapidly over open flats and slopes, pausing only briefly to take nectar in their search for receptive females. Mated pairs have been found from 1035 to 1630 hr; males are the carriers. In Iowa, a female was observed laying eggs at 1530 hr. Flower visitation takes place from 0845 to 1605 hr. Freshly emerged males imbibe moisture at mud or damp sand. In the tropics this species emigrates to higher elevations at the onset of the dry season. The winter form spends most of its life resting on low vegetation.

Broods. There are three generations in the southern states, two reproductive broods in late May–early June and mid-July–August and the winter form from late September to late April of the succeeding year. The last brood does not become reproductive until spring, after a long reproductive diapause.

Early Stages. The yellow-green spindle-shaped eggs are laid on young host leaves, as is true for other *Colias* species. The caterpillar is green and covered with blackish or black setiferous tubercles (*chalazae*). The caterpillars are extremely variable, being either unmarked or bearing cross-bands or longitudinal lines of yellow and black.

Adult Nectar Sources. Dog Faces have been observed nectaring at alfalfa, verbena, coreopsis, and shrub houstonia.

Caterpillar Host Plants. A variety of small-leaved plants in the pea family are selected, especially species of indigo bush (*Dalea*). In western Iowa, prairie clovers (*Pentalostemon alba* and *P. purpurea*) were the primary hosts. False indigo (*Amorpha fruticosa*), clovers (*Trifolium*), soybean (*Glycine*) and alfalfa (*Medicago sativa*) are also utilized.

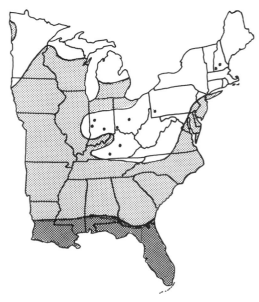

27. *Colias cesonia*

Giant Brimstone
Anteos maerula (Fabricius), 1775
PLATE 8 · FIGURE 45 · MAP 28

Synopsis. This very large sulphur is a very rare stray in Mississippi, southern Florida, and the Keys. It ranges from the Antilles and northern Mexico south through Central America to Peru. In its native range it flies powerfully, high above the ground, and may be seen or collected with ease only when the males congregate at damp spots or when visiting flowers. The Giant Brimstone's caterpillars are known to feed on various species of *Cassia*.

Cloudless Sulphur
Phoebis sennae eubule (Linnaeus), 1767
PLATE 8 · FIGURES 16 & 17 · MAP 29

Etymology. The genus is named after Phoebe, daughter of Gaea and sister of Apollo. The species name is derived from senna, the common name for one of the *Cassia* host plants.
Synopsis. The Cloudless Sulphur is the common large yellow sulphur of most of our southeastern United States and the American Tropics. It sometimes undertakes large emigrations during the fall (August–November)
Butterfly. Male forewing: $\bar{X}$ = 3.4 cm, range 3.2–3.5 cm; female forewing: $\bar{X}$ = 3.3 cm, range 3.1–3.5 cm. The male Cloudless Sulphur is the only large sulphur that combines pure-yellow wings dorsally and a discal spot on the ventral hind wing. The Large Orange Sulphur (*Phoebis agarithe maxima*) is solid orange dorsally, the Orange-barred Sulphur (*Phoebis philea*) has red-orange markings on the dorsal wings, and the Statira Sulphur (*Aphrissa statira floridensis*) has paler yellow outer portions of its wings and lacks the discal spot on the ventral hind wing. Both yellow and white forms of the female may be seen. Furthermore, the Cloudless Sulphur is seasonally dimorphic. The summer-form male has almost obsolete markings on the ventral hind wing, while the winter form is larger and has accentuated dark markings ventrally.
Range. This large yellow butterfly is resident from southern Georgia and Florida west through the Gulf states and Texas to southern California. From there it ranges south through the Antilles, Mexico, and Central America to Argentina. In the eastern United States, it regularly emigrates north to Delaware and southern Illinois. It is sporadic farther north but has reached most northern states and southern Canada on rare occasions.
Habitat. The Cloudless Sulphur is found in a variety of open disturbed situations, including beaches, parks, open scrub, abandoned corn fields, and roadsides.

28. *Anteos maerula*

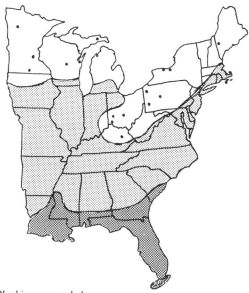

29. *Phoebis sennae eubule*

Life History. *Behavior.* Males patrol all day from early morning. They usually fly rapidly within a few meters of the ground. Mated pairs have been seen from 1045–1200 hr, with the male as carrier. Oviposition was observed in coastal Virginia at 1120 hr, the female depositing single white eggs on young host terminals. The butterfly sometimes undertakes large movements or emigrations, which occur chiefly in the fall and are usually in a southeast direction. During these movements, the butterflies fly 2–4 m above the ground and move in a direct course. From about September to March, the winter-form adults probably undergo a reproductive diapause. Freshly emerged males often gather in swarms at wet mud or sand. This species has been reported to roost communally at night.

Broods. There are probably only two broods of the summer form and one of the winter form. The species is found all year in coastal South Carolina, Georgia, Florida, and Mississippi. In other southern states it usually colonizes by June and may bring off one complete brood. Farther north it does not appear until August or September and usually does not reproduce.

Early Stages. Development is direct. None of the immature stages can overwinter or bypass unfavorable periods. The yellow-green egg is spindle-shaped. The caterpillar may be either yellow or green, with small black setiferous tubercles (chalazae). It has a yellow stripe with blue points bordering its upper edge along each side. The chrysalis is colored variously from green to bright pink, with a yellow lateral stripe bordered below in green, and yellow antennal sheaths.

Adult Nectar Sources. The Cloudless Sulphur and its kin visit a variety of long-tubed flowers and may fly as high as 15 m to take advantage of flowering trees. Preferred nectar plants include lantana, bougainvilla, hibiscus, and cordia. In coastal Virginia the species was observed visiting several species of wild morning glories (*Ipomoea* and *Meremia*), while cardinal flower was visited in Tennessee.

Caterpillar Host Plants. All known hosts are various sennas (*Cassia*) of the Cesalpinia family. These include *C. chamaecrista*, *C. obtusifolia*, and *C. fasciculata* in the eastern states. Many other sennas are utilized in the Tropics. Reports of clover are probably erroneous.

Orange-Barred Sulphur
Phoebis philea (Johansson), 1763
PLATE 8 · FIGURE 48 · MAP 30

Synopsis. This beautiful tropical butterfly was not a resident of the eastern states until it successfully colonized Florida around 1928. It has a high coursing flight, but the adults may be observed closely when they visit flowers or moist sand.

Butterfly. Male forewing: $\overline{X}$ = 4.1 cm, range 3.7–4.5 cm; female forewing: $\overline{X}$ = 3.8 cm, range 3.0–4.5 cm. The male may be identified easily by its vertical red-orange bar, which terminates on the dorsal forewing costa, and the broad red-orange band on the outer margin of the dorsal hind wing. The female is broadly suffused with magenta pink ventrally and has both white and yellow forms. In addition, there are seasonal forms. The summer males have the ventral hind wing surface relatively clear, whereas the larger winter males have the ventral hind wing heavily marked with small red spots. The Florida butterflies are of the typical Central American subspecies.

Range. The Orange-barred Sulphur is resident from peninsular Florida and northern Mexico south through Cuba, Hispaniola, and Central America to southern Brazil. The species occasionally establishes temporary breeding populations in Georgia, but elsewhere in the East it is an extremely rare vagrant. Most records occurred in the 1930s shortly after the species first entered Florida; they extended sporadically as far north as Wisconsin and Maine.

Habitat. This species is found in Tropical and Subtropical Life Zones in open lowland situations. It frequents forest edges, canopy, and city gardens. Occasionally, it is found in open fields.

30. *Phoebis philea*

Life History. *Behavior*. The males not only patrol near ground level, but also frequently fly high along forest edges and over the canopy. The freshly emerged males take moisture from wet mud or sand. The Orange-barred Sulphur is less emigratory than the Cloudless Sulphur.

Broods. The species occurs all year in southern Florida, but the number of generations there is uncertain. Klots (1951) reports at least two generations, and three broods are likely.

Early Stages. The caterpillar is yellow-green, with many black setiferous tubercles. Laterally, it has blackish and yellow bands. The latter contain reddish black spots ringed with white.

Adult Nectar Sources. Adults visit long-tubed flowers, such as bougainvilla and hibiscus.

Caterpillar Host Plants. The Orange-barred Sulphur larva feeds almost exclusively on sennas (*Cassia*). In Florida, *Cassia bicapsularis* is a known host, and many other *Cassia* species as well as royal poinciana (*Poinciana pulcherima*) are selected in other parts of its range.

Large Orange Sulphur
Phoebis agarithe maxima (Neumoegen), 1891
PLATE 9 · FIGURES 49–51 · MAP 31

Synopsis. This large orange tropical butterfly is resident within our area only in southern Florida. Unlike the other two resident *Phoebis*, its caterpillars feed only on plants in the mimosa family.

Butterfly. Male forewing: $\overline{X}$ = 3.6 cm, range 3.2–3.7 cm; female forewing: $\overline{X}$ = 3.6 cm, range 3.1–3.9 cm. The male is the only pierid in our area that is entirely yellow-orange dorsally. The female is similar to that of the Cloudless Sulphur, but has several blurred bars on the outer forewings. The female is dimorphic, as in other *Phoebis*, with both white and orange forms. Moreover, there are two seasonal forms, with the winter form more heavily marked ventrally. The winter form probably undergoes a reproductive diapause.

Range. The Large Orange Sulphur is resident in peninsular Florida and southern Texas, whence it extends south through the Antilles, Mexico, and Central America to Peru. Strays, primarily from Texas, have occurred as far north as Wisconsin and southern Maine.

Life History. *Behavior*. Adult behavior is similar to that of the Cloudless Sulphur, except that the Large Orange Sulphur is less migratory.

Early Stages. The caterpillar is green, with a yellow lateral line edged below with black.

Adult Nectar Sources. This tropical butterfly has nectar choices almost identical to those of the Cloudless Sulphur. In Florida it has been seen sipping nectar at bougainvilla, hibiscus, lantana, turk's cap, shepherd's needle, and rose periwinkle.

Caterpillar Host Plants. The Large Orange Sulphur will feed only on woody plants of the mimosa family. Known host genera are *Pithecellobium* and *Inga*. *Pithecellobium guadalupense* and *P. unguiscati* are hosts in Florida (Dyar 1900).

31. *Phoebis agarithe maxima*

Statira Sulphur
Aphrissa statira floridensis (Neumoegen), 1891
MAP 32

Synopsis. This species launches great migratory flights over the ocean from northern South America. The Florida population represents a separate subspecies not known to be migratory.

Butterfly. Male forewing: $\overline{X}$ = 3.2 cm, range 2.7–3.4 cm; female forewing: $\overline{X}$ = 3.4 cm, range 3.2–3.6 cm. The male has the dorsal surface of both wings divided by a ridged scale line. Internally to the line, the wings are lemon yellow, externally a paler yellow. The apical margin is thinly edged with black and the ventral wings are pale yellow. The female differs from that of the Cloudless Sulphur by having wing margins edged with black dorsally, and by the small, solid-black forewing cell spot. *Aphrissa* is sometimes treated as a subgroup of *Phoebis*. There may well be seasonal forms, but they have not been documented.

Range. The Statira Sulphur occurs from southern Florida and southern Texas south through the lowland tropics of the Antilles, Mexico, and Central America to Argentina. Individuals almost never stray from the southern Florida population, although this species has been found in Georgia on at least one occasion.

Habitat. This is a species of open tropical scrub habitat, and it is restricted to the Tropical Life Zone.

Life History. *Broods.* The Statira Sulphur is bivoltine in coastal southern Florida (with flights from June to mid-September and November through early February).

Early Stages. The last-instar caterpillar is pale orange, with a faint greenish tinge, a blue-black subspiracular band, and a pale-orange head.

Adult Nectar Sources. Kimball (1965) reports that adults visit the red-orange flowers of scarlet bush. In Costa Rica, the adults nectar at cordias and several pasture composites.

Caterpillar Host Plants. *Dalbergia exastophyllum* and *Calliandra* are both reported as hosts in Florida. In the Caribbean, Riley (1975) reports that senna (*Cassia*) and Spanish lime (*Melicoccus bijugatus*) are selected. If all these reports are correct, the Statira Sulphur has a very broad host range, since all three leguminous families, Cesalpiniaceae, Fabaceae, and Mimosaceae, are represented.

32. *Aphrissa statira floridensis*

Orbed Sulphur
Aphrissa orbis (Poey), 1832

MAP 33

Synopsis. This species is native only to Cuba and Hispaniola. In 1973, a stray male was collected on Big Pine Key, Florida. The male is a very pale lemon yellow with a large orange splotch covering the basal third of the dorsal forewing.

Lyside
Kricogonia lyside (Godart), 1819

PLATE 9 · FIGURE 52 · MAP 34

Etymology. The generic name is derived from the Greek *krikos,* "ring, circle," and *goneia,* "generation." The species name is not classical in derivation.

Synopsis. Lyside belongs to a small tropical American genus, the species and proper names of which have long been confused. It sometimes undertakes huge emigrations, particularly in northern Mexico and southern Texas. Its status in the eastern United States is poorly known.

Butterfly. Male forewing: $\bar{X}$ = 2.2 cm, range 1.8–2.7 cm; female forewing: $\bar{X}$ = 2.4 cm, range 1.8–2.7 cm. This little butterfly may be distinguished by the abrupt angle of the forewing apex, yellow basal area of the forewing, and a satiny sheen of the ventral hind-wing surface. Males usually have a vertical black bar at the dorsal hind-wing apical angle. Both sexes are highly variable, and both white and yellow female forms exist. The butterfly is probably also seasonally dimorphic.

Range. Lyside extends from southern Florida, where it is a rare resident, and southern Texas south through the Antilles, Mexico, and Central America to Venezuela. It has strayed to western Missouri.

Habitat. Lyside is at home in lowland tropical scrub or seasonally dry forests.

Life History. *Behavior.* This species periodically undertakes huge emigrations, especially in late summer or fall. The reasons behind these movements are not well understood. *Broods.* The number of generations is unknown. Lyside may be found in southern Florida from May through August. *Early Stages.* Caterpillars feed at night and hide in bark crevices by day. The mature caterpillar is dull green with dorsal and lateral gray or silvery lines; the dorsal line is bordered by chocolate brown. The sides are variegated golden yellow and brown (Wolcott 1927).

Adult Nectar Sources. In southern Florida, Lyside has been observed visiting shepherd's needle and black mangrove.

Caterpillar Host Plants. Lignum vitae (*Guaiacum sanctum*) is the probable larval host in southern Florida. *Porliera angustifolia* is a known host in Texas. Both plants belong to the Zygophyllaceae family.

Barred Sulphur
Eurema daira (Godart), 1819
PLATE 9 · FIGURES 53 & 54 · MAP 35

Etymology. A *daira* is a tambourine used in Islamic countries.

Synopsis. This small butterfly thrives in disturbed fields and overgrown pastures and may be the most abundant tropical American butterfly. It has different seasonal forms with strikingly different potential longevities. The seasonal forms have been mistakenly referred to as separate species.

Butterfly. Male forewing: $\overline{X} = 1.6$ cm, range 1.5–1.8 cm; female forewing: $\overline{X} = 1.7$ cm, range 1.5–1.9 cm. The male Barred Sulphur is similar to the Dainty Sulphur but is larger, has more quadrate forewings, and lacks a hind-wing black costal bar. Dorsally, male Barred Sulphurs differ from females in having a black band along the forewing inner margin and a black band along the hind-wing outer margin. The yellow area of the male forewing is highly reflective of ultraviolet light. There are two seasonal forms: the summer (wet-season) form is smaller, having satiny white ventral hind wings, while the winter (dry-season) form is larger, has reduced black markings above, and has a brick-red ventral hind wing (Mather 1956). In addition, the latter form always has two small black dots in the center of the ventral hind wing. Besides the sexual and seasonal dimorphism, there is extensive geographic variation. Most populations in our area are of the yellow hind-wing form; however, some white hind-wing individuals in southern Florida indicate genetic influx from the Antilles (Clench 1970; Smith et al. 1982).

Range. The Barred Sulphur occurs from the southeastern United States, especially the coastal plain, and southern Texas south throughout the Caribbean, Mexico, and Central America to Argentina. Resident populations occur in Alabama, Florida, Georgia, Louisiana, Mississippi, and South Carolina. Emigrants rarely wander north to Virginia.

33. *Aphrissa orbis*

34. *Kricogonia lyside*

Habitat. The Barred Sulphur favors abandoned pastures, open pine woods, or coastal dunes where its hosts grow. The species occurs in Tropical, Subtropical, and Lower Austral Life Zones where winter minimum temperatures rarely reach freezing.

Life History. *Behavior.* Male Barred Sulphurs patrol in open areas in search of females during warm daylight hours (0800–1400 hr). Upon locating a female, the male flutters until she alights. Then he lands to one side of her, disjoints one forewing, and moves it up and down in front of her. If she is receptive, the male curves his abdomen laterally to couple. Most matings occur in early afternoon (1200–1430 hr). Oviposition occurs between 0830 and 1330, with a strong peak between 1000 and 1200 hr. Ovipositing females lay a single egg on a young host terminal, then move on, often taking nectar before laying another egg. When females encounter each other, they display a weak form of interaction, circling around each other briefly before parting. Freshly emerged males often throng at damp soil along roads or streams. The winter-form individuals often form loose aggregations at forest edges or along streams, where they rest in the shade, visiting flowers each day to maintain their energy reserves. At the onset of winter (or dry season) conditions, Barred Sulphurs often undertake dense emigrations. In the tropics, emigrations often take the butterflies upslope, where cooler, moister conditions prevail and nectar sources are more abundant. Winter-form individuals may live as long as 4 or 5 months, while summer-form individuals live an average of only 2 days and a maximum of 9 days (Opler unpublished). *Broods.* The number of generations has not been documented, but summer-form broods are continuous from mid-April to mid-October in Mississippi, and may number four or five. There is only one generation of the winter form (mid-October to March). The latter individuals are in reproductive diapause until spring.

Early Stages. The larvae feed on host leaves. The mature caterpillar is light green above and translucent green below. There are a pale lateral stripe and a dark green dorsal line (Haskins 1933).

Adult Nectar Sources. Barred Sulphurs visit a tremendous variety of flowers, but those most frequented are joint vetch (their host), *Sida* species, and herbaceous composites, such as shepherd's needle.

Caterpillar Host Plants. Joint vetches (*Aeschynomene*, including *A. viscidula*) are the principal hosts. Also utilized are pencil flower (*Stylosanthes biflora*) and other fabaceous legumes.

35. *Eurema daira*

Boisduval's Sulphur
Eurema boisduvaliana (Felder & Felder), 1865
MAP 36

Synopsis. Boisduval's Sulphur is an extremely rare find in the lower Florida Keys, where it has been recorded only twice. It is rare even in the Antilles, and is common only in Mexico and Central America. The male superficially resembles the Mexican Sulphur but is yellow and has less deeply indented black forewing markings. Its early stages are unreported.

Mexican Sulphur
Eurema mexicana (Boisduval), 1836
MAP 37

Etymology. The species name is derived from Mexico, where this species is common.

Synopsis. The Mexican Sulphur is not resident anywhere in the eastern United States, but is a periodic immigrant from

Texas and northern Mexico. It was a particularly common immigrant during the late 1920s and early 1930s.

Butterfly. Male forewing: $\overline{X}$ = 2.4 cm, range 2.1–2.9 cm; female forewing: $\overline{X}$ = 2.2 cm, range 2.0–2.3 cm. Dorsally, the Mexican Sulphur is cream white with wide, deeply indented, black margins, giving it a bit of "dog's head" appearance. There are very short triangular "tails" at the outer angle of the hind wing.

Range. The Mexican Sulphur is resident from Texas and northern Mexico south through Mexico and Central America. It extends north most years into the Southwest deserts and Great Plains states. It is a rare late-summer (September–October) immigrant to Arkansas, Illinois, Mississippi, and Michigan.

Habitat. The Mexican Sulphur favors dry flats and desert scrub in the northern portion of its range but is found in moister, more densely overgrown, second growth farther south in Latin America.

Caterpillar Host Plant. The host is reported to be *Cassia*.

Little Sulphur
Eurema lisa (Boisduval & LeConte), 1829
PLATE 10 · FIGURES 55 & 56 · MAP 38

Synopsis. The Little Sulphur is the most widespread of the small sulphurs (genus *Eurema*) in the eastern United States. Even though it cannot survive freezing winters, it reoccupies much of the North by late summer each year. It does this so predictably that some have wondered whether a separate winter-hardy sibling might exist. It is the most abundant "yellow" in parts of the South.

Butterfly. Male forewing: $\overline{X}$ = 1.8 cm, range 1.5–1.9 cm; female forewing: $\overline{X}$ = 1.8 cm, range 1.6–2.0 cm. Dorsally, the Little Sulphur has scalloped black margins that are reduced on the female. The rare Jamaican Sulphur (*Eurema nise*) has narrow black borders, lacks the small discal black spot on the dorsal forewing, and has a plainer yellow ventral hind wing. A white or pale-yellow female form is sometimes common. Seasonal forms have not been described.

Range. The species is resident in relatively frost-free portions of the southeastern states, southern Texas, and Mexico south through the Antilles and Central America to Costa Rica. The Little Sulphur invades and establishes temporary populations each summer as far north as Maine, New York, Ontario, and Michigan.

Habitat. The Little Sulphur is found in dry sandy fields, abandoned corn fields, roadsides, along railroad tracks, and occasionally in open woods. The species is resident in Tropical to Subtropical Life Zones but invades the Upper Austral and Transition Life Zones each summer.

Life History. *Behavior.* Males patrol during warm daylight hours. Mating pairs have been observed from 1055 to 1445 hr, with the male as carrier. Newly emerged males take moisture at damp spots along roads or streams. The Little

36. *Eurema boisduvaliana*

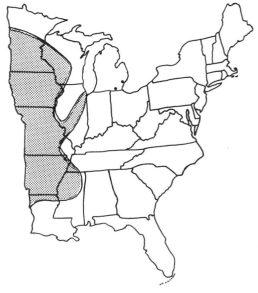

37. *Eurema mexicana*

Sulphur regularly emigrates long distances and millions have been seen in flights over the Atlantic and Caribbean.

Broods. There are four or five reproductive generations each summer, and probably a long overwintering generation of diapausing adults. The species is found all year in Florida and during most months in Georgia, Louisiana, and Mississippi. In more northern states the species appears in late spring to midsummer and flies until late September or October.

Early Stages. The mature caterpillar is slender and grass green, with either one or two lateral white lines.

Adult Nectar Sources. In the North, the Little Sulphur nectars at small-flowered composites, such as asters and goldenrods.

Caterpillar Host Plants. The favored larval hosts are sennas (*Cassia*), including wild sensitive plant (*C. nictitans*) and partridge pea (*C. fasciculata*). Other reported host plants, such as sensitive plant (*Mimosa pudica*), hog peanut (*Amphicarpa*), and clovers (*Trifolium*), are doubtful hosts and require confirmation.

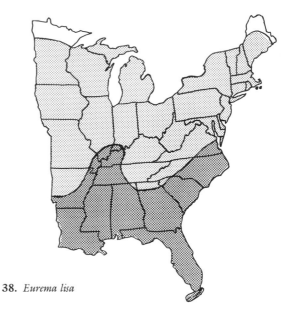

38. *Eurema lisa*

Chamberlain's Sulphur
Eurema chamberlaini (Butler), 1897
MAP 39

Synopsis. This small sulphur is endemic to the Bahama Islands and is similar to the Jamaican Sulphur (*Eurema nise*), from which it differs by being more orange and by having a triangular patch of white scales at the base of the ventral forewing. It was found in Dade County, Florida, during March, 1963. Its early stages are unknown.

Jamaican Sulphur
Eurema nise (Cramer), 1775
PLATE 10 · FIGURE 57 · MAP 40

Etymology. *Nisa* is Latin, a female proper name. *Nise* is Latin, an archaic form of nisi, "neither."

Synopsis. The Jamaican Sulphur is restricted to southern Florida, where it is sporadic but occasionally common. It is probably overlooked because of its close resemblance to the Little Sulphur (*Eurema lisa*). It is common and widespread in the lowland tropics.

Butterfly. Male forewing: $\bar{X}$ = 1.5 cm, range 1.4–1.6 cm; female forewing: $\bar{X}$ = 1.5 cm, range 1.4–1.6 cm. The distinguishing features of this species are discussed under the Little Sulphur. Wet- and dry-season forms exist, but white female forms are absent.

Range. The Jamaican Sulphur's range extends from extreme southern Florida and the Texas Gulf Coast south through Jamaica, Mexico, and Central America to Argentina.

Habitat. The Jamaican Sulphur is most often seen at the brushy edge of forests, in second growth, and even in pastures. It is limited to Tropical and Subtropical Life Zones.

Life History. *Broods.* The Jamaican Sulphur has been found in southern Florida from May through August. If resident, the species must occur all year as an adult; a diapausing winter (dry-season) form may be seen less often. The number of broods is unknown.

Early Stages. The mature caterpillar is green, covered with short, fine, white setae, and has a whitish lateral line.

Adult Nectar Sources. In Costa Rica, this species nectars at flowers of small composites, *Sida* species, and *Cordia* species.

Caterpillar Host Plant. The only recorded larval host is sensitive plant (*Mimosa pudica*).

Shy Sulphur
Eurema messalina blakei (Maynard), 1891

PLATE 10 · FIGURE 58

Etymology. Messalina was the spouse of the Roman Emperor Claudius. The subspecies was named in honor of Charles Alfred Blake (1834–1903), a student of both Lepidoptera and Hymenoptera.

Synopsis. This Antillean butterfly was recorded in southern Florida in the 1880s, and there is only one authentic specimen. Dorsally, the male is cream white with black margins. Males also have a distinct pink bar on the ventral forewing inner margin that fades after death. The female is similar but its black hind-wing band is reduced to a small smudge. The species is restricted to the Antilles, favoring brushy areas and has a life history similar to that of the Barred Sulphur. The hosts are tick trefoils (*Desmodium*).

Bush Sulphur
Eurema dina helios (M. Bates), 1934

PLATE 10 · FIGURES 59 & 60 · MAP 41

Etymology. The species name is derived from the Greek *diné*, "whirlwind."

Synopsis. The Bush Sulphur is unusual in that its larvae feed on members of the Simarouba family instead of legumes. It is a rare resident of southern Florida.

Butterfly. Male forewing: $\bar{X}$ = 1.7 cm, range 1.5–2.0 cm; female forewing: $\bar{X}$ = 1.9 cm, range 1.7–2.2 cm. Dorsally, the male Bush Sulphur is orange-yellow, with the outer and costal margins narrowly edged in black. Ventrally, it is a relatively unmarked yellow. The female is yellow above and below, with a black forewing apex and a red splotch on the ventral hind-wing outer margin. Dry-season (winter) and wet-season (summer) forms occur. The former is larger, with the male more orange dorsally, and the black margins slightly more extensive.

Range. The Bush Sulphur ranges from southern Florida, southern Texas, and southern Arizona south through the Antilles, Mexico, and Central America to Panama. It was first found in Florida in 1962.

Habitat. The Bush Sulphur is found in brushy fields or in open forest, often in wet areas. It is found in Subtropical or Tropical Life Zones with a pronounced dry season.

Life History. *Broods.* There are probably two or three wet-season generations and one dry-season generation of adults in reproductive diapause.

Adult Nectar Sources. Lantana and small-flowered composites are preferred in Central America.

Caterpillar Host Plants. Hosts are restricted to trees and shrubs of the Simarouba family. The species has a close association with Mexican alvaradoa (*Alvaradoa amorphoides*) in Florida and is known to feed on bitterbush (*Picramnia*) in Central America.

39. *Eurema chamberlaini*

Sleepy Orange
Eurema nicippe (Cramer), 1779

PLATE 11 · FIGURES 61 & 62 · MAP 42

Etymology. Nicippe was a Roman poet (third century B.C.) who wrote epigrams (short satirical pieces).

Synopsis. The Sleepy Orange is a familiar southern butterfly of open pine woods, fields, and roadsides. It occasionally joins other species in massive emigrations and may be seen mud-puddling in large groups or "clubs."

Butterfly.   Male forewing: $\overline{X}$ = 2.4 cm, range 1.8–2.6 cm; female forewing: $\overline{X}$ = 2.3 cm, range 1.6–2.6 cm. The Sleepy Orange is most similar to the Orange Sulphur, from which it differs by having a more irregular dorsal black marginal band that extends halfway along the forewing costa. In addition, the ventral hind wing is mottled and lacks the circular cell spots of the Orange Sulphur. A yellow form is occasionally seen. The Sleepy Orange is seasonally dimorphic (Mather and Mather 1958). The summer form has a yellow ventral hind wing with or without brown or tan marks, and the diapausing winter form has a brick-red, brown, or tan ventral hind wing, without yellow.

Range.   The Sleepy Orange is resident in nonmountainous areas of the eastern United States south of 40° latitude and along the United States–Mexican border in the Southwest. From these areas it ranges south to the northern Antilles, Mexico, and northern Central America. In the East it wanders north (usually in late summer) to Massachusetts, Ontario, New York, Michigan, and Wisconsin. It may occasionally produce a brood or two at these latitudes, but, of course, cannot survive winters there (Hessel 1956).

Habitat.   Low-lying areas, including open river woodland, pine flats, open fields, and roadsides in the Tropical and Lower Austral Life Zones are the usual haunts of the Sleepy Orange.

Life History.   *Behavior.* Males patrol in search of females during most warm daylight hours. One mated pair was seen at 1120 hr, the male flying with a freshly emerged female. Freshly emerged males often visit muddy spots along roads or streams to take up moisture. Occasionally, groups of a hundred or more are seen. Until spring arrives, the winter-form individuals usually rest quietly on low, shaded vegetation and occasionally take nectar. In addition, in some circumstances they may take part in large, mixed-species emigrations. Winter males may have a lifespan of 5 or 6 months, while summer-form individuals may live only about a week.

Broods. In areas where the Sleepy Orange is resident there are probably three or four generations of the summer form and only one of the diapausing winter form. In Mississippi the summer form occurs from May to October and the winter form from October to April.

Early Stages. The mature larva is slender, green, and covered with many short, fine setae. It has a lateral white and yellow stripe subtended by black. There are both black and green forms of the chrysalis (Evans 1958). The life cycle requires 30 days.

Adult Nectar Sources.   Sleepy Orange adults utilize a variety of flowers, including shepherd's needle.

Caterpillar Host Plants.   Larvae feed primarily on sennas (*Cassia*), e.g., *C. fasciculata*, *C. occidentalis*, *C. marilandica*, and *C. bicapsularis*. Reports of clover (*Trifolium*) are based only on laboratory findings.

40. *Eurema nise*

41. *Eurema dina helios*

Dainty Sulphur
Nathalis iole (Boisduval), 1836
PLATE 11 · FIGURE 63 · MAP 43

Etymology. The species is named after Iole, daughter of Eurytus, King of Oechalia.

Synopsis. The Dainty Sulphur has a combination of characters that indicate it has no close relatives within the pierid family. It is probably quite primitive. The species is an excellent colonist, and despite its small size and frail appearance, may extend its range by hundreds of miles in a single season. It appears to have reached Florida, possibly from the Bahamas, around 1913.

Butterfly. Male forewing: $\overline{X} = 1.3$ cm, range 1.2–1.5 cm; female forewing: $\overline{X} = 1.4$ cm, range 1.3–1.5 cm. The Dainty Sulphur is similar to the Barred Sulphur (*Eurema daira*), but the former is consistently smaller, has more pointed forewings, and has a black bar along the costal margin of the dorsal hind wing. Clench (1976a) points out that, in life, the males have a bright orange oval patch at the base of the costal margin on the dorsal hind wing that fades to dull yellow within 2 months after death. This color change may be due to a heat-labile pteridine, one of the pigments responsible for most of the white, yellow, and orange colors found among pierid butterflies (see Introduction). The Dainty Sulphur has two seasonal forms, one with fairly clear yellow hind wings beneath and the other with greatly increased black scaling below. Douglas and Grula (1978) have found that these two forms are produced by differing day-length regimes experienced by the caterpillars. The clear-yellow form is produced when the photoperiod is long (16 hr) and the dark form when

the photoperiod is short (10 hr). In nature the dark form flies during the spring and fall, while the clear-yellow form flies in summer. The dark form can gain heat more rapidly when temperatures are cool, through lateral basking, and thus can more rapidly attain flight temperature. This phenomenon of seasonally variable forms is an example of the *seasonal polyphenism* described by Shapiro (1976). All-white or all-orange individuals of this butterfly are occasionally found.

Range. The Dainty Sulphur occurs in southern Georgia, Florida, the Bahamas, Jamaica, and Hispaniola, having invaded Florida about 1913 (Clench 1976b). A second, more extensive, distributional area extends from the Gulf Coast and Mexico northward through the Great Plains and Southwest sporadically to southwestern Ontario, Virginia, Manitoba, Wyoming, and Washington. The butterfly cannot successfully overwinter in areas with regular cold winters. The species regularly recolonizes the Midwest from the South each summer, reaching areas more than a thousand kilometers from its areas of permanent residence. By contrast, the southeastern Antilles-derived population is much more sedentary and does not undertake northward emigrations.

Habitat. The Dainty Sulphur favors dry open areas, including sandy coastal flats, dry hillsides, and weedy fields or road edges. It is resident in Tropical and Lower Austral Life Zones, but may establish temporary populations in Transition Life Zone areas.

Life History. *Behavior.* The male Dainty Sulphur patrols

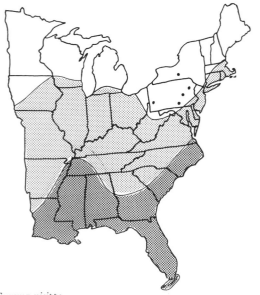

42. *Eurema nicippe*

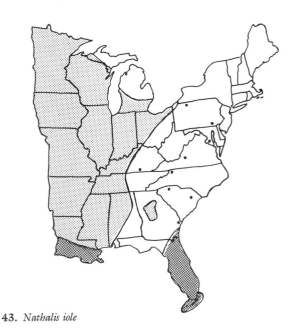

43. *Nathalis iole*

close to the ground in search of females during most daylight hours. One mating pair was found at 1550 hr, with the male as the active carrier. All yellows and sulphurs perch with their wings closed and warm up by orienting their wings perpendicular to the sun's rays (lateral basking).

Broods. The number of generations is indefinite. Douglas and Grula (1978) report a 5-week generation time, and four broods are reported for southern California. The species is reported to be abundant throughout the year in Florida, but it occurs for no more than 6 months in any other eastern state. Extreme dates for several states range as follows: Georgia—May 6 to November 29, Louisiana—July 8 to October 14, Illinois—July 4 to November 13, and Michigan—July 20 to September 13.

Early Stages. The larva is dark green, with setae and paired forward-directed reddish tubercles on the thoracic segments. It has a purple dorsal stripe and lateral fused black and yellow lines. The chrysalis lacks the pointed frontal projection of all other North American pierids, and is slung from a girdle, as are all our papilionids and pierids.

Adult Nectar Sources.   Labrador tea has been reported to be visited in Minnesota.

Caterpillar Host Plants.   The Dainty Sulphur caterpillar feeds only on members of the aster family, particularly on fetid marigold (*Dyssodia*), shepherd's needle (*Bidens pilosa*), sneezeweed (*Helenium*), and cultivated marigold (*Tagetes*). Reports on chickweed (*Stellaria media*) are suspicious, since this plant is a member of the pink family.

Family Lycaenidae: Gossamer Wings

The cosmopolitan Lycaenidae family is probably the most diverse and is second only to the skippers in number of species in our area. Most species are small and the eyes are usually emarginate (indented) near the antennae. The front legs of the males are somewhat reduced, although the females' legs are fully developed.

Both perching and patrolling male-location systems are used, although the latter is more or less restricted to the blues. Courtship, mating, and oviposition usually occur in the afternoon. The butterflies perch with closed or open wings depending upon the subgroup. Many metalmarks perch upside down on the undersides of leaves. The proboscises of most species are short, and floral nectar is the predominant adult food. Only male blues take moisture at wet sand or mud with any regularity.

The caterpillars feed on a variety of plants, often selecting flowers or fruits instead of leaves. Harvester caterpillars feed on aphids, an unusual diet for a butterfly. The eggs are echinoid (urchin-shaped) and are usually laid singly. The caterpillars have small retracted heads and dorsoventrally flattened sluglike bodies, which are often covered with short, fine, dense hair. Green is a frequent body color and there are often diagonal white or yellow chevrons and stripes. Many species have secretory glands on the abdominal dorsum that produce a sugary solution, and many species are tended by ants in a presumed mutualistic association. Lycaenid caterpillars often burrow into their host substrate and are difficult to locate. The chrysalids are usually brown and are rounded at either end. Winter is usually passed by the egg, by partially grown larva (rarely), or by the pupa.

Subfamily Lycaeninae: Harvesters, Coppers, Hairstreaks, and Blues

The Lycaeninae subfamily is distinguished from the metalmarks by a number of structural features. In our area, four tribes occur and are best treated as separate units.

Harvesters (Tribe Gerydini)

The harvesters are a small tribe represented in North America by a single species.

Harvester
Feniseca tarquinius (Fabricius), 1793

PLATE 11 · FIGURE 64 · MAP 44

Etymology. *Feniseca* means "mower" or "person who mows" (Latin). The species is named after Lucius Tarquinius Superbus, semilegendary Etruscan king of Rome.

Synopsis. The Harvester is unusual in that its caterpillars feed on aphids instead of the usual vegetarian fare of their brethren. The adult has an extremely short proboscis, which it uses to feed on aphid "honeydew." Its closest relatives are found in southeast Asia and Africa.

Butterfly. Male forewing: $\bar{X} = 1.43$ cm, range 1.32–1.55 cm. This butterfly has an angled or pointed forewing apex and areas of orange surrounded by black on the dorsal wing surfaces. Ventrally, the hind wing is brownish purple with circles of white scaling. The orange areas are more extensive on individuals of the spring generation.

Range. The Harvester occurs from Nova Scotia west across southern Ontario to Minnesota, thence south to the Gulf states, central Florida, and central Texas.

Habitat. Deciduous or mixed forests, particularly along streams, are the home of this butterfly.

Life History. *Behavior.* Male Harvesters perch at the edge of sunlit leaves 2 to 8 m above the ground. They fly out at other males but usually return to the same vicinity, landing on the center of a leaf, then walking to its edge. Freshly emerged males are occasionally seen at wet spots on roads or along streams.

Broods. The number of generations is not well documented, but there appear to be two in such areas as northern Michigan and Maine, and three everywhere to the south. The statement of three generations is based upon flight dates from several states, and is made in spite of the Clarks' (1951) declaration of up to eight broods in Virginia and the extremely brief larval development time (see below).

Early Stages. The caterpillars feed under a clump of wooly aphids and a silken web. Their complete development is brief, requiring only 7 to 11 days (Clark, 1926). The caterpillar is green-brown with faint olive stripes and has long white hairs emanating from the intersegmental areas (joints). The pupa is brown to brown-green with an arrangement of irregular dark patches resembling a monkey's face.

Adult Food. The Harvester's proboscis is proportionately shorter than that of any other butterfly. Adults feed primarily on aphid honeydew but are also seen at animal droppings.

Caterpillar Hosts. The caterpillars are carnivorous. They feed on wooly aphids (*Prociphilus, Neoprociphilus, Schizoneura,* and *Pemphigus*), which in turn feed principally on alders but also on witch hazel, beech, ash, wild currant, and hawthorn.

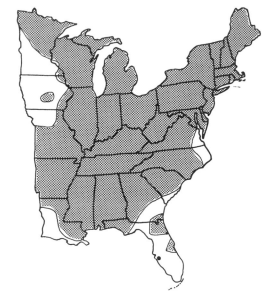

44. *Feniseca tarquinius*

Coppers (Tribe Lycaenini)

Coppers are predominately north temperate (or holarctic) in occurrence. The adults are small to medium and are often strongly sexually dimorphic. The males usually have iridescent red or purple dorsal wing surfaces. Specialized scent patches, or androconia, are absent.

Most males perch with partially outspread wings, although in a few species they engage in patrolling. Adults feed only on floral nectar and very rarely visit moist spots for water.

The caterpillars feed on plants in several families, but those in the buckwheat and rose families are most frequently consumed. The eggs are usually laid under host leaves, and the first-instar larvae winter inside the egg capsule (chorion).

American Copper
Lycaena phlaeas (Linnaeus), 1761
PLATE 11 · FIGURES 65 & 66 · MAP 45

Etymology. The genus is probably named after Lycaon, king of the Arcadian town Lykosury (Greek myth).

Synopsis. The American Copper is one of our most beautiful common butterflies, with its iridescent orange-red forewings. Although the species is called the American Copper and was named subspecies *americana,* our eastern populations probably were introduced from Scandinavia during the colonial period.

Butterfly. Male forewing: $\overline{X}$ = 1.3 cm, range 1.2–1.4 cm; female forewing: $\overline{X}$ = 1.5 cm, range 1.3–1.6 cm. No other eastern butterfly shares the spotted orange-red forewing and

gray-black red-margined hind wing. In New England, the form "fasciata" has elongated black spots on the forewing, and may sometimes be common. Early spring butterflies are a lighter red. Our butterflies are most similar to those found in Scandinavian countries.

Range. The American Copper ranges from Nova Scotia west across southern Canada and the Great Lakes states to North Dakota. It ranges south to Georgia (mountains), Tennessee, and Arkansas. Native single-brooded populations occur in the western mountains and the Arctic.

Habitat. The butterfly is always found in habitats disturbed by man, including old fields, pastures, vacant lots, and landfills.

Life History. *Behavior.* The males perch atop grass or on tall composites in their open habitat and fly at and interact actively with almost any passing insect. Mating occurs all day (0845–1430 hr), with the female acting as the carrier.

Broods. In the northern portion of our area this copper is bivoltine (June–early July and August–September) and probably has three broods everywhere to the south, e.g., Virginia (mid-April through May, mid-June through July, and mid-August through September). In most areas individuals may be found until late October.

Early Stages. The pale-green eggs are laid singly on host leaves or stems. The young caterpillars chew holes in the underside of young host leaves and later make longitudinal channels. Development takes about three weeks and the chrysalis may be found under leaves or rocks. Winter is passed in the pupal stage. The caterpillars are covered with short hairs and are variably colored rose-red to green. There is a red dorsal stripe on some caterpillars. The chrysalis is light brown, is tinged with pale yellow-green, and is spotted with black.

Adult Nectar Sources. In the spring, common buttercup seems to be preferred, while white clover, yarrow, ox-eye daisy, and butterflyweed are chosen in early summer. Later, various composites are visited in late summer and fall.

Caterpillar Host Plants. Sheep sorrel (*Rumex acetosella*) is the primary host almost everywhere in the East, while curled dock (*Rumex crispus*) is fed on occasionally. Both plants are weeds introduced from Europe. This lends strong circumstantial support to the idea that the copper is also an introduction.

Great Copper
Lycaena xanthoides dione (Scudder), 1869
PLATE 12 · FIGURES 67 & 68 · MAP 46

Etymology. The species name is derived from the Greek term *xanthos,* "yellow."

Synopsis. Basically a grassland butterfly, the Great Copper is found in both the Great Plains states and valley grasslands of California.

Butterfly. Male forewing: $\overline{X}$ = 2.0 cm, range 1.7–2.1 cm; female forewing: $\overline{X}$ = 2.0 cm, range 2.0–2.2 cm. This is our only "copper" with navy-gray dorsal wing surfaces.

Range. The Great Copper ranges from Illinois, western Wisconsin, Minnesota, and Manitoba west to Alberta, west-ern Montana, and central Colorado. It is found southward to Oklahoma and Missouri. The species may be expected to spread eastward and southward in the future. The typical subspecies occurs in California.

Habitat. This copper is found in native prairies, pastures, and weedy fields.

Life History. *Behavior.* Males perch on tall vegetation or flowers holding their wings partly spread while awaiting the appearance of receptive females (Scott and Opler 1975). They also patrol back and forth on some occasions. Male

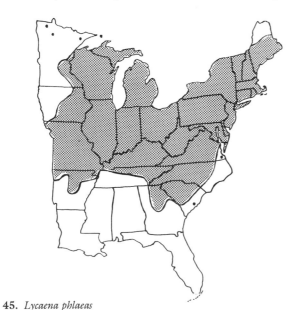

45. *Lycaena phlaeas*

46. *Lycaena xanthoides dione*

—male interactions often result in long chases, with one male returning to the vicinity of the original perch. Mating occurs in the morning (0940–1120 hr). Most females mate only once, but a few mate twice in their lifetime. The adults live as long as 11 days.

Broods. The butterfly is univoltine, with a flight each summer. Illinois flight dates range from June 9 to July 30.

Early Stages. Eggs are laid singly on the underside of host leaves and feeding takes place the following spring. The caterpillar is green, with a darker dorsal stripe or with a red dorsal stripe surrounded by yellow-green. The chrysalis is brown-pink and is dappled with many black marks.

Adult Nectar Sources. Great Coppers are avid flower visitors. In Iowa, they were seen in numbers at dogbane, common milkweed, Sullivant's milkweed, butterflyweed, and large houstonia.

Caterpillar Host Plants. Broad dock (*Rumex obtusifolius*), an alien weed, is one recorded host, but the native prairie host plant is not recorded.

Bronze Copper
Lycaena hyllus (Cramer), 1775
PLATE 12 · FIGURE 69 · MAP 47

Etymology. The species is named for Hyllus, son of Herakles (myth).

Synopsis. This large copper is strikingly dimorphic, the male being iridescent orange-purple dorsally, and the female having a pale-orange, black-dotted forewing and dark-gray hind wing.

Butterfly. Male forewing: $\overline{X} = 1.8$ cm, range 1.7–1.8 cm; female forewing: $\overline{X} = 2.0$ cm, range 1.7–2.2 cm. Ventrally, the forewings are pale orange with a white outer margin and the hind wings are matte white with a broad orange submarginal band. Ventrally both wings have many round black dots. This combination separates the Bronze Copper from any other eastern butterfly.

Range. The Bronze Copper ranges from Newfoundland and the Maritime Provinces west across southern Canada to Alberta, eastern Montana and central Colorado. It ranges south to Maryland, West Virginia, Mississippi (once), and Arkansas.

Habitat. This copper is found in low, wet meadows, marshes, and pond edges, generally on neutral to alkaline soils.

Life History. *Behavior.* Males perch on low vegetation in the vicinity of their hosts. They will sun with wings spread wide (180°), an unusual habit for coppers. Mating occurs in the afternoon (1215–1735 hr). In late afternoon they perch with closed wings oriented perpendicular to the sun's rays.

Broods. In the northern portions of its range, the Bronze Copper is bivoltine, e.g., in New York's Finger Lakes region, June 17–July 14 and August 5–September 22. Farther south, as in Maryland and Pennsylvania, where the butterflies emerge as early as May 17 and are on the wing as late as November 3, there is also a rudimentary third brood.

Early Stages. The caterpillar is bright yellow-green with a blackish dorsal stripe, while the chrysalis is light orange-brown with darker blotches.

Adult Nectar Sources. Bronze Coppers seldom visit flowers, but red clover and blackberry are two plants at whose flowers these butterflies nectar.

Caterpillar Host Plants. Water dock (*Rumex orbiculatus*) is the native host throughout most of the Bronze Copper's range, but curled dock (*Rumex crispus*) is also selected with some frequency.

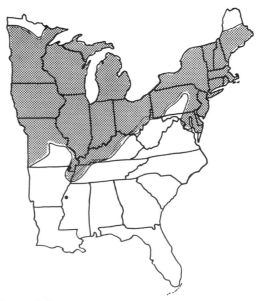

47. *Lycaena hyllus*

Bog Copper
Lycaena epixanthe (Boisduval & LeConte), 1833

PLATE 12 · FIGURES 70 & 71 · MAP 48

Synopsis. This small butterfly is limited to acid bogs, not the fresh marshes or marsh meadows favored by the Bronze Copper. Its flight is low and weak among the cranberries it favors as a host.

Butterfly. Male forewing: $\overline{X}$ = 1.1 cm, range 1.0–1.2 cm; female forewing: $\overline{X}$ = 1.1 cm, range 1.1–1.2 cm. Dorsally, the male has a blue-purple iridescent gloss, while the female is mouse gray. Ventrally, the butterflies vary from cream yellow to gray-white and have round black spots. There are several clinal subspecies. Populations of some bogs have individuals marked differently from those of the other sites, due to the poor dispersal of the copper and isolation of the bogs. Most populations are very small, consisting of less than a hundred butterflies.

Range. The Bog Copper occurs from Newfoundland and the Maritime Provinces west to Minnesota and south to New Jersey, western Maryland, and southern Michigan.

Habitat. This mothlike butterfly is limited to acid bogs with cranberries in the Transition (rarely) and Canadian Life Zones.

Life History. *Broods.* The Bog Copper is univoltine. Flight dates in Michigan's upper peninsula range from July 10 to August 21, while in New York the butterflies may be found from June 21 to August 1.

Adult Nectar Sources. The Bog Copper rarely visits flowers.

Caterpillar Host Plants. The only known hosts are cranberries (*Vaccinium macrocarpon* and *V. oxycoccos*). Eggs are laid singly under leaves and overwinter until the following spring.

48. *Lycaena epixanthe*

Dorcas Copper
Lycaena dorcas Kirby, 1837

MAP 49

Etymology. The species is derived from Dorcas, a Christian female disciple who made coats and garments for the poor.

Synopsis. This butterfly is so closely related to the Purplish Copper that some individuals cannot be distinguished. It differs in minor pattern differences, brood number, and hosts.

Butterfly. Male forewing: $\overline{X}$ = 1.3 cm, range 1.3–1.4 cm; female forewing: $\overline{X}$ = 1.3 cm, range 1.2–1.4 cm. The Dorcas Copper differs from the Purplish in having a reduced, orange submarginal hind-wing band and in having more pronounced black spotting ventrally. In females, the orange-yellow areas on all wings dorsally tend to be reduced or absent. The highly restricted subspecies *claytoni* occurs in central Maine and may be endangered.

Range. The Dorcas Copper ranges from Newfoundland and Labrador west across Canada to Washington and British Columbia, and north to Northwest Territories, the Yukon, and northern Alaska. It is found south to central Maine (subspecies *claytoni*), northern Ohio, Michigan, northern Wisconsin, Minnesota, and then throughout the Rocky Mountains to northern New Mexico (Ferris 1977).

Habitat. In our area the Dorcas Copper is found in upland fields and on the fringes of sphagnum bogs in the vicinity of its host, shrubby cinquefoil.

Life History. *Broods.* The Dorcas Copper is univoltine throughout its range. The Maine population has a flight period from July 20 to August 25, while in southern Michigan the butterfly flies from June 19 to August 15. In northern Michigan individuals may be seen as late as September 21.

Behavior. Dorcas Coppers remain in the vicinity of their hosts and do not range widely. Eggs are laid singly on the underside of host leaflets and fall to the ground in the autumn. The young caterpillars must find their way back to the host leaves the following April. There are five instars, and development is usually complete by mid-June or early July (Newcomb 1909).

Caterpillar Host Plant. In the East the only documented host is shrubby cinquefoil (*Potentilla fruticosa*), although other plants (primarily cinquefoils) are used elsewhere in its range.

Purplish Copper
Lycaena helloides (Boisduval), 1852
MAP 50

Etymology. The species is named after Helle, a maiden who in fleeing from her stepmother Ino, fell into the Dardanelles (Hellespont) and drowned (Greek myth). The name helloides means "similar to helle," i.e., after the European *Lycaena helle*.

Synopsis. The Purplish Copper is a more southern, multiple-brooded relative of the Dorcas Copper that probably became a separate species during the Ice Ages.

Butterfly. Male forewing: $\overline{X}$ = 1.4 cm, range 1.3–1.5 cm, female forewing: $\overline{X}$ = 1.6 cm, range 1.5–1.7 cm. This copper may be distinguished from the Dorcas by its more extensive submarginal orange band on the male's dorsal hind wings, its more pale orange on the female dorsally, and its less extensive black marks on the ventral wing surfaces. It also differs in habitat, behavior, number of broods, and hosts. Early spring individuals are darker below.

Range. The Purplish Copper is found from southern Ontario westward in a gradually widening arc encompassing the upper Midwest, northern prairie states, nondesert portions of the western states, and southern Canada. It is now expanding its range into Missouri and Illinois.

Habitat. This species is usually found in disturbed habitats in our area. These include roadsides, open fields, and wet meadows.

Life History. *Behavior.* Males usually patrol extensively but occasionally perch in their attempts to locate suitable females. The species wanders more widely than does the Dorcas Copper.

Broods. In Illinois and Michigan the butterfly is bivoltine, although as many as seven broods may be produced in the lowlands of California. In Illinois the butterflies are found from May 19 to mid-July and from mid-August to October 21. The second generation is more abundant.

Early Stages. Single eggs are laid in litter near the host, on young shoots, or under leaves. In the summer, development is direct to the chrysalis. It has been reported that the pupa

49. *Lycaena dorcas*

50. *Lycaena helloides*

overwinters, but winter is probably passed in the egg stage at many localities. The caterpillar is green, with many lateral, oblique, yellow lines and is covered with many short, whitish hairs (Coolidge 1924).

Caterpillar Host Plants. The Purplish Copper rarely selects cinquefoils, but usually oviposits on or near various docks (*Rumex*) and knotweeds (*Polygonum*). Reports of other hosts must be viewed with suspicion.

Hairstreaks (Tribe Theclini)

Hairstreaks are small butterflies whose tropical species are often iridescent blue above, while their more temperate species have duller colors. There are usually one or two long hairlike projections from the trailing hind-wing margin and often a small red and black eyespot ("thecla spot") at the hind-wing anal angle. The males usually have scent patches or brands on the forewing disc.
Males of most species perch during late afternoon while awaiting the passage of receptive mates. Mated pairs of many species have been found at night. The butterflies usually perch with closed wings, and lateral basking is universal. The butterflies often perch high in trees, and their flight is rapid and erratic.

Atala
Eumaeus atala florida Röber, 1926
PLATE 12 · FIGURE 72 · MAP 51

Etymology. The genus is named after Eumaeus, the faithful swineherd of Odysseus. Atala was the heroine, the daughter of an American Indian chief, in a romantic novel by Chateaubriand.

Synopsis. Atala is a tropical butterfly endemic to the northern Caribbean. It is very rare and was believed to be extirpated in our area at one time (Rawson 1961). Recently discovered colonies are seriously threatened by development (Baggett 1982). It is our only butterfly that uses a cycad as a caterpillar host.

Butterfly. Male forewing: $\overline{X}$ = 2.1 cm, range 1.9–2.3 cm; female forewing: $\overline{X}$ = 2.2 cm, range 2.0–2.4 cm. There is no other United States butterfly with which Atala might be easily confused, except possibly the Great Purple Hairstreak (*Atlides halesus*). Atala is sexually dimorphic; dorsally, the male's wings are covered with iridescent green, while the females have blue iridescence restricted to the basal areas.

Range. Atala occurs or occurred in southeastern Florida (Broward, Dade, and Monroe Counties), the Bahamas (Abaco, Andros, Grand Bahama), and Cuba.

Habitat. In Florida, Atala is limited to shaded tropical hardwood hammocks, and the butterflies rarely stray to adjacent open areas. However, in the Bahamas their habitat is open Caribbean pine forest, and the butterflies are frequently seen in the open.

Life History. *Behavior.* The butterflies spend much of their time in dense shade perching on top of or beneath leaves or on twigs or branches. When perched, the Atalas are amazingly docile. They may be easily picked up or repeatedly touched. Their flight is slow, direct, and fluttering; it is reminiscent of some day-flying moths. Courtship is very similar to that of some Danaine nymphalids, e.g., the Queen (*Danaus gilippus*), in which the male hovers in front of the female while fanning his abdominal hair tufts with his wings.

Broods. In Florida, Atala is found all year except in October but is most abundant during early summer.

Early Stages. The eggs are laid in batches on young leaf shoots and the resultant caterpillars feed openly in groups.

51. *Eumaeus atala florida*

The mature caterpillar is brilliant red, with the body segments each "humped up" and bearing transverse rows of four tubercles. Each tubercle in turn bears a tuft of setae. In addition, there are two dorsal rows of bright yellow spots.
Adult Nectar Sources. Intensive studies by Jason Wein-traub have shown lantana to be the preferred nectar source for adults, while shepherd's needle, periwinkle, and wild coffee flowers are also visited.
Caterpillar Host Plants. The only native host in Florida and the Bahamas is coontie (*Zamia pumila*).

Great Purple Hairstreak
Atlides halesus (Cramer), 1777
PLATE 13 · FIGURES 73 & 74 · MAP 52

Etymology. The genus name is possibly an incorrect formation for Atlantides, meaning "the race of Atlas." The species is named after Stephen Hales, an English botanist.
Synopsis. The Great Purple Hairstreak is one of the most stunning eastern butterflies, having brilliant blue, iridescent upper wing surfaces. It is our only butterfly whose caterpillars feed on mistletoe.
Butterfly. Male forewing: $\bar{X}$ = 1.8 cm, range 1.4–1.9 cm; female forewing: $\bar{X}$ = 2.1 cm, range 2.0–2.4 cm. Among eastern butterflies, only the Atala (*Eumaeus atala florida*) also has a bright red abdomen, but Atala lacks any hint of tails, in opposition to this species.
Range. The Great Purple Hairstreak ranges from New Jersey, Kentucky, and southern Illinois south to Florida and the Gulf, usually in low-lying areas. The species also ranges west to central California and thence south to Guatemala.
Habitat. In the East, the Great Purple Hairstreak is usually found in association with moist deciduous or mixed forest.
Life History. *Behavior.* Males perch at the tops of trees or other prominences in the afternoon awaiting receptive females. At rest, the butterfly "rubs" its hind wings back and forth. This presumably draws the attention of the would-be predator, who assumes that the tails are antennae. When the attacker goes after the "false head," he ends up with only a bit of hind wing and the butterfly escapes (Robbins 1981). This behavior is shown by most hairstreaks and many blues.
Broods. There are three generations in the southeastern states, although the adults may be found during every month in Florida. In Virginia the broods are timed as follows: late March–early April, July, and mid-August–late November.
Early Stages. The caterpillar pupates under loose bark or in crevices at the base of trees containing its mistletoe host. The mature caterpillar is green and densely covered with short, fine, orange hairs. The chrysalid is robust, chestnut brown, heavily mottled with black, and is covered with short yellow-orange hairs.
Adult Nectar Sources. Adults feed at wild plum, shepherd's needle, sweet pepperbush, goldenrod, and Hercules club as favorites.
Caterpillar Host Plant. This lovely butterfly's larvae feed only on mistletoe (*Phoradendron*).

52. *Atlides halesus*

Maesites Hairstreak
Chlorostrymon maesites (Herrich-Schäffer), 1864
MAP 53

Etymology. The genus name is derived from the greek *chloros*, "green," and *Strymon*, a river in Macedonia.
Synopsis. This Antillean endemic is a jewel, having green below and iridescent blue-purple above in the male. It usually perches high above the ground and is rarely encountered.

Butterfly. Male forewing: $\bar{X}$ = 1.0 cm, range 1.0–1.1 cm; female forewing: $\bar{X}$ = 1.0 cm, range 1.0–1.1 cm. Within its range, the Maesites is the only small hairstreak with green ventral wing surfaces. Dorsally, the male is an iridescent blue-purple and the female a duller gray-blue.

Range. The Maesites Hairstreak is restricted to a few southern Florida Keys, as well as several of the larger Caribbean Islands. Maesites is possibly conspecific with the Telea Hairstreak (*Chlorostrymon telea*), which ranges from southern Texas to South America.

Habitat. Maesites Hairstreak is associated with evergreen or semideciduous forest and hammocks within seasonally dry Tropical Life Zone areas.

Life History. *Behavior.* The males perch 2 or 3 m above the ground on sunlit tree foliage, rendering it difficult to observe or capture them.

Broods. There are probably two or three generations each year, with peak numbers found in late May to mid-June. In the Florida Keys, there are records from December to early February, April, and late May to the end of July.

Adult Nectar Sources. In Florida, adults have been found on flowers of Brazilian pepper, Guamachil, ape's earring, and shepherd's needle. In Central America the closely related Telea Hairstreak is fond of mango flowers.

Caterpillar Host Plant. The caterpillars have been raised in the laboratory on woman's tongue (*Albizia lebbeck*), a tropical mimosoid legume (R. Boscoe, unpublished). Other legumes are probably used in nature.

53. *Chlorostrymon maesites*

St. Christopher's Hairstreak
Chlorostrymon simaethis (Drury), 1770

MAP 54

Synopsis. This widespread tropical hairstreak has been found in extreme southern Florida since February 1974 and is now resident there. Presumably it strayed from Cuba. This butterfly is found in seasonally dry, lowland, tropical forests from the Caribbean and southern Texas south to South America. Adults are usually found on small white flowers of tropical trees and vines. The caterpillars feed on balloon vine (*Cardiospermum halicacabum*) in Florida, and are known to use many other hosts elsewhere in the species' range.

Alcestis Hairstreak
Phaeostrymon alcestis (Edwards), 1871

Synopsis. This southwestern butterfly was found once in Barry Country, Missouri. It may be distinguished from the Northern Hairstreak (*Fixsenia ontario*) by the bold white discal marks on both wings ventrally. It normally ranges from Kansas west to northeastern New Mexico and south to Texas and southeastern Arizona. The hairstreak has a single annual flight in late May and June. Its caterpillar host is western soapberry (*Sapindus saponaria* var. *drummondii*).

Coral Hairstreak
Harkenclenus titus (Fabricius), 1793

PLATE 13 · FIGURE 75 · MAP 55

Etymology. The genus name is an anagram of *Harry Kendon Clench* (1925–1980), a curator at the Carnegie Museum and lycaenid specialist. The species name is that of a Roman emperor.

Synopsis. The Coral Hairstreak is a widespread North American species whose larvae feed on wild plums. The adults have an inordinate predilection for butterflyweed flowers.

Butterfly. Male forewing: $\bar{X}$ = 1.5 cm, range 1.2–1.6 cm; female forewing: $\bar{X}$ = 1.7 cm, range 1.5–1.8 cm. The adult, which lacks tails and has a submarginal band of red-orange spots, can be confused with only the tailed Acadian Hairstreak. The Coral Hairstreak is extremely variable geographically, and there are several named clinal subspecies of dubious merit.

Range. The Coral Hairstreak occurs from Maine west through southern Ontario, the Great Lakes States, and southern Manitoba to southern British Columbia and thence south to the Florida panhandle, the Gulf states, southern Texas, central New Mexico, and northeastern California.

Habitat. The Coral Hairstreak requires open areas in or near scrubby woodland. It has been found in city parks, barrens, neglected pastures, and sparsely wooded hillsides.

Life History. *Behavior.* Males perch on hilltops or small trees on flats from about 1000 to 1900 hr and are extremely aggressive toward other males and small insects. During late afternoon they orient themselves laterally to the sun's rays. Mated pairs have been seen from 1200 to 1400 hr, but mating may occur later as well. Females lay their eggs on host twigs.

Broods. The Coral Hairstreak is univoltine in early to mid-summer, with emergence progressively earlier southward. In Michigan's upper peninsula, flight dates range from July 24 to August 23; in Virginia, from June 20 to July 30; and in Georgia, from May 14 to July 6.

Early Stages. The deep green eggs overwinter on host twigs and hatch during the following spring, after which development is direct. The caterpillars feed on host flowers and young fruit, and are usually tended by ants, which seek the sugar segretions produced by the caterpillar's abdominal glands (Harvey and Webb 1981). The mature caterpillar has a black head with a dull white streak across the lower front. Its body is dull green with a yellow tint anteriorly and dull pink patches on the thoracic dorsum. The pupa is pale brown with many dark brown or black speckles and is also covered with many short brown hairs.

Adult Nectar Sources. The majority of flower visits are to butterflyweed, but dogbane, common milkweed, swamp milkweed, New Jersey tea, meadowsweet, white sweet clover, and shrub houstonia receive some visits.

Caterpillar Host Plants. The usual hosts are wild cherry (*Prunus serotina*), wild plum (*Prunus virginiana* and *P. angustifolia*. Chokeberry (*Aronia melanocarpa*) is a host in some areas.

54. *Chlorostrymon simaethis*

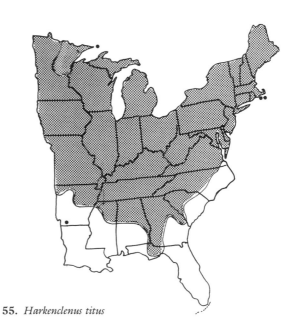

55. *Harkenclenus titus*

Acadian Hairstreak
Satyrium acadica (Edwards), 1862

PLATE 13 · FIGURE 76 · MAP 56

Etymology. The genus name is derived from the mythical Greek *Satyros,* a class of small woodland deities. The species name is derived from Acadia, the original name for northern New England and Nova Scotia.

Synopsis. The Acadian Hairstreak, which is pale gray beneath, is a denizen of willow thickets along streams and in marshes. It is a close relative of the western Sylvan Hairstreak (*Satyrium sylvinus*).
Butterfly. Male forewing: $\overline{X}$ = 1.5 cm, range 1.4–1.6 cm; female forewing: $\overline{X}$ = 1.6 cm, range 1.4–1.7 cm. The Acadian Hairstreak may be distinguished by its gray ventral wing surfaces with their postmedian row of round black dots.
Range. The species ranges from Nova Scotia west to southern British Columbia and south to New Jersey, Maryland, the upper Midwest, Colorado, and Idaho.
Habitat. This species is found along willow-lined streams and in marshes with low willows. It is a Transition Life Zone species.
Life History. *Behavior.* Males perch on willows and other nearby vegetation from 1 to 2 m above ground during the early morning and late afternoon (0730–0900 and 1730–1900 hr). Males perch parallel to the afternoon sun and usually sit at a slight tilt to maximize their exposure. Mated pairs have been observed at 1045 hr and in the evening from 1800–2000 hr.
Broods. The Acadian Hairstreak is univoltine. Flight dates are progressively earlier southward. They range from July 24 to August 6 in Michigan's upper peninsula and late June to early July in the Delaware Valley. In Illinois there are records from June 14 to August 22.
Early Stages. The mature larva is broad and green and has two longitudinal yellow stripes on each side, between which are placed rows of oblique yellow or white bars. The head is

brown. The pupa is dull yellow-brown with brown-black spotting.
Adult Nectar Sources. Dogbane, butterflyweed, common and swamp milkweeds, New Jersey tea, meadowsweet, and thistles are visited.
Caterpillar Host Plants. Larval hosts are limited to willows, including black willow (*Salix nigra*) and silk willow (*Salix sericea*).

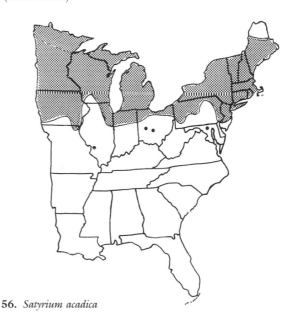

56. *Satyrium acadica*

Edwards' Hairstreak
Satyrium edwardsii (Grote & Robinson), 1867
PLATE 13 · FIGURE 77 · MAP 57

Etymology. The species was named in honor of William Henry Edwards (1822–1909), sometimes considered the father of American butterfly study (see Introduction).
Synopsis. Edwards' Hairstreak is a butterfly of oak thickets and barrens and may appear sporadically in other forested habitats.
Butterfly. Male forewing: $\overline{X}$ = 1.5 cm, range 1.4–1.6 cm; female forewing: $\overline{X}$ = 1.5 cm, range 1.4–1.7 cm. The Edwards' Hairstreak is similar to the Banded Hairstreak but may be distinguished ventrally by the postmedian line of separate, oval, black spots on the wings.
Range. This species extends from the Maritime Provinces west to Manitoba and south to Georgia, Missouri, and Texas. It is very rare and sporadic south of Pennsylvania.
Habitat. This species is found in oak thickets within sand barrens, shale barrens, limestone ridges, or other rocky habitats. The species is almost entirely limited to the Transition Life Zone.

Life History. *Behavior.* The males are fond of perching on oak twigs and leaves, and may be seen interacting during most daylight hours (0900–1800 hr).
Broods. Edwards' Hairstreak is univoltine. Flight dates range from June 20 to July 24 in Michigan and from May 13 to July 6 in Georgia.
Early Stages. The mature larva has a black head with a narrow white band on the front. Its body is brown, with paler, dull, yellow-brown markings and is covered with black tubercles bearing short brown hairs. The pupa is dull yellow-brown with many darker spots.
Adult Nectar Sources. Adults visit milkweeds, dogbane, goldenrod, meadowsweet, staghorn sumac, New Jersey tea, and white sweet clover in their efforts to obtain energy sources.
Caterpillar Host Plants. Scrub oak (*Quercus ilicifolia*) is the usual host, but black oak (*Quercus velutina*) is also fed upon.

Banded Hairstreak
Satyrium calanus (Hübner), 1809

PLATE 14 · FIGURES 79 & 80 · MAP 58

Etymology.   The species is named for Calanus, an ancient East Indian gymnosophist.

Synopsis.   The Banded Hairstreak is our most common eastern hairstreak, although years of abundance may be followed by years of great scarcity. It is distinguished from the Hickory Hairstreak only with difficulty.

Butterfly.   Male forewing: X̄ = 1.5 cm, range 1.5–1.7 cm; female forewing: X̄ = 1.5 cm, range 1.4–1.7 cm. Ventrally, the Banded Hairstreak has relatively continuous unbroken postmedian lines. In the Edwards' Hairstreak (*Satyrium edwardsii*), these lines are broken into separate ovoid spots, and in the Hickory Hairstreak (*Satyrium caryaevorum*) they are widely spaced and irregular. Typical *calanus* from Florida intergrades with subspecies *falacer* in coastal Georgia and South Carolina. The former is slightly larger (forewing length up to 1.8 cm), has a dorsal orange spot near the tails, and has longer tails. Evidence indicates there may be two species representing this complex in Florida (D. Baggett, personal communication).

Range.   The Banded Hairstreak occurs from Nova Scotia west to central Manitoba and south to central Florida, the Gulf Coast, Texas, and central Colorado.

Habitat.   The Banded Hairstreak occurs in a variety of forested situations and adjacent edges and fields. Its only requirements are the presence of suitable hosts and nearby nectar sources. It is reasonably tolerant of urban areas, having been found commonly in several New York City parks.

Life History.   *Behavior.* Males perch on shrubs or low tree branches on hilltops or in forest glades during most daylight hours (0730–1800 hr). Their interactions with other males are particularly strong in early morning and late afternoon. *Broods.* This butterfly is univoltine, the flight dates becoming progressively earlier southward. In Michigan's upper peninsula, flight dates extend from July 6 to August 20; in Virginia they range from June 7 to August 3. In southeastern Georgia typical *calanus* flights occur from May 8 to 29, and in Florida they begin in April.

Early Stages. The pale-green eggs are laid on host twigs in summer but do not hatch until the following spring. The young larvae feed on catkins at first and then switch their attention to young leaves, upon which they complete their development. The caterpillar is green, marked with faint white oblique lines, and is covered with brownish pubescence. A brown form of the caterpillar is also found. It usually has a series of brown trapezoidal marks running up the dorsum.

Adult Nectar Sources.   With its short proboscis, the butterfly favors flowers with short tubes. Dogbane and common milkweed are preferred, but New Jersey tea, small-flowered dogwood, yarrow, staghorn sumac, meadowsweet, white sweet clover, and chinquapin are also visited avidly.

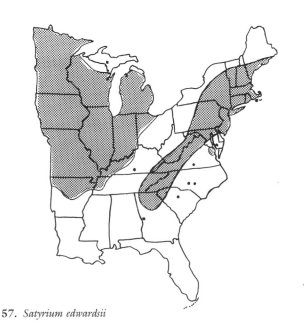

57. *Satyrium edwardsii*

58. *Satyrium calanus*

Caterpillar Host Plants. Larval hosts include white oak (*Quercus alba*), chestnut oak (*Quercus prinus*), bluejack oak (*Quercus incana*), turkey oak (*Quercus laevis*), butternut (*Ju-glans cinerea*), Texas black walnut (*Juglans microcarpa*), and hickories (*Carya*). Hawthorn (*Crataegus*) has been reported but is not a likely host.

Hickory Hairstreak
Satyrium caryaevorum (McDunnough), 1942
MAP 59

Etymology. The species name is derived from *Carya*, the generic name for hickory, and the Latin verb *vorare*, "to eat."

Synopsis. The Hickory Hairstreak is extremely cyclic, experiencing periodic boom years. It is very similar to the Banded Hairstreak, and a few individuals may be indistinguishable on the basis of superficial characters.

Butterfly. Male forewing: $\bar{X}$ = 1.5 cm, range 1.4–1.6 cm; female forewing: $\bar{X}$ = 1.5 cm, range 1.4–1.6 cm. In most cases the Hickory Hairstreak may be distinguished from the Banded Hairstreak (*Satyrium calanus*) by the ventral postmedian line, which is double and filled in with dark brown; while that of the ventral hind wing is offset. Klots (1960) gives a detailed description of the Hickory Hairstreak's features.

Range. This hairstreak ranges from Vermont and Connecticut west through southern Quebec and Ontario to Minnesota and Iowa, thence south patchily to Arkansas and Georgia.

Habitat. This localized hairstreak is found in association with deciduous forest and second growth within the Transition Life Zone. It is usually found only where there are richer soils, whereas the Banded Hairstreak is most abundant on poor rocky or sandy soils (R. Robbins, personal communication).

Life History. *Behavior.* Males begin perching later in the morning than males of the Banded Hairstreak, i.e., after 0900 hr, and they usually select perches very high up in trees.

Broods. The Hickory Hairstreak is univoltine. It usually has a flight in late June and early July, but the butterflies may be seen into early August in northern New York (July 1–August 10).

Early Stages. The mature caterpillar is yellow-green, with its dorsal area bounded by two even, parallel white lines. It has oblique yellow-white lines and a lateral white line running around most of its body. The chrysalid is brown, speckled with black.

Adult Nectar Sources. Dogbane, common milkweed, staghorn sumac, white sweet clover, and New Jersey tea are all visited with some frequency.

Caterpillar Host Plants. Larval hosts are predominantly hickories. Shagbark hickory (*Carya ovata*) and pignut hickory (*Carya glabra*) are known hosts, while butternut (*Juglans cinerea*), red oak (*Quercus rubra*), chestnut (*Castanea dentata*), white ash (*Fraxinus americana*), and black ash (*Fraxinus nigra*) have also been reported.

59. *Satyrium caryaevorum*

King's Hairstreak
Satyrium kingi (Klots & Clench), 1952
MAP 60

Etymology. The species was named in honor of Mr. Harold L. "Verne" King, its discoverer. Mr. King, a retired executive, has spent many years collecting hairstreaks and metalmarks in Latin America.

Synopsis. King's Hairstreak is a denizen of shaded swamp woods and hammocks along the south Atlantic and Gulf Coast.

Butterfly. Male forewing: $\overline{X}$ = 1.6 cm, range 1.4–1.7 cm; female forewing: $\overline{X}$ = 1.6 cm, range 1.4–1.6 cm. King's Hairstreak is characterized by the displaced outer hind-wing margin just above the tails and the ventral spot bands. Its closest relative is the Striped Hairstreak (Klots and Clench 1952).

Range. This rare hairstreak occurs from coastal Maryland southward to northern Florida and east along the Gulf Coast to Mississippi. It occurs primarily on the coastal plain but is also found in low mountains and piedmont in the Gulf states.

Habitat. King's Hairstreak is found in hardwood hammocks, near wooded streams, and at the edge of swamps. Its habitats are within the Lower Austral Life Zone.

Life History. *Behavior.* Adults rarely venture our of their shaded forest habitat.

Broods. Like others in the genus, King's Hairstreak is univoltine. In coastal North Carolina, flights occur from May 31 to June 6, and in coastal Georgia from May 11 to June 1.

Early Stages. Eggs are laid on host twigs and hatch the following spring. The young caterpillars bore into leaf buds and later feed on young expanding leaves. Unlike most hairstreak larvae, they do not eat the flowers or fruits. The caterpillars are similar to those of the Striped Hairstreak but are a lighter green.

Adult Nectar Sources. King's Hairstreak has been found at Allegheny Chinquapin flowers in North Carolina and Georgia and at blooms of sourwood in southeastern Virginia.

Caterpillar Host Plants. The only larval host is horse sugar or common sweetleaf (*Symplocos tinctoria*), which is the only native member of the sweetleaf family (Symplocaceae) in North America.

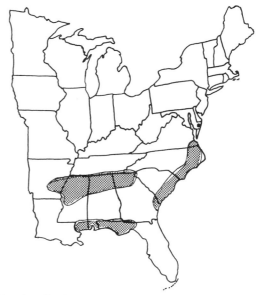

60. *Satyrium kingi*

Striped Hairstreak
Satyrium liparops (LeConte), 1833
PLATE 13 · FIGURE 78 · MAP 61

Etymology. The species name is possibly derived from *liparos,* Greek for oily, and *opis,* meaning nymph or naiad, an epithet of Diana.

Synopsis. Through much of its range, the Striped Hairstreak is often lost among the myriad Banded Hairstreaks (*Satyrium calanus*), which it superficially resembles, although it may be common in southern New England.

Butterfly. Male forewing: $\overline{X}$ = 1.4 cm, range 1.3–1.5 cm; female forewing: $\overline{X}$ = 1.4 cm, range 1.3–1.7 cm. Ventrally, this species may be distinguished by the widely spaced, broken white lines. Some populations have orange patches on the dorsal surfaces, and several geographic subspecies are named.

Range. The Striped Hairstreak extends from the Maritime Provinces west to central Manitoba and south to central Florida, the Gulf states, and Texas. It is very rare in the South.

Habitat. The Striped Hairstreak is found at edges and openings in deciduous or mixed deciduous woods of various stages, including young second growth. It may also be found in shaded swamps and bogs or in acid barrens.

Life History. *Behavior.* Males perch on bushes or tree foliage within 2 m of the ground during most daylight hours (0730–1830). They spend a great amount of time walking on foliage or flowers.

Broods. The Striped Hairstreak is univoltine. Flight dates are progressively earlier as one proceeds to the south. In Michigan's upper peninsula they fly from July 6 to August 6, in Illinois from June 10 to August 12, and in coastal Georgia and central Florida from May 6 to 28.

Early Stages. Eggs are laid singly on host twigs but do not hatch until the following spring. The young caterpillar bores into host buds and later eats the young leaves. Flowers and young fruit may be eaten as well. The caterpillar is light green, with a dark-green dorsal stripe and several oblique yellow-green stripes along each side. The pupa is a dull, dusky, yellow-brown, with a reddish tinge, and is dotted with reddish brown.

Adult Nectar Sources. Favored nectar sources are common milkweed, dogbane, staghorn sumac, New Jersey tea, white sweet clover, goldenrod, meadowsweet, viburnum, and chinquapin. Other plants are visited rarely.

Caterpillar Host Plants. Plants in the heath (Ericaceae) and rose (Rosaceae) families are utilized most often. Heath family plants include blueberry (*Vaccinium arboreum* and *V. corymbosum*) and flame azalea (*Rhododendron calendulaceum*). Rose family hosts include wild plum and cherry (*Prunus*), hawthorn (*Crataegus*), shadbush (*Amelanchier*), blackberry (*Rubus*), apples (*Malus*), and chokeberry (*Aronia*). There are also reports for oak, willow, and American hornbeam (*Carpinus caroliniana*).

61. *Satyrium liparops*

Light-banded Hairstreak
Tmolus azia (Hewitson), 1873

MAP 62

Synopsis. This tropical hairstreak was discovered in Dade County, Florida, in 1974. It was probably introduced from Central or South America. It is a small, dull species whose caterpillars feed on plants in the mimosa family such as wild tamarind (*Lysiloma bahamensis*) and Mimosa. In Florida it flies from late April to September. The adults visit shepherd's needle and palmetto for nectar.

Red-banded Hairstreak
Calycopis cecrops (Fabricius), 1793

PLATE 14 · FIGURE 81 · MAP 63

Etymology. The genus name is derived from a formation combining *calyx* ("cup" or "flower bud") and *opis,* meaning nymph or naiad, an epithet for Diana. The species is named after *Kekrops,* the first king of Attica and founder of Athens (Greek).

Synopsis. The Red-banded Hairstreak is a northern representative of a tropical American group that has many similar species. The female lays eggs on the underside of dead leaves lying near potential hosts.

Butterfly. Male forewing: $\overline{X}$ = 1.3 cm, range 1.1–1.4 cm; female forewing: $\overline{X}$ = 1.2 cm, range 1.1–1.3 cm. This hairstreak's small size, ventral red postmedian band, and gray-black wings with dorsal basal iridescent blue should serve to distinguish it.

Range. The Red-banded Hairstreak ranges from Long Island south along the Atlantic coastal plain to Florida and then west to coastal Texas. It ranges north along the Mississippi River drainage to Ohio. Vagrants have turned up in Michigan and northern Illinois.

Habitat. This butterfly is most common in overgrown fields with tangled small trees and shrubs and in hammocks behind coastal dunes.

Life History. *Behavior.* Males perch on low trees and shrubs, particularly in the afternoon. Freshly emerged males occasionally sit on moist sand or mud along trails to sip moisture.

Broods. Two generations seem to be the rule throughout this hairstreak's range, except in Florida, where adults have been found almost every month. In northern Virginia, the broods extend from late April to the end of June, and from the last week of July to early October. A few individuals found at the end of October and early November represent a partial third generation.

Early Stages. The female lays single eggs under leaves on the ground during the afternoon (1400–1510 hr), and the young caterpillars must find their way to appropriate host plants. The caterpillars are grey-black, an unusual color for this family, and develop very slowly. Winter may be passed by caterpillars or the chrysalis.

Adult Nectar Sources. Favored flowers include wild cherry, dogbane, yarrow, common milkweed, sumac, sweet pepperbush, New Jersey tea, and tickseed sunflower.

Caterpillar Host Plants. Larval hosts include staghorn sumac (*Rhus typhina*), dwarf sumac (*Rhus copallina*), wax myrtle (*Myrica cerifera*), and several oaks.

Dusky Blue Hairstreak
Calycopis isobeon (Butler & Druce), 1872

Synopsis. The tropical Dusky Blue Hairstreak was collected twice in Mississippi. It ranges from coastal Texas south through Mexico and Central America to Panama. It differs from its close relative, the Red-banded Hairstreak (*Calycopis cecrops*), in having a broader ventral postmedian red band on the hind wing than on the forewing. This species should be looked for in coastal Louisiana.

Olive Hairstreak
Mitoura gryneus (Hübner), 1819
PLATE 14 · FIGURE 82 · MAP 64

Etymology. The species is named after Grynea, a town in Mysia, northwest Asia Minor.

Synopsis. When present, this attractive butterfly may easily be found by tapping the branches of red cedar trees, its host. It has greatly benefited by man's disturbance of the eastern forests, since its host thrives in overgrazed pastures and abandoned farmlands.

Butterfly. Male forewing: $\overline{X}$ = 1.4 cm, range 1.3–1.4 cm; female forewing: $\overline{X}$ = 1.4 cm, range 1.4–1.5 cm. The Olive Hairstreak is extremely similar to Hessel's Hairstreak. It is more yellow-green on the ventral hind wing, and its last white mark on the postmedian band is concave above and convex below. In addition, the two white marks of the ventral forewing's postmedian band are lined up, not offset as they

62. *Tmolus azia*

63. *Calycopis cecrops*

are on Hessel's Hairstreak. Individuals of the spring generation have much orange above, while those emerging in the summer are usually black-brown above. The subspecies *sweadneri* occurs in Florida.

Range.   The Olive Hairstreak occurs from central New England west through southern Ontario to southern Minnesota and eastern Nebraska, thence south to central Florida, the Gulf states, and Texas.

Habitat.   The Olive Hairstreak is associated with small to medium-sized red cedars in old fields, serpentine barrens, and bluffs.

Life History.   *Behavior.*   Both sexes perch on red cedar trees and occasionally visit flowers. Males perch high and interact primarily in the afternoon. A mated pair was found perched on red cedar in the early afternoon (1345 hr). Oviposition has been reported near midday (1130 hr). Adults live about a week.

Broods.   The Olive Hairstreak is a spring univoltine species with sporadic emergences later in the year. The later emergences peak about 2 months after the first brood but may continue sporadically until near the end of the summer. In northern New York emergence does not occur until May 20, while at the other extreme, in Florida, the first generation emerges in late February or early March. The gap between the generations becomes wider as one travels south.

Early Stages.   The larvae feed on young host leaves. They are deep green, with oblique whitish subdorsal bars along each side. The pupa is a rich brown, with head and thorax a faint luteous green. There are black-brown vermicular blotches, especially on the abdomen.

Adult Nectar Sources.   In Virginia, winter cress is visited by the spring generation, while common milkweed, dog-bane, and wild carrot are visited by summer adults. In Florida, shepherd's needle and wild plum are visited. Elsewhere, butterflyweed, white sweet clover, dogwood, and houstonia are selected.

Caterpillar Host Plant.   The only natural host is red cedar (*Juniperus virginiana*). Although females will not oviposit on white cedar (*Chamaecyparis thyoides*), the caterpillars will feed upon it and develop normally when transferred (Remington and Pease 1955).

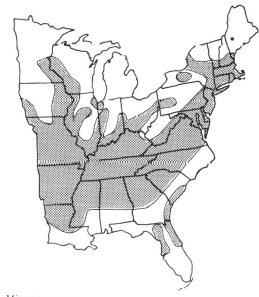

64.   *Mitoura gryneus*

Hessel's Hairstreak
Mitoura hesseli Rawson & Ziegler, 1950
PLATE 14 · FIGURE 83 · MAP 65

Etymology.   This species was named in honor of its discoverer, Sidney A. Hessel (1907–1974) of Connecticut, who was a banker by profession and a lepidopterist by avocation.

Synopsis.   Hessel's Hairstreak is virtually identical to the Olive Hairstreak, but it feeds only on white cedar, a valuable timber tree. Cutting of the cedars and draining of swamps have reduced this species' occurrence markedly.

Butterfly.   Male forewing: $\bar{X}$ = 1.2 cm, range 1.2–1.3 cm; female forewing: $\bar{X}$ = 1.3 cm, range 1.2–1.3 cm. The characteristics of this species are detailed by Rawson and Ziegler (1950). Dorsally, the summer individuals are darker than those of the spring brood. There are scattered brown patches on the lower surface of the hind wings. Florida individuals are very large and have two well-developed tails on each hind wing.

Range.   Hessel's Hairstreak is a denizen of the coastal plain and occasionally the Piedmont; it ranges from southern New Hampshire to northern Florida.

Habitat.   Hessel's Hairstreak is found in or adjacent to white cedar trees in acid bogs or swamps, which usually occur in the midst of pine barrens or sandy pine forests.

Life History.   *Behavior.*   Most reproductive activities occur on the host white cedar trees. Freshly emerged individuals may be found on damp earth along roads through swamps. The butterflies are exceptionally docile when they visit flowers (Beck 1983).

Broods. Hessel's Hairstreak is univoltine (May 16–31) at the northern end of its range, but it has two broods as it ranges southward: New Jersey, April 29–May 24, July 23–24; Virginia to North Carolina, April 13–20, July 18–28.

Early Stages. In the spring, eggs are laid on young host terminals, and the larvae develop directly to the pupal stage. Some pupae produce a second brood in midsummer, but most wait until the following spring. The mature larvae are dark blue-green with oblique white bars along the sides. The pupa is dark brown. The life cycle requires 48 days (Rawson et al. 1951; Gifford and Opler 1983).

Adult Nectar Sources. Flowers visited include sand myrtle, shadbush, highbush blueberry, sweet pepperbush, buttonbush, dogbane, and swamp milkweed.

Caterpillar Host Plant. The only natural host is white cedar (*Chamaecyparis thyoides*). In the laboratory, some larvae will accept red cedar (*Juniperus virginiana*) and develop normally on it.

Brown Elfin
Incisalia augustus (Kirby), 1837
PLATE 14 · FIGURE 84 · MAP 66

Etymology. The generic name is derived from the Latin *incisum* (part), while the species epithet is a Latin masculine proper name.

Synopsis. The Brown Elfin is the most plainly marked elfin. It is also the most widespread, ranging from northern Alaska to northern Mexico and from coast to coast. Some consider the western forms to be a separate species.

Butterfly. Male forewing: $\overline{X}$ = 1.2 cm, range 1.1–1.3 cm; female forewing: $\overline{X}$ = 1.2 cm, range 1.1–1.2 cm. The Brown Elfin lacks taillike extensions, is brown dorsally (reddish in females), and chestnut brown ventrally, with an irregular dark basal area. Freshly emerged individuals have an iridescent green sheen over the ventral wing surfaces. The subspecies *croesioides* occurs over most of the eastern area.

Range. The species extends from Newfoundland north and west to Alaska, thence south to northern Georgia, northern portions of the Gulf states, central Arizona, and northern Baja California.

Habitat. In the East, the Brown Elfin occurs primarily in areas with acidic soil, most often in pine-oak-heath barrens, which may be sand, serpentine, or shale barrens. This elfin also occurs in acid bogs. The Brown Elfin is found primarily in Transition and Canadian Life Zones but also occurs in the Upper Austral.

Life History. *Behavior.* Males perch during most warm daylight hours (0830–1700) on projecting twigs of shrubs or on the ground in small clearings. Mating occurs from 1100 to 1600 hr. Powell (1968) describes perching and mating be-

65. *Mitoura hesseli*

66. *Incisalia augustus*

havior. In New Jersey, we observed oviposition in a Labrador tea flower head at 1015 hr. In New England, one marked adult survived at least 23 days (R. K. Robbins, unpublished).

Broods. The Brown Elfin is univoltine, with emergence occurring progressively earlier southward. Examples of flight-period extremes are May 19 to July 16 in Michigan's upper peninsula, April 8 to May 3 in New York's Finger Lakes, and March 21 to April 19 in Georgia.

Early Stages. Eggs are laid in host flower buds, and upon hatching the young caterpillars bore into flowers, where they feed. Later they feed on young fruits. There are three molts, and the caterpillar requires about 22 days to complete development. Pupation occurs in sand or litter at the base of the host, and the adult emerges the following spring. The caterpillar is bright green, with a yellow-green dorsal stripe and oblique lateral stripes and dashes. The chrysalid is light brown and covered with short hairs (Cook 1906–7).

Adult Nectar Sources. Adults visit blueberry, willow, spicebush, wild plum, footsteps of spring, and winter cress.

Caterpillar Host Plants. Larval hosts are primarily members of the heath family in the eastern states, e.g., sugar huckleberry (*Vaccinium vacillans*), leatherleaf (*Chamaedaphne*), bearberry (*Arctostaphylos uva-ursi*), huckleberry (*Gaylussacia*), and Labrador tea (*Ledum groenlandicum*). The western populations of the Brown Elfin are much more catholic in their choice of larval hosts.

Hoary Elfin
Incisalia polios Cook & Watson, 1907
PLATE 15 · FIGURE 85 · MAP 67

Etymology. The species name *polios* means gray in Greek.

Synopsis. The Hoary Elfin is widespread but occurs in relatively few localized colonies. Where it occurs, it is one of the first spring butterflies.

Butterfly. Male forewing: $\overline{X}$ = 1.2 cm, range 1.1–1.3 cm; female forewing: $\overline{X}$ = 1.2 cm, range 1.0–1.3 cm. Ventrally, the Hoary Elfin may be distinguished by its small size, white postmedian line on the forewing, and the blue-gray clouding along the outer margins of both wings, narrowly on the forewing and broadly on the hind wing.

Range. This elfin occurs from Nova Scotia and Maine west and north through southern Quebec and Ontario to central Alaska, thence south to the mountains of Virginia, lakeside dunes of northern Indiana and Illinois, Minnesota, the southern Rocky Mountains, and coastal Oregon (Ferris and Fisher 1973).

Habitat. In the eastern states the Hoary Elfin favors sunny, relatively dry slopes in sandy pine barrens, adjacent to acid bogs, lakeside dunes, rocky slopes and ridges, or forest edges. It is limited to Transition and Canadian Life Zone habitats.

Life History. *Behavior.* Males perch most of the day in clearings at woodland edges or in small sunny bogs. They fly close to the ground and often rest on their host plant.

Broods. The Hoary Elfin is a spring univoltine, with flight occurring generally from late April through May, but occasionally extending into June. The earliest record is of an April 5th specimen from Virginia.

Early Stages. Eggs are laid at the base of flower buds and on leaves of the host. The egg is a flattened sphere with a central depression, and the first-instar larva is pale green with long hairs. About 27 days are required to complete development. The second- and third-instar caterpillars are yellow-green with dorsal and lateral rosy stripes and shading. The final-(fourth) instar caterpillar is green. The chrysalis is dark brown.

Adult Nectar Sources. Adults visit the flowers of leatherleaf, willow, wild strawberry, and pyxie.

Caterpillar Host Plants. Bearberry (*Arctostaphylos uva-ursi*) is the only documented larval host, but trailing arbutus (*Epigaea repens*) has been implicated by association at some localities.

67. *Incisalia polios*

Frosted Elfin
Incisalia irus (Godart), 1824

PLATE 15 · FIGURE 86 · MAP 68

Etymology. The species is named after Irus (or Iros), an arrogant beggar in the house of Odysseus, King of Ithaca (Homerian epic, Greek).

Synopsis. The Frosted Elfin occurs in very localized colonies and feeds only on several kinds of legumes with inflated pods.

Butterfly. Male forewing: $\overline{X}$ = 1.3 cm, range 1.2–1.5 cm; female forewing: $\overline{X}$ = 1.3 cm, range 1.2–1.5 cm. The Frosted Elfin is very similar to Henry's Elfin, but has a long stigma on the male forewing and a "thecla" spot on the ventral hind wing. The subspecies *hadros,* which occurs in the extreme southwestern portion of our area, is larger and has blurred or absent ventral markings. It is sometimes considered a separate species.

Range. The Frosted Elfin ranges from southern Maine west to Michigan and central Wisconsin, then south to Georgia, northern portions of the Gulf states, and eastern Texas. It always occurs in extremely local colonies.

Habitat. The Frosted Elfin is found on or close to its hosts in open woods, woodland edges, old fields, or pine-oak and oak-heath scrub.

Life History. *Broods.* Like all its relatives, the Frosted Elfin is univoltine. Adults fly in spring and flight extremes vary from May 2 to June 5 in Michigan and from March 14 to May 17 in Georgia.
Early Stages. The caterpillars feed on flowers and developing pods. Pupation occurs in a loose cocoon in litter at the base of the host.

Caterpillar Host Plants. Larval hosts are limited to legumes (Fabaceae family) with inflated pods. The two hosts reported most often are wild indigo (*Baptisia tinctoria*) and lupine (*Lupinus perennis*), but blue false indigo (*Baptisia australis*) and rattlebox (*Crotalaria sagittalis*) are also selected at times.

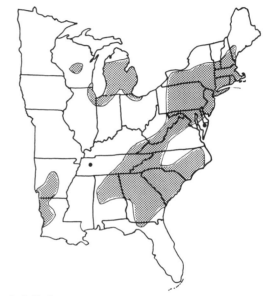

68. *Incisalia irus*

Henry's Elfin
Incisalia henrici (Grote & Robinson), 1867

PLATE 15 · FIGURE 87 · MAP 69

Etymology. The species name is derived from the Latin *Henrici,* for Henry's.

Synopsis. Henry's Elfin was misplaced for 40 years after its description, since biologists of the day used its Latin name for the Hoary Elfin, which had not yet been described. This species uses very different host plants in different parts of its range.

Butterfly. Male forewing: $\overline{X}$ = 1.2 cm, range 1.2–1.3 cm; female forewing: $\overline{X}$ = 1.3 cm, range 1.2–1.4 cm. Henry's Elfin may be distinguised by its long, taillike hind-wing projections, lack of gray scaling along the outer margin of the

ventral hind wing, and the lack of a stigma on the male's forewing. Northern individuals are much darker than southern ones, and in populations from North Carolina's Outer Banks most individuals have strong green highlights ventrally.

Range. Henry's Elfin occurs from Nova Scotia west through southern Canada to Wisconsin, thence south throughout most of our area to central Florida, the Gulf Coast, and Texas.

Habitat. This species, like most elfins, is most often found in barrens with acidic, sandy or rocky soils. Since it feeds on very different plants at different sites (see below), it may reasonably be expected in very different situations. It is often found in at least somewhat open areas but usually in the vicinity of pine or pine-oak woodland.

Life History. *Behavior.* Males often perch and interact on their chosen hosts during most warm daylight hours. One mated pair was observed at 1030 hr, while an ovipositing female was seen at 1130 hr. Oviposition behavior is stereotyped depending on the local host. In the case of redbud, eggs are laid on the flowers and buds, but on American holly, a single egg is laid directly on the center of an old host leaf just prior to bud break. Males often visit mud or damp sand for moisture. The butterflies rarely move far from their hosts.
Broods. Henry's Elfin is univoltine, and its flight occurs progressively earlier as one goes southward. For example, in Michigan, flight dates range from May 4 to 28, in Illinois from March 30 to May 9, and in Mississippi from February 17 to April 6.
Early Stages. The young caterpillars feed on buds and young leaves. The larvae develop directly in about a month and pupate, probably in litter at the base of the host (Gifford and Opler 1983). Emergence occurs the following spring. The caterpillars are light green, with lighter green bars and stripes, or red-brown with lighter red-brown stripes (Cook 1907).
Adult Nectar Sources. In sites where redbud is the larval host, the plant's flowers are also the prime nectar source. In other cases, flowers visited are usually not the host's. In coastal North Carolina, where hollies are selected for egg-laying, willow flowers constitute the prime nectar supply, and in Texas, flowers of *Forstiera* are used where Texas persimmon is the caterpillar's food. Wild plum and hawthorn flowers are utilized in northern Florida.
Caterpillar Host Plants. In many locations, e.g., Florida, Illinois, Pennsylvania, and Virginia, redbud (*Cercis can-*

adensis) seems to be the sole, or at least primary, host. Elsewhere, fleshy fruited trees in several unrelated families are selected. As mentioned above, Texas persimmon (*Diospyros texana*) is the host in central Texas, while in Florida and North Carolina at least three hollies are used: dahoon (*Ilex cassine*), American holly (*I. opaca*), and yaupon (*I. vomitoria*). Blueberries and huckleberries (*Vaccinium*) seem to be hosts in such diverse areas as Florida and Pennsylvania, while maple-leaf viburnum (*Viburnum acerifolium*) is a host in Michigan. Wild plum (*Prunus*) has been suggested but not confirmed.

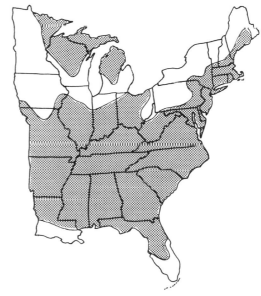

69. *Incisalia henrici*

Bog Elfin
Incisalia lanoraieensis Sheppard, 1934
MAP 70

Etymology. The species name is derived from the type locality, Lanoraie, Quebec.
Synopsis. The Bog Elfin is limited to black spruce bogs in northern New England and southeastern Canada. It is the smallest hairstreak in our area.
Butterfly. Male forewing: $\overline{X}$ = 1.1 cm, range 1.0–1.1 cm; female forewing: $\overline{X}$ = 1.0 cm, range 1.0–1.1 cm. The Bog Elfin is always much smaller than the Pine Elfin, has a reduced or smudged pattern ventrally, and has a black submarginal spot at the anal angle on the ventral hind wing.
Range. The Bog Elfin occurs in southeastern Canada

(southern Quebec and New Brunswick), Maine, and northern New Hampshire.
Habitat. The Bog Elfin is found only in black spruce –tamarack–sphagnum bogs.
Life History. *Broods.* The single annual flight extends from May 18 to June 9.
Early Stages. The first-instar larvae mine young spruce needles; later, larvae continue their feeding externally.
Caterpillar Host Plant. The only larval host is black spruce (*Picea mariana*).

PLATE 1

FIGURE 1. Black Swallowtail (*Papilio polyxenes asterius*), mature caterpillar, Fairfax County, Virginia.

FIGURE 4. Variegated Fritillary (*Euptoieta claudia*), mature caterpillar, Putnam County, Georgia.

FIGURE 2. Giant Swallowtail (*Papilio cresphontes*), mature caterpillar feeding on citrus, Indian River County, Florida.

FIGURE 5. Florida Leaf Wing (*Anaea floridalis*), mature caterpillar, Big Pine Key, Florida.

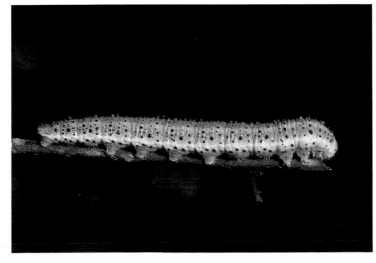

FIGURE 3. Olympia Marble (*Euchloe olympia*), mature caterpillar, Pendleton County, West Virginia.

FIGURE 6. Silver-spotted Skipper (*Epargyreus clarus*), mature caterpillar, Fairfax County, Virginia.

PLATE 2

FIGURE 7. Pipe Vine Swallowtail (*Battus philenor*), ♂, dorsal basking, Bath County, Virginia.

FIGURE 10. Zebra Swallowtail (*Eurytides marcellus*), ♂, spring form, Montgomery County, Maryland.

FIGURE 8. Pipe Vine Swallowtail, ♂, at moisture, Rockingham County, Virginia.

FIGURE 11. Zebra Swallowtail, ♀, spring form, Montgomery County, Maryland.

FIGURE 9. Polydamas Swallowtail (*Battus polydamas lucayus*), taking nectar, Martinique.

FIGURE 12. Zebra Swallowtail, ♂, summer form, at moisture, Montgomery County, Maryland.

PLATE 3

FIGURE 13. Black Swallowtail (*Papilio polyxenes asterius*), ♂, taking nectar at thistle, Montgomery County, Maryland.

FIGURE 16. Tiger Swallowtail (*Papilio glaucus*), ♂, at moisture, Montgomery County, Maryland.

FIGURE 14. Black Swallowtail, ♀, dorsal basking, Fairfax County, Virginia.

FIGURE 17. Tiger Swallowtail, ♀, yellow form, taking nectar at thistle, Fairfax County, Virginia.

FIGURE 15. Black Swallowtail, ♂, at moisture, Fauquier County, Virginia.

FIGURE 18. Tiger Swallowtail, ♀, black form, taking nectar at swamp milkweed, Montgomery County, Maryland.

PLATE 4

FIGURE 19. Giant Swallowtail (*Papilio cresphontes*), ♂, dorsal basking, Volusia County, Florida.

FIGURE 22. Spicebush Swallowtail, ♂, rare blue form, taking nectar at multiflora rose, Montgomery County, Maryland.

FIGURE 20. Schaus' Swallowtail (*Papilio aristodemus ponceanus*), ♂, dorsal basking, Florida Keys.

FIGURE 23. Spicebush Swallowtail, ♀, dorsal basking, Pendleton County, West Virginia.

FIGURE 21. Spicebush Swallowtail (*Papilio troilus*), ♂, at moisture and dorsal basking, Fairfax County, Virginia.

FIGURE 24. Spicebush Swallowtail, ♂, taking moisture, Fairfax County, Virginia.

PLATE 5

FIGURE 25. Bahaman Swallowtail (*Papilio andraemon bonhotei*), ♂, dorsal basking, Bahamas.

FIGURE 28. Florida White (*Appias drusilla neumoegenii*), mated pair, Upper Matecumbe Key, Florida.

FIGURE 26. Palamedes Swallowtail (*Papilio palamedes*), ♂, dorsal basking, Suffolk County, Virginia.

FIGURE 29. Checkered White (*Pontia protodice*), ♂, taking nectar at aster, Orange County, Virginia.

FIGURE 27. Palamedes Swallowtail, ♂, resting, Suffolk County, Virginia.

FIGURE 30. Checkered White, ♀, Collier County, Florida.

PLATE 6

FIGURE 31. Mustard White (*Pieris napi*), Eagle Summit, Alaska.

FIGURE 34. European Cabbage Butterfly, ♀, taking nectar at wild oregano, Fairfax County, Virginia.

FIGURE 32. West Virginia White (*Pieris virginiensis*), ♂, Frederick County, Maryland.

FIGURE 35. Olympia Marble (*Euchloe olympia*), ♂, dorsal basking, Alleghany County, Maryland.

FIGURE 33. European Cabbage Butterfly (*Pieris rapae*), ♂, taking nectar at winter cress, Fairfax County, Virginia.

FIGURE 36. Olympia Marble, ibid.

PLATE 7

Figure 37. Great Southern White (*Ascia monuste phileta*), ♂, taking nectar at citrus, Indian River County, Florida.

Figure 40. Clouded Sulphur (*Colias philodice*), ♂, taking moisture, Montgomery County, Maryland.

Figure 38. Falcate Orange-tip (*Anthocharis midea*), ♂, taking nectar at violet, Anne Arundel County, Maryland.

Figure 41. Orange Sulphur (*Colias eurytheme*), ♀, taking nectar at goldenrod, Montgomery County, Maryland.

Figure 39. Falcate Orange-tip, ♀, taking nectar at winter cress, Fairfax County, Virginia.

Figure 42. Orange Sulphur, ♀, white form, taking nectar at tickseed sunflower, Fairfax County, Virginia.

PLATE 8

FIGURE 43. Pink-edged Sulphur (*Colias interior*), ♂, at moisture, Grant County, West Virginia.

FIGURE 46. Cloudless Sulphur (*Phoebis sennae eubule*), ♀, Caroline County, Virginia.

FIGURE 44. Dog Face (*Colias cesonia*), ♂, taking nectar at alfalfa, Sioux County, Iowa.

FIGURE 47. Cloudless Sulphur, ♂, Collier County, Florida.

FIGURE 45. Giant Brimstone (*Anteos maerula*), resting, Trinidad. Note background resemblance.

FIGURE 48. Orange-barred Sulphur (*Phoebis philea*), ♂, at moisture, Trinidad.

PLATE 9

FIGURE 49. Large Orange Sulphur (*Phoebis agarithe maxima*), ♂, taking nectar at bougainvilla, Upper Matecumbe Key, Florida.

FIGURE 52. Lyside (*Kricogonia lyside*), ♂, taking nectar, Yucatan, Mexico.

FIGURE 50. Large Orange Sulphur, ♂, Collier County, Florida.

FIGURE 53. Barred Sulphur (*Eurema daira*), ♂, winter or dry-season form, Dade County, Florida.

FIGURE 51. Large Orange Sulphur, ♀, white form, Collier County, Florida.

FIGURE 54. Barred Sulphur, ♂, summer or wet-season form, Bahamas.

PLATE 10

FIGURE 55. Little Sulphur (*Eurema lisa*), ♂, Collier County, Florida.

FIGURE 58. Shy Sulphur (*Eurema messalina blakei*), ♀, Bahamas.

FIGURE 56. Little Sulphur, ♀, resting on partridge pea, its host plant, Sioux County, Iowa.

FIGURE 59. Bush Sulphur (*Eurema dina helios*), ♂, at moisture, Bahamas.

FIGURE 57. Jamaican Sulphur (*Eurema nise*), ♀, Bahamas.

FIGURE 60. Bush Sulphur, ♀, ibid.

PLATE 11

FIGURE 61. Sleepy Orange (*Eurema nicippe*), ♂, at moisture, Bahamas.

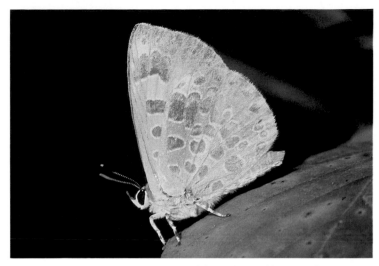

FIGURE 64. Harvester (*Feniseca tarquinius*), ♂, perching, Fairfax County, Virginia.

FIGURE 62. Sleepy Orange, ♀, taking nectar at shepherd's needle, Alachua County, Florida.

FIGURE 65. American Copper (*Lycaena phlaeas*), ♂, perching, Pendleton County, West Virginia.

FIGURE 63. Dainty Sulphur (*Nathalis iole*), summer form, Stock Island, Florida.

FIGURE 66. American Copper, ♂, taking nectar at butterflyweed, Fairfax County, Virginia.

PLATE 12

Figure 67. Great Copper (*Lycaena xanthoides dione*), ♂, taking nectar at butterflyweed, Dickinson County, Iowa.

Figure 70. Bog Copper (*Lycaena epixanthe*), ♂, Bradford County, Pennsylvania.

Figure 68. Great Copper, ♂, ibid.

Figure 71. Bog Copper, ♂, ibid.

Figure 69. Bronze Copper (*Lycaena hyllus*), ♂, Caroline County, Maryland.

Figure 72. Atala (*Eumaeus atala florida*), Dade County, Florida.

PLATE 13

FIGURE 73. Great Purple Hairstreak (*Atlides halesus*), ♂, taking nectar at boneset, Suffolk County, Virginia.

FIGURE 74. Great Purple Hairstreak, ♀, taking nectar at Hercules club, Suffolk County, Virginia.

FIGURE 75. Coral Hairstreak (*Satyrium titus*), ♀, taking nectar at mountain mint, Fairfax County, Virginia.

FIGURE 76. Acadian Hairstreak (*Satyrium acadica*), ♂, perching, Sioux County, Iowa.

FIGURE 77. Edwards' Hairstreak (*Satyrium edwardsii*), ♂, perching, Grant County, West Virginia.

FIGURE 78. Striped Hairstreak (*Satyrium liparops*), taking nectar at New Jersey tea, Fairfax County, Virginia.

PLATE 14

FIGURE 79. Banded Hairstreak (*Satyrium calanus falacer*), ♂, perching, Montgomery County, Maryland.

FIGURE 82. Olive Hairstreak (*Mitoura gryneus*), resting on red cedar, its host, Fairfax County, Virginia.

FIGURE 80. Banded Hairstreak, ♀, taking nectar at ox-eye daisy, Fairfax County, Virginia.

FIGURE 83. Hessel's Hairstreak (*Mitoura hesseli*), ♀, nectaring at buttonbush, Suffolk County, Virginia.

FIGURE 81. Red-banded Hairstreak (*Calycopis cecrops*), taking nectar at sweet pepperbush, Suffolk County, Virginia.

FIGURE 84. Brown Elfin (*Incisalia augustus*), ♀, lateral basking, Anne Arundel County, Maryland.

PLATE 15

FIGURE 85. Hoary Elfin (*Incisalia polios*), ♂, Burlington County, New Jersey.

FIGURE 86. Frosted Elfin (*Incisalia irus*), ♂, Fairfax County, Virginia.

FIGURE 87. Henry's Elfin (*Incisalia henrici*), perching, Anne Arundel County, Maryland.

FIGURE 88. Pine Elfin (*Incisalia niphon*), nectaring at blueberry, Fairfax County, Virginia.

FIGURE 89. Southern Hairstreak (*Fixsenia favonius*), ♂, perching, Indian River County, Florida.

FIGURE 90. Northern Hairstreak (*Fixsenia ontario*), ♂, nectaring, Houston County, Alabama.

PLATE 16

FIGURE 91. White M Hairstreak (*Parrhasius m-album*), Indian River County, Florida.

FIGURE 94. Bartram's Hairstreak (*Strymon acis bartrami*), nectaring at croton, its host plant, Big Pine Key, Florida.

FIGURE 92. Gray Hairstreak (*Strymon melinus*), nectaring at Queen Anne's lace, Fairfax County, Virginia.

FIGURE 95. Columella Hairstreak (*Strymon columella modesta*), nectaring at lantana, Big Pine Key, Florida.

FIGURE 93. Martialis Hairstreak (*Strymon martialis*), nectaring at lantana, Big Pine Key, Florida.

FIGURE 96. Early Hairstreak (*Erora laeta*), nectaring at fleabane, Harlan County, Kentucky.

PLATE 17

FIGURE 97. Eastern Pygmy Blue (*Brephidium isophthalma pseudofea*), nectaring at lippia, Indian River County, Florida.

FIGURE 100. Miami Blue (*Hemiargus thomasi bethunebakeri*), nectaring at seaside heliotrope, Key Largo, Florida.

FIGURE 98. Cassius Blue (*Leptotes cassius theonus*), ♂, nectaring at lippia, Monroe County, Florida.

FIGURE 101. Ceraunus Blue (*Hemiargus ceraunus antibubastus*), ♂, Monroe County, Florida.

FIGURE 99. Marine Blue (*Leptotes marina*), ♀, perching on mesquite, one of its caterpillar hosts, Coconino County, Arizona.

FIGURE 102. Reakirt's Blue (*Hemiargus isola*), nectaring at horehound, Coconino County, Arizona.

PLATE 18

FIGURE 103. Eastern Tailed Blue (*Everes comyntas*), ♂, nectaring at Venus' looking glass, Montgomery County, Maryland.

FIGURE 106. Appalachian Blue (*Celastrina neglectamajor*), ♂, nectaring at tooth-wort, Frederick County, Maryland.

FIGURE 104. Eastern Tailed Blue, ♂, ibid.

FIGURE 107. Silvery Blue (*Glaucopsyche lygdamus*), ♂, at moisture (Spring Azure at right), Allegheny County, Maryland.

FIGURE 105. Spring Azure (*Celastrina ladon*), ♂, form "violacea," at moisture, Fairfax County, Virginia.

FIGURE 108. Silvery Blue, ♂, ibid.

PLATE 19

FIGURE 109. Dusky Blue (*Celastrina ebenina*), ♂, Highland County, Virginia.

FIGURE 112. Northern Metalmark, ♂, ibid.

FIGURE 110. Little Metalmark (*Calephelis virginiensis*), ♀, taking nectar, Brevard County, Florida.

FIGURE 113. Gulf Fritillary (*Agraulis vanillae*), ♀, taking nectar at shepherd's needle, Monroe County, Florida.

FIGURE 111. Northern Metalmark (*Calephelis borealis*), taking nectar, Bath County, Virginia.

FIGURE 114. Gulf Fritillary, ♂, Big Pine Key, Florida.

PLATE 20

FIGURE 115. Eastern Snout Butterfly (*Libytheana bachmanii*), taking nectar at goldenrod, Fairfax County, Virginia.

FIGURE 116. Eastern Snout Butterfly, pale form, ibid.

FIGURE 117. Eastern Snout Butterfly, dark form, Montgomery County, Maryland.

FIGURE 118. Zebra (*Heliconius charitonius tuckeri*), taking nectar at shepherd's needle, Key West, Florida.

FIGURE 119. Variegated Fritillary (*Euptoieta claudia*), ♀, taking nectar at boneset, Montgomery County, Maryland.

FIGURE 120. Variegated Fritillary, ♂, taking nectar at red clover, Putnam County, Georgia.

PLATE 21

FIGURE 121. Diana (*Speyeria diana*), ♂, taking nectar at swamp milkweed, Bath County, Virginia.

FIGURE 124. Diana, ♀ ibid.

FIGURE 122. Diana, ♂, ibid.

FIGURE 125. Great Spangled Fritillary (*Speyeria cybele*), ♂, taking nectar at New Jersey tea, Fairfax County, Virginia.

FIGURE 123. Diana, ♀, taking nectar at common milkweed, Bath County, Virginia.

FIGURE 126. Great Spangled Fritillary, ♂, taking nectar at common milkweed, Montgomery County, Maryland.

PLATE 22

FIGURE 127. Aphrodite Fritillary (*Speyeria aphrodite*), ♂, dorsal basking, Grant County, West Virginia.

FIGURE 128. Aphrodite Fritillary, ♀, taking nectar at dogbane, Orange County, Virginia.

FIGURE 129. Aphrodite Fritillary, ♀, ibid.

FIGURE 130. Regal Fritillary (*Speyeria idalia*), ♂, dorsal basking, Orange County, Virginia.

FIGURE 131. Regal Fritillary, ♀, ibid.

FIGURE 132. Regal Fritillary, ♂, ibid.

PLATE 23

FIGURE 133. Mountain Silver-spot (*Speyeria atlantis*), ♂, dorsal basking, Grant County, West Virginia.

FIGURE 136. Bog Fritillary, taking nectar at bistort, Pyrenees, France.

FIGURE 134. Mountain Silver-spot, ♂, ibid.

FIGURE 137. Silver-bordered Fritillary (*Boloria selene*), dorsal basking, Pyrenees, France.

FIGURE 135. Bog Fritillary (*Boloria eunomia*), ♂, Pyrenees, France.

FIGURE 138. Silver-bordered Fritillary, ibid.

PLATE 24

FIGURE 139. Julia (*Dryas iulia largo*), ♂, taking nectar at shepherd's needle, Key Largo, Florida.

FIGURE 142. Purple Lesser Fritillary (*Boloria titania*), Italian Alps.

FIGURE 140. Meadow Fritillary (*Boloria bellona*), ♂, Dickinson County, Iowa.

FIGURE 143. Gorgone Checkerspot (*Chlosyne gorgone*), ♂, dorsal basking, Sioux County, Iowa.

FIGURE 141. Meadow Fritillary, mating pair, Bath County, Virginia.

FIGURE 144. Gorgone Checkerspot, ♂, perching position, ibid.

PLATE 25

FIGURE 145. Silvery Checkerspot (*Chlosyne nycteis*), dorsal basking, Montgomery County, Maryland.

FIGURE 148. Harris' Checkerspot, ♂, taking nectar at spreading dogbane, Ontario, Canada.

FIGURE 146. Silvery Checkerspot, ♂, ibid.

FIGURE 149. Phaon Crescent (*Phyciodes phaon*), ♀, taking nectar at shepherd's needle, Alachua County, Florida.

FIGURE 147. Harris' Checkerspot (*Chlosyne harrisii*), ♂, dorsal basking, Ontario, Canada.

FIGURE 150. Phaon Crescent, mating pair, ibid.

PLATE 26

FIGURE 151. Cuban Crescent (*Phyciodes frisia*), ♀, dorsal basking, Monroe County, Florida.

FIGURE 154. Question Mark (*Polygonia interrogationis*), ♂, summer form, dorsal basking, Montgomery County, Maryland.

FIGURE 152. Pearl Crescent (*Phyciodes tharos*), ♂, spring form, dorsal basking, Fairfax County, Virginia.

FIGURE 155. Question Mark, ♂, summer form, ibid.

FIGURE 153. Pearl Crescent, ♀, spring form, ibid.

FIGURE 156. Question Mark, ♀, winter form, Fairfax County, Virginia.

PLATE 27

FIGURE 157. Northern Pearl Crescent (*Phyciodes pascoensis*), ♂ dorsal basking, Pendleton County, West Virginia.

FIGURE 160. Baltimore, ♂, perching, Pendleton County, West Virginia.

FIGURE 158. Northern Pearl Crescent, ♀, ibid.

FIGURE 161. Gray Comma (*Polygonia progne*), ♂, perching, Pendleton County, West Virginia.

FIGURE 159. Baltimore (*Euphydryas phaeton*), ♂, dorsal basking, Pendleton County, West Virginia.

FIGURE 162. Gray Comma, ♂, ibid.

PLATE 28

FIGURE 163. Comma (*Polygonia comma*), ♂, summer form, dorsal basking, Montgomery County, Maryland.

FIGURE 166. Satyr (*Polygonia satyrus*), ♂, dorsal basking, Ontario, Canada.

FIGURE 164. Comma, ♂, winter form, Fairfax County, Virginia.

FIGURE 167. Mourning Cloak (*Nymphalis antiopa*), ♀, dorsal basking, Fairfax County, Virginia.

FIGURE 165. Comma, ♂, ibid.

FIGURE 168. Mourning Cloak, ♂, taking nectar at common milkweed, Sioux County, Iowa.

PLATE 29

FIGURE 169. Milbert's Tortoise Shell (*Nymphalis milberti*), ♂, dorsal basking, Ontario, Canada.

FIGURE 172. Painted Lady (*Vanessa cardui*), ♀, taking nectar at cosmos, Fairfax County, Virginia.

FIGURE 170. American Painted Lady (*Vanessa virginiensis*), ♀, taking nectar at common milkweed, Montgomery County, Maryland.

FIGURE 173. Painted Lady, ibid.

FIGURE 171. American Painted Lady, ♂, taking nectar at marigold, Fairfax County, Virginia.

FIGURE 174. Caribbean Buckeye (*Junonia evarete genoveva*), ♂, dorsal basking, Monroe County, Florida.

PLATE 30

FIGURE 175. Red Admiral (*Vanessa atalanta rubria*), ♂, taking nectar at shepherd's needle, Key West, Florida.

FIGURE 178. Buckeye, ♂, at moisture, Fairfax County, Virginia.

FIGURE 176. Red Admiral, ♂, ibid.

FIGURE 179. Malachite (*Siproeta stelenes*), dry-season form, Costa Rica.

FIGURE 177. Buckeye (*Junonia coenia*), ♂, taking nectar at tickseed sunflower, Fairfax County, Virginia.

FIGURE 180. Malachite, ibid.

PLATE 31

FIGURE 181. White Peacock (*Anartia jatrophae guantanamo*), ♂, dorsal basking, Nassau, Bahamas.

FIGURE 184. Red-spotted Purple (*Limenitis arthemis astyanax*), ♂, dorsal basking, Montgomery County, Maryland.

FIGURE 182. White Admiral (*Limenitis arthemis arthemis*), ♂, dorsal basking, Ontario, Canada.

FIGURE 185. Red-spotted Purple, ♂, feeding at dead copperhead, Fairfax County, Virginia.

FIGURE 183. White Admiral, ♂, ibid.

FIGURE 186. Ruddy Dagger Wing (*Marpesia petreus*), ♀, taking nectar at shepherd's needle, Monroe County, Florida.

PLATE 32

FIGURE 187. Viceroy (*Limenitis archippus*), ♂, perching, Montgomery County, Maryland.

FIGURE 190. Florida Purple Wing, ♂, resting.

FIGURE 188. Viceroy, ♂, perching, Fairfax County, Virginia.

FIGURE 191. Hackberry Butterfly (*Asterocampa celtis*), ♂, Fairfax County, Virginia.

FIGURE 189. Florida Purple Wing (*Eunica tatila tatilista*), ♂, dorsal basking, Upper Matecumbe Key, Florida.

FIGURE 192. Hackberry Butterfly, ♂, ibid.

PLATE 33

FIGURE 193. Florida Leaf Wing (*Anaea floridalis*), ♂, dorsal basking, Big Pine Key, Florida.

FIGURE 196. Tawny Emperor (*Asterocampa clyton*), ♂, dorsal basking, Loudoun County, Virginia.

FIGURE 194. Florida Leaf Wing, ♀, ibid.

FIGURE 197. Tawny Emperor, ♀, dorsal basking, Sioux County, Iowa.

FIGURE 195. Florida Leaf Wing, ♀, resting, note background resemblance.

FIGURE 198. Tawny Emperor, ♀, ibid.

PLATE 34

FIGURE 199. Pearly Eye (*Enodia portlandia*), Suffolk County, Virginia.

FIGURE 202. Creole Pearly Eye (*Enodia creola*), ♂, Suffolk County, Virginia.

FIGURE 200. Northern Pearly Eye (*Enodia anthedon*), ♂, Montgomery County, Maryland.

FIGURE 203. Creole Pearly Eye, ♀, resting on switch cane, its host plant, Suffolk County, Virginia.

FIGURE 201. Northern Pearly Eye, dorsal basking, ibid.

FIGURE 204. Appalachian Eyed Brown (*Satyrodes appalachia*), mated pair, Fairfax County, Virginia.

PLATE 35

FIGURE 205. Smoky Eyed Brown (*Satyrodes fumosa*), dorsal basking, Dickinson County, Iowa.

FIGURE 208. Carolina Satyr (*Hermeuptychia hermes sosybius*), Suffolk County, Virginia.

FIGURE 206. Smoky Eyed Brown, ibid.

FIGURE 209. Georgia Satyr (*Neonympha areolatus*), Gates County, North Carolina.

FIGURE 207. Gemmed Satyr (*Cyllopsis gemma*), Putnam County, Georgia.

FIGURE 210. Inornate Ringlet (*Coenonympha inornata*), taking nectar at ox-eye daisy, Ontario, Canada.

PLATE 36

FIGURE 211. Little Wood Satyr (*Megisto cymela*), ♀, dorsal basking, Fairfax County, Virginia.

FIGURE 214. Common Wood Nymph, ♀, ibid.

FIGURE 212. Little Wood Satyr, mated pair, ibid.

FIGURE 215. Common Wood Nymph (*Cercyonis pegala nephele*), Dickinson County, Iowa.

FIGURE 213. Common Wood Nymph (*Cercyonis pegala pegala*), ♂, Fairfax County, Virginia.

FIGURE 216. Jutta Arctic (*Oenis jutta*), Fairbanks, Alaska.

PLATE 37

FIGURE 217. Monarch (*Danaus plexippus*), ♂, taking nectar at common milkweed, Fairfax County, Virginia.

FIGURE 220. Queen (*Danaus gilippus*), ♀, taking nectar at shepherd's needle, Key West, Florida.

FIGURE 218. Monarch, ♀, taking nectar at cosmos, Fairfax County, Virginia.

FIGURE 221. Silver-spotted Skipper (*Epargyreus clarus*), nectaring at zinnia, Fairfax County, Virginia.

FIGURE 219. Monarch, ♂, taking nectar at New York ironweed, Montgomery County, Maryland.

FIGURE 222. Silver-spotted Skipper, mated pair, Montgomery County, Maryland.

PLATE 38

FIGURE 223. Mangrove Skipper (*Phocides pigmalion okeechobee*), nectaring at shepherd's needle, Upper Matecumbe Key, Florida.

FIGURE 226. Long-tailed Skipper, Stock Island, Florida.

FIGURE 224. Hammock Skipper (*Polygonus leo*), nectaring at *Sida*, Big Pine Key, Florida.

FIGURE 227. Dorantes Skipper (*Urbanus dorantes*), nectaring at shepherd's needle, Key West, Florida.

FIGURE 225. Long-tailed Skipper (*Urbanus proteus*), Costa Rica.

FIGURE 228. Gold-banded Skipper (*Autochton cellus*), dorsal basking, Fairfax County, Virginia.

PLATE 39

FIGURE 229. Hoary Edge (*Achalarus lyciades*), ♂, dorsal basking, Montgomery County, Maryland.

FIGURE 232. Southern Cloudy Wing, ♀, ibid.

FIGURE 230. Hoary Edge, nectaring at common milkweed, ibid.

FIGURE 233. Northern Cloudy Wing (*Thorybes pylades*), dorsal basking, Orange County, Virginia.

FIGURE 231. Southern Cloudy Wing (*Thorybes bathyllus*), ♀, nectaring at teasel, Montgomery County, Maryland.

FIGURE 234. Northern Cloudy Wing, ♂, nectaring at viper's bugloss, Pendleton County, West Virginia.

PLATE 40

FIGURE 235. Southern Sooty Wing (*Staphylus hayhurstii*), ♀, nectaring at marigold, Fairfax County, Virginia.

FIGURE 238. Dreamy Dusky Wing (*Erynnis icelus*), ♂, perching, Fairfax County, Virginia.

FIGURE 236. Florida Dusky Wing (*Ephyriades brunnea floridensis*), ♂, Key West, Florida.

FIGURE 239. Sleepy Dusky Wing (*Erynnis brizo*), ♀, nectaring at blueberry, Spotsylvania County, Virginia.

FIGURE 237. Florida Dusky Wing, ♀, Big Pine Key, Florida.

FIGURE 240. Horace's Dusky Wing (*Erynnis horatius*), ♀, nectaring at narrow-leaved mountain mint, Fairfax County, Virginia.

PLATE 41

FIGURE 241. Juvenal's Dusky Wing (*Erynnis juvenalis*), ♂, dorsal basking, Montgomery County, Maryland.

FIGURE 244. Mottled Dusky Wing (*Erynnis martialis*), ♀, Allegheny County, Maryland.

FIGURE 242. Juvenal's Dusky Wing, ♀, nectaring at winter cress, Fairfax County, Virginia.

FIGURE 245. Wild Indigo Dusky Wing (*Erynnis baptisiae*), ♂, Fairfax County, Virginia.

FIGURE 243. Juvenal's Dusky Wing, mating pair, Montgomery County, Maryland.

FIGURE 246. Grizzled Skipper (*Pyrgus centaureae wyandot*), at moisture, Allegheny County, Maryland.

PLATE 42

FIGURE 247. Checkered Skipper (*Pyrgus communis*), ♂, dorsal basking, Montgomery County, Maryland.

FIGURE 250. Tropical Checkered Skipper, ♀, nectaring at shepherd's needle, Alachua County, Florida.

FIGURE 248. Checkered Skipper, nectaring at blue mistflower, Montgomery County, Maryland.

FIGURE 251. Common Sooty Wing (*Pholisora catullus*), ♀, Fairfax County, Virginia.

FIGURE 249. Tropical Checkered Skipper (*Pyrgus oileus*), ♂, nectaring, Trinidad.

FIGURE 252. Swarthy Skipper (*Nastra lherminier*), ♂, Fairfax County, Virginia.

PLATE 43

FIGURE 253. Clouded Skipper (*Lerema accius*), nectaring at shepherd's needle, Alachua County, Florida.

FIGURE 256. Southern Skipperling (*Copaeodes minima*), Collier County, Florida.

FIGURE 254. Least Skipper (*Ancyloxipha numitor*), ♂, Montgomery County, Maryland.

FIGURE 257. European Skipper (*Thymelicus lineola*), ♂, Jefferson County, West Virginia.

FIGURE 255. Powesheik Skipper (*Oarisma powesheik*), ♂, nectaring at purple coneflower, Dickinson County, Iowa.

FIGURE 258. Leonard's Skipper (*Hesperia leonardus*), ♂, nectaring at New York ironweed, Fairfax County, Virginia.

PLATE 44

FIGURE 259. Fiery Skipper (*Hylephila phyleus*), ♂, nectaring at swamp milkweed, Calvert County, Maryland.

FIGURE 262. Ottoe Skipper (*Hesperia ottoe*), ♂, perched on skeleton plant, Sioux County, Iowa.

FIGURE 260. Fiery Skipper, ♀, nectaring at shepherd's needle, Key West, Florida.

FIGURE 263. Ottoe Skipper, ♀, ibid.

FIGURE 261. Fiery Skipper, ♀, nectaring at sweet pepperbush, Gates County, North Carolina.

FIGURE 264. Cobweb Skipper (*Hesperia metea*), ♂, perching, Burlington County, New Jersey.

PLATE 45

Figure 265. Meske's Skipper (*Hesperia meskei*), ♀, Big Pine Key, Florida.

Figure 268. Peck's Skipper, nectaring at red clover, ibid.

Figure 266. Dakota Skipper (*Hesperia dacotae*), ♀, nectaring at purple coneflower, Dickinson County, Iowa.

Figure 269. Tawny-edged Skipper (*Polites themistocles*), ♂, perching, Grant County, West Virginia.

Figure 267. Peck's Skipper (*Polites coras*), ♂, perching, Fairfax County, Virginia.

Figure 270. Tawny-edged Skipper, ♀, nectaring at purple coneflower, Dickinson County, Iowa.

PLATE 46

FIGURE 271. Indian Skipper (*Hesperia sassacus*), ♂, perching, Pendleton County, West Virginia.

FIGURE 274. Cross Line Skipper, ♂, perching, ibid.

FIGURE 272. Cross Line Skipper (*Polites origenes*), ♂, nectaring at red clover, Montgomery County, Maryland.

FIGURE 275. Whirlabout (*Polites vibex*), ♀, nectaring at lantana, Alachua County, Florida.

FIGURE 273. Cross Line Skipper, ♀, Fairfax County, Virginia.

FIGURE 276. Whirlabout, ♂, nectaring at shepherd's needle, ibid.

PLATE 47

FIGURE 277. Long Dash (*Polites mystic*), ♂, nectaring at sow thistle, Grant County, West Virginia.

FIGURE 280. Broken Dash (*Wallengrenia otho*), ♀, Monroe County, Florida.

FIGURE 278. Long Dash, ♀, Dickinson County, Iowa.

FIGURE 281. Northern Broken Dash (*Wallengrenia egeremet*), ♂, perching, Fairfax County, Virginia.

FIGURE 279. Long Dash, nectaring at hedge nettle, Sullivan County, Pennsylvania.

FIGURE 282. Northern Broken Dash, ♂, perching, Suffolk County, Virginia.

PLATE 48

FIGURE 283. Little Glassy Wing (*Pompeius verna*), ♂, perching, Montgomery County, Maryland.

FIGURE 286. Satchem, ♀, nectaring at zinnia, ibid.

FIGURE 284. Little Glassy Wing, ♀, nectaring at common milkweed, ibid.

FIGURE 287. Satchem, ♀, ibid.

FIGURE 285. Satchem (*Atalopedes campestris*), ♂, perching, Calvert County, Maryland.

FIGURE 288. Arogos Skipper (*Atrytone arogos*), ♂, nectaring at purple coneflower, Dickinson County, Iowa.

PLATE 49

FIGURE 289. Delaware Skipper (*Atrytone logan*), ♂, perching, Pendleton County, West Virginia.

FIGURE 292. Northern Golden Skipper (*Poanes hobomok*), ♂, perching, Pendleton County, West Virginia.

FIGURE 290. Mulberry Wing (*Poanes massasoit*), ♂, Dorchester County, Maryland.

FIGURE 293. Northern Golden Skipper, ♂, ibid.

FIGURE 291. Mulberry Wing, nectaring at red clover, ibid.

FIGURE 294. Aaron's Skipper (*Poanes aaroni*), ♂, nectaring at marsh fleabane, New Kent County, Virginia.

PLATE 50

FIGURE 295. Southern Golden Skipper (*Poanes zabulon*), ♂, perching, Fairfax County, Virginia.

FIGURE 298. Yehl Skipper (*Poanes yehl*), ♂, perching, Suffolk County, Virginia.

FIGURE 296. Southern Golden Skipper, ♂, ibid.

FIGURE 299. Yehl Skipper, ♀, ibid.

FIGURE 297. Southern Golden Skipper, ♀, nectaring at red clover, ibid.

FIGURE 300. Palmetto Skipper (*Euphyes arpa*), ♂, nectaring at blue flag, Brevard County, Florida.

PLATE 51

FIGURE 301. Broad-winged Skipper (*Poanes viator*), ♀, Calvert County, Maryland.

FIGURE 302. Broad-winged Skipper, ♀, nectaring at pickerelweed, New Kent County, Virginia.

FIGURE 303. Dion Skipper (*Euphyes dion*), ♂, perching, Fairfax County, Virginia.

FIGURE 304. Black Dash (*Euphyes conspicua*), ♂, perching, Cumberland County, Pennsylvania.

FIGURE 305. Black Dash, ♂, ibid.

FIGURE 306. Black Dash, ♀, ibid.

PLATE 52

FIGURE 307. Two-spotted Skipper (*Euphyes bimacula*), ♂, perching, Dickinson County, Iowa.

FIGURE 310. Dun Skipper, ♀, Fairfax County, Virginia.

FIGURE 308. Two-spotted Skipper, ♀, nectaring at hedge nettle, Sullivan County, Pennsylvania.

FIGURE 311. Dusted Skipper (*Atrytonopsis hianna*), ♂, perching, Fairfax County, Virginia.

FIGURE 309. Dun Skipper (*Euphyes ruricola metacomet*), ♂, nectaring at dogbane, Pendleton County, West Virginia.

FIGURE 312. Dusted Skipper, ♀, nectaring at Japanese honeysuckle, Fairfax County, Virginia.

PLATE 53

FIGURE 313. Monk (*Asbolis capucinus*), ♂, Key West, Florida.

FIGURE 316. Carolina Roadside Skipper, ♂, ibid.

FIGURE 314. Lace-winged Roadside Skipper (*Amblyscirtes aesculapius*), ♂, perching, Suffolk County, Virginia.

FIGURE 317. Carolina Roadside Skipper, ♂, Suffolk County, Virginia.

FIGURE 315. Carolina Roadside Skipper (*Amblyscirtes carolina*), ♂, perching, Gates County, North Carolina.

FIGURE 318. Roadside Skipper (*Amblyscirtes vialis*), ♂, nectaring at mint, Allegheny County, Maryland.

PLATE 54

FIGURE 319. Eufala Skipper (*Lerodea eufala*), Key West, Florida.

FIGURE 322. Salt Marsh Skipper, ♂, New Kent County, Virginia.

FIGURE 320. Twin Spot Skipper (*Oligoria maculata*), ♂, Collier County, Florida.

FIGURE 323. Ocola Skipper (*Panoquina ocola*), nectaring at shepherd's needle, Alachua County, Florida.

FIGURE 321. Salt Marsh Skipper (*Panoquina panoquin*), ♀, nectaring at golden aster, Sussex County, Delaware.

FIGURE 324. Giant Yucca Skipper (*Megathymus yuccae*), ♂, Indian River County, Florida.

Western Banded Elfin
Incisalia eryphon (Boisduval), 1852
MAP 72

Synopsis. The Western Banded Elfin is extremely similar to the Pine Elfin, and may usually be distinguised by its lack of a midcell bar on the ventral forewing and the more highly angled submarginal chevrons on the ventral hind wing. It has been found locally in Maine, Minnesota, and Michigan. In Maine the butterflies were found in a black spruce–sphagnum bog perching on young spruce trees. In western North America the caterpillars feed only on leaves of hard pine. Its life history is simlar to the Pine Elfin's. In the East, the single brood of adults flies from May 18 to June 9.

Southern Hairstreak
Fixsenia favonius (J. E. Smith), 1797
PLATE 15 · FIGURE 89 · MAP 73

Etymology. The species is named after *favonius,* the western spring wind (Latin).

Synopsis. The Southern Hairstreak is closely related to the Northern Hairstreak, with whose range it overlaps in southeastern Georgia. This species probably became separate from the Northern Hairstreak when Florida was an isolated island.

Butterfly. Male forewing: $\overline{X}$ = 1.4 cm, range 1.2–1.6 cm; female forewing: $\overline{X}$ = 1.6 cm, range 1.5–1.8 cm. The large dorsal red-orange patches and the large orange submarginal ventral hind-wing patch serve to distinguish this hairstreak.

Range. The Southern Hairstreak occurs in southeastern Georgia and most of peninsular Florida.

Habitat. Oak hammocks in the Tropical and Lower Austral Life Zones are the home for this Floridian endemic.

Life History. *Broods.* There is one annual generation, which flies from May 11 to June 6 in Georgia and from mid-March to early May in peninsular Florida.

Early Stages. In a painting by the pioneer John Abbot, the mature caterpillar is light green with a narrow, dorsal, dark-green stripe, oblique, green, lateral stripes, and a yellow lateral stripe along the spiracular area.

Adult Nectar Sources. Chinquapin, white sweet clover, viburnum, blueberry, sweetspire, and shepherd's needle flowers are visited for nectar.

Caterpillar Host Plants. The hosts are believed to be various oaks, particularly live oak (*Quercus virginiana*).

72. *Incisalia eryphon*

73. *Fixsenia favonius*

Northern Hairstreak
Fixsenia ontario (Edwards), 1868
PLATE 15 · FIGURE 90 · MAP 74

Etymology. The species name is based on the Canadian Province of the same name.

Synopsis. The Northern Hairstreak is rare and very patchy in its occurrence. It is usually found near small oaks and is always a good find.

Butterfly. Male forewing: $\overline{X}$ = 1.3 cm, range 1.1–1.5 cm; female forewing: $\overline{X}$ = 1.4 cm, range 1.3–1.5 cm. This rare, sporadic butterfly has orange patches on the dorsal wing surfaces and otherwise might be confused with the Gray Hairstreak, although it is browner ventrally and has a jagged postmedian line near the anal hind-wing angle. There are several unnamed geographic variants in the East, as well as three named western subspecies. It is very closely related to the Southern Hairstreak (*Fixsenia favonius*), and could be conspecific with it.

Range. The Northern Hairstreak occurs from southern Ontario and Quebec south and east to New York and Massachusetts and south and west through Michigan, Illinois, Kansas, and Texas to Arizona. The species ranges south to northern Florida and the Gulf states, but does not occur in southern Florida. Within this large range its occurrence is extremely patchy.

Habitat. The Northern Hairstreak occurs on wooded coastal dunes, pine barrens, shale barrens, open oak woods, pine-oak woods, and occasionally in rich deciduous woods if they are on sandy soils.

Life History. *Behavior.* Males perch and fly in oak thickets, while females rest on oak leaves. Adults are found most commonly at flowers in late afternoon (1630–1800 hr).

Broods. The Northern Hairstreak is univoltine. In the Delaware Valley, it flies during late June and early July, but it flies much earlier in the South. Flight dates of April 26 to May 27 pertain in Georgia. Generally, it can be found 1 to 2 weeks earlier than the *Satyrium* hairstreaks.

Early Stages. Eggs are laid on host twigs and do not hatch until the following spring when buds are swelling. The young larvae bore into individual flowers and feed on pollen. After the third molt, the larvae feed on young leaves. The larvae are pale green (Gifford and Opler 1983).

Adult Nectar Sources. Flowers of dogbane, New Jersey tea, common milkweed, farkleberry, blueberry, white sweet clover, and wild quinine are visited by this rare species.

Caterpillar Host Plants. Oaks are the only known hosts. White oak (*Quercus alba*), post oak (*Q. stellata*), live oak (*Q. virginiana*), and laurel oak (*Q. laurifolia*) have been documented. Bear oak (*Q. ilicifolia*) is also a likely host in the Appalachian barrens.

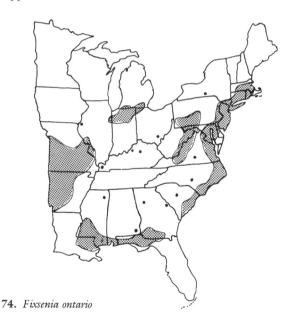

74. *Fixsenia ontario*

White M Hairstreak
Parrhasius m-album (Boisduval & LeConte), 1833
PLATE 16 · FIGURE 91 · MAP 75

Etymology. *Albus* is Latin for "white."

Synopsis. Although common is the Southeast, the White M Hairstreak is of regular but sporadic occurrence in the North. Its life history is little known considering it is such a widespread species.

Butterfly. Male forewing: $\overline{X}$ = 1.6 cm, range 1.5–1.7 cm; female forewing: $\overline{X}$ = 1.7 cm, range 1.6–1.9 cm. The com-

bination of iridescent blue on the dorsal surface and the gray ventral surface with costal white dot and "white M" in postmedian line on the hind wings should serve to distinguish this lovely insect.

Range. The White M Hairstreak occurs from Massachusetts and New Jersey west to Iowa and Kansas and south to Florida and the Gulf Coast. Away from the Deep South, it is most common along the Appalachian chain. Other sub-species occur in Mexico, Central America, and Venezuela.

Habitat. The White M Hairstreak occurs in or near broad-leaf woodlands from the Tropical through Transition Life Zones. Various habitats include scrubby woods, swamp forest, woodland edges, and dense upland hardwood forest.

Life History. *Broods.* The White M Hairstreak has three generations throughout most of our area, but possibly only two in some northern areas and four in Florida. In Virginia it flies from mid-April to early October, and in Georgia from February 10 to October 2.

Early Stages. The mature caterpillar is green with a dark dorsal stripe and seven oblique dark green stripes on each side of the body (Clench 1961).

Adult Nectar Sources. The White M Hairstreak favors a variety of flowers, including viburnum, sumac, common milkweed, goldenrod, sweet pepperbush, sourwood, dogwood, wild plum, poinsettia, and lantana.

Caterpillar Host Plants. In the eastern states, oaks are apparently the sole hosts. Live oak (*Quercus virginiana*) and others are selected.

Gray Hairstreak
Strymon melinus (Hübner), 1818
PLATE 16 · FIGURE 92 · MAP 76

Etymology. The species name is derived from the Latin *melinus*, "quince yellow."

Synopsis. The Gray Hairstreak is the most widespread hairstreak in North America. Its caterpillars feed on the widest array of plants of any butterfly and occasionally cause economic injury to cotton and bean crops.

Butterfly. Male forewing: $\overline{X} = 1.5$ cm, range 1.3–1.6 cm; female forewing: $\overline{X} = 1.5$ cm, range 1.3–1.6 cm. The steel-gray dorsal wing surface, the pale-gray ventral wings, the red-orange spot above the tail on the dorsal hind-wing surface, and the orange-margined postmedian line should serve to distinguish the Gray Hairstreak. The species is both seasonally and geographically variable. Spring individuals are generally small and very dark gray below. Later, the butterflies are larger and paler gray ventrally.

Range. The Gray Hairstreak occurs from Nova Scotia west to central British Columbia and south through the entire United States and the seasonally dry lowlands of Mexico and Central America to South America. It does not occur in the West Indies. It is a rare immigrant northward.

Habitat. The Gray Hairstreak may be found everywhere in nonforested areas from Tropical through Transition Life Zones. It is especially common in weedy disturbed areas.

Life History. *Behavior.* Males perch on shrubs or small trees from 1300 hr to just before dusk. In spring they perch close to the ground, but move higher as the year progresses. After an encounter with another individual, the male often returns to his favorite perch. Mated pairs have been seen only at night (2210–2245 hr). Oviposition has been observed at 1500 hr.

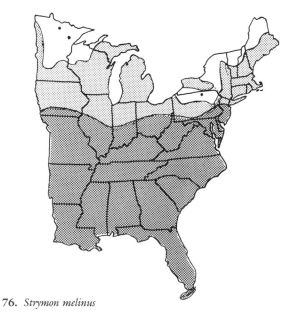

75. *Parrhasius m-album*

76. *Strymon melinus*

Broods. The Gray Hairstreak is probably bivoltine in northern areas, e.g., Maine (May 21–August 31), but has three or four flights in the central United States. It flies as early as mid-February and as late as the end of October in Mississippi. The timing of the generations is not well synchronized, possibly due to differences in developmental rates on different hosts.

Early Stages. The young caterpillars bore into the flowers and fruits of their hosts. Older larvae may feed on leaves. The caterpillars are extremely variable, being green, yellow, or red-brown, with oblique markings of various colors, and covered with short yellow-brown hairs. The pupa is pale brown and is marked with dark-brown speckles.

Adult Nectar Sources. The Gray Hairstreak nectars at a variety of flowers, including winter cress, dogbane, milkweeds, mints, white sweet clover, tick trefoils, and goldenrod.

Caterpillar Host Plants. The larvae feed on flowers and fruits of an almost endless list of plants in many families. Pea and mallow family species are most often selected, including tick trefoils (*Desmodium*), vetches (*Vicia*), beans (*Phaseolus*), bush clover (*Lespedeza*), clovers (*Trifolium*), mallow (*Malva*), and cotton (*Gossypium*).

Martialis Hairstreak
Strymon martialis (Herrich-Schäffer), 1864
PLATE 16 · FIGURE 93 · MAP 77

Etymology. The species is named for M. Valerius Martialis, a first-century (A.D.) Roman poet.

Synopsis. Martialis Hairstreak is another West Indian butterfly closely related to Bartram's Hairstreak.

Butterfly. Male forewing: $\overline{X}$ = 1.4 cm, range 1.3–1.6 cm; female forewing: $\overline{X}$ = 1.4 cm, range 1.3–1.6 cm. This species is very similar to Bartram's Hairstreak, but has large blue patches dorsally and lacks the two white basal spots on the ventral hind wing.

Range. Martialis Hairstreak occurs in southern Florida and the West Indies south to Jamaica.

Habitat. Adults frequent fields and open sunny areas adjacent to the coastline in the Tropical Life Zone.

Life History. *Broods.* In Florida, the butterfly is found from February through August, and sporadically in October and December. The number of generations is indefinite.

Early Stages. The mature caterpillar is dull green, and its dorsal portion is covered with short white hairs.

Adult Nectar Sources. The adults nectar at shepherd's needle, lantana, lippia, Brazilian pepper, bay cedar, and tournefortia.

Caterpillar Host Plants. Larval hosts have been documented as Florida trema (*Trema micrantha*) and bay cedar (*Suriana martiana*).

77. *Strymon martialis*

Bartram's Hairstreak
Strymon acis bartrami (Comstock & Huntington), 1943
PLATE 16 · FIGURE 94 · MAP 78

Etymology. The species is named after either a river in Sicily (Ovidius) or one of the Cyclades (Plinius). The subspecies is named after one of the Bartrams, a family of botanists from Philadelphia. John Sr. (1699–1777) wrote "Antiquities of Florida," while his son William (1739–1823) traveled extensively.

Synopsis. This Antillean endemic hairstreak is most common on Big Pine Key in Florida, where it is threatened by development. Each island's population is different and several subspecies are named.

Butterfly. Male forewing: $\bar{X}$ = 1.2 cm, range 1.1–1.3 cm; female forewing: $\bar{X}$ = 1.2 cm, range 1.1–1.3 cm. This butterfly is characterized by its ventral pattern. On the hind wing there are two basal spots, a bold black and white linear submarginal band and a large red-orange patch subtending a white submarginal band. The very long tails are recurved. Bartram's Hairstreak is a subspecies endemic to Florida.

Range. The species is found in southern Florida and the West Indies east to Dominica.

Habitat. In Florida the species is most common on Big Pine Key, where it occurs on narrow-leafed croton bushes within openings of pine forest.

Life History. *Broods.* There appear to be three or four generations in Florida, where the species flies from February through May, July–August, and October–November.

Adult Nectar Sources. The adults nectar at flowers of shepherd's needle and narrow-leafed croton.

Caterpillar Host Plants. Larvae feed on narrow-leafed croton (*Croton linearis*).

Columella Hairstreak
Strymon columella modesta (Maynard), 1873

PLATE 16 · FIGURE 95 · MAP 79

Etymology. The species is named after the first-century (A.D.) Latin writer L. Junius Columella.

Synopsis. The Columella is a small tropical hairstreak of weedy fields and pastures. It is undistinguished in its appearance.

Butterfly. Male forewing: $\bar{X}$ = 1.2 cm, range 1.0–1.4 cm; female forewing: $\bar{X}$ = 1.2 cm, range 1.1–1.4 cm. The Columella Hairstreak is distinguished by the two round black spots on the costal margin of the ventral hind wing, and by the absence of any dorsal orange marks. Dorsally, the female has an area of dull blue-gray near the tails.

Range. This species occurs from peninsular Florida, southern Texas, and southern California south through the West Indies, Mexico, and Central America to Columbia and Venezuela.

Habitat. The Columella Hairstreak is limited to weedy fields, second growth, and pastures in seasonally dry Tropical lowlands.

Life History. *Broods.* In Florida, the Columella Hairstreak has been found in every month except October. The number of separate broods is unknown.

Early Stages. Development is direct in all stages. The egg is a pea-green flattened sphere and hatches in about 7 days. The mature caterpillar is green with a darker green dorsal line and

78. *Strymon acis bartrami*

79. *Strymon columella modesta*

a dirty white cervical shield. The body is covered with translucent white tubercles bearing short chestnut-brown hairs. The pupa is pink buff with scattered brown specks and a greenish dorsal band (Comstock and Dammers 1935).

Adult Nectar Sources.   Small-flowered composites, croton, and various trees with small white flowers are visited.
Caterpillar Host Plants.   Larval hosts are various mallows. *Abutilon permolle* and *Suriana maritima* are hosts in Florida.

Disguised Hairstreak
Strymon limenia (Hewitson), 1868
MAP 80

Synopsis.   The Disguised Hairstreak, normally found only in the West Indies, has been found on several of the Florida Keys in recent years. It is extremely similar to the Columella Hairstreak, but may be distinguished by its more elongate forewings and the small red-orange spot at the anal angle of the dorsal hind wing. It is often found on shepherd's needle flowers with the Columella. So far there are flight records for April, May, and December. Its early stages are not known.

Early Hairstreak
Erora laeta (Edwards), 1862
PLATE 16 · FIGURE 96 · MAP 81

Etymology.   The genus name is possibly derived from Aurora, the dawn goddess of Roman myth. The species name is derived from the Latin *laetus,* "happy" or "fat."
Synopsis.   Although occasionally common, the Early Hairstreak is sporadic, localized, and highly sought by lepidopterists. It is closely related to the Arizona Hairstreak (*Erora quaderna*) of the Southwest and Mexico.

Butterfly.   Male forewing: $\overline{X}$ = 1.1 cm; female forewing: $\overline{X}$ = 1.1 cm, range 1.0–1.1 cm. This species lacks tails and has a gray-green ventral hind wing with submedian and submarginal rows of small red spots. The female has much blue dorsally, while the male has only a small blue band along the outer margin.

80. *Strymon limenia*

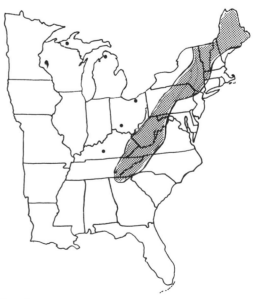

81. *Erora laeta*

Range. The Early Hairstreak occurs from Nova Scotia and New Brunswick west to northern Michigan. It ranges southwestward along the Appalachians to northern Georgia.

Habitat. The Early Hairstreak is found associated with deciduous woods in the Transition and Canadian Life Zones. Dirt roads through beech-maple forest or open ridge-tops are situations where it is most often encountered.

Life History. *Behavior.* Males of *Erora quaderna*, a related species, perch high in trees along ridge-tops in late afternoon. The Early Hairstreak probably does the same. Adults of both sexes may be found on moist soil along streams or on dirt roads through forests (D. Bowers 1978).

Broods. In the north (Michigan, New York, and Maine), the species has one full generation in late April to early June and a partial second generation in July or early August. Farther south (Virginia, Kentucky, and North Carolina), there appear to be two broods and a partial third, i.e., in April–early May, late June–early July, and late August or early September.

Early Stages. The pale-green eggs are laid singly. The mature caterpillar is pale green to rusty brown with reddish, brown and dark-green patches. The pupa is rust speckled with dark brown (Klots and dos Passos 1982).

Adult Nectar Sources. Adults have been found nectaring at hardtack, fleabane, and ox-eyed daisy.

Caterpillar Host Plants. Beaked hazel (*Corylus cornuta*) and beech (*Fagus grandifolia*) are the only known hosts.

Fulvous Hairstreak
Electrostrymon angelia (Hewitson), 1874

MAP 82

Synopsis. The Fulvous Hairstreak is another West Indian species and was first found in Florida during 1973. It may be distinguished by the dorsal surfaces, which are largely orange, bordered with black, and the brown ventral surfaces, which have an irregular white median line. It has been found in southern Florida and the Keys from January to mid-November. The adults perch on leaves in broken sunlight or shade at the edge of hardwood hammocks. It visits the flowers of shepherd's needle, sea grape, and Brazilian pepper. Its caterpillars feed on Brazilian pepper (*Schinus terebenthefolius*), which is a member of the cashew family.

82. *Electrostrymon angelia*

Blues (Tribe Plebejini)

Like coppers, blues are predominantly north temperate. Many more species occur in the Old World than in the New World. The adults are small and are almost always sexually dimorphic. The males are usually extensively iridescent blue above, while the females have more limited blue areas, if any.

The males usually employ the patrolling mate-location strategy and visit moist spots freely when newly emerged. Adults visit flowers for nectar.

Most species limit their choice of larval host plants to a single family, often legumes. The caterpillars often feed on host flowers or young seeds. Many blue caterpillars have secretory glands on the dorsal surface of the tenth segment and are tended by ants in exchange for the sugary secretion they produce. It is thought that the ants reduce the chances that small parasitic wasps or flies will lay their eggs on the caterpillars.

Western Pygmy Blue
Brephidium exilis (Boisduval), 1852
MAP 83

Synopsis.   The Western Pygmy Blue is a small butterfly found sparingly in Louisiana, Arkansas, and Missouri. It ranges through the Antilles and is common along the coast and in arid alkaline habitats in the western states and northern Mexico. Unlike the Eastern Pygmy Blue, it is not closely tied to tidal habitat. Its hosts are various saltbush species (*Atriplex*) and pigweed (*Chenopodium album*). The blue-green eggs are laid on host flower buds at midday and the resultant caterpillars are yellow-green and covered with small brown tubercles surmounted by minute white processes.

Eastern Pygmy Blue
Brephidium isophthalma pseudofea (Morrison), 1873
PLATE 17 · FIGURE 97 · MAP 84

Etymology.   The species name is derived from the Greek *isophthalma*, "having the same eyes."

Synopsis.   This species and the Western Pygmy Blue (*Brephidium exilis*) are the smallest butterflies in our area. They are so small that one must look twice before realizing that the tiny object is actually a butterfly.

Butterfly.   Male forewing: $\overline{X}$ = 0.9 cm, range 0.8–0.9 cm; female forewing: $\overline{X}$ = 1.0 cm, range 0.9–1.1 cm. The two pygmy blues may be distinguished from other blues by their small size, red-brown ventral wing surfaces, and the submarginal row of six metallic black spots on the ventral hind wing. The Eastern Pygmy Blue may be distinguished from the Western Pygmy Blue by its more uniformly brown ventral wing surfaces and the entirely brown fringes.

Range.   This tiny insect occurs from coastal South Carolina south along both coasts of Florida and west along the immediate Gulf Coast to Louisiana and occasionally Texas.

Habitat.   Eastern Pygmy Blues are always close to saltwater, being found on tidal flats and coastal strand habitat in the Lower Austral Life Zone (Heppner 1974).

Life History.   *Behavior.*   The flight is slow and weak. Males patrol back and forth within inches of their host plant. Courtship takes place in late afternoon.

83. *Brephidium exilis*

84. *Brephidium isophthalma pseudofea*

Broods. The number of generations is not documented. In Florida these blues may be found all year, while in Georgia they have been seen from May 19 to August 15.

Early Stages. The egg is pale blue-green with a raised white network, while the larva is yellow-green, pale green, or green with white-tipped brown tubercles. There is a yellow-white dorsal line and a bright yellow lateral band. The pupa is variable, but is usually light yellow-brown with darker brown dots.

Adult Nectar Sources. Pygmy Blues nectar at palmetto palm and saltwort flowers in Florida.

Caterpillar Host Plants. The primary host is annual glasswort (*Salicornia bigelovii*) and possibly woody glasswort (*Salicornia virginica*). Saltwort (*Batis maritima*) has been suggested but not confirmed.

Cassius Blue
Leptotes cassius theonus (Lucas), 1857
PLATE 17 · FIGURE 98 · MAP 85

Etymology. The species name is a proper Roman name, as for example that given the conspirator against Julius Caesar. The subspecies is named after Theon of Smyrna, a Greek mathematician, musician, and astronomer.

Synopsis. The Cassius Blue is a common butterfly in residential areas in the southern two-thirds of Florida. Its females are more brightly colored than the males, a reversal of the usual situation in lycaenids.

Butterfly. Male forewing: $\overline{X}$ = 1.2 cm, range 1.1–1.5 cm; female forewing: $\overline{X}$ = 1.1 cm, range 0.9–1.3 cm. The Cassius Blue is distinguished from other blues by the fine barring ventrally and by the females, which have much white dorsally. In Missouri and Arkansas, where either this species or the Marine Blue may occur, the former may be distinguished by the white on the females, the pure blue dorsal wing surfaces on the male, and the white along the anal margins of the male's hind wings.

Range. The Cassius Blue occurs from Florida and southern Texas south through the Antilles, Mexico, and Central America to Argentina. It occasionally strays north to South Carolina, Arkansas and Missouri. The Cassius Blues that appear west of the Mississippi represent the subspecies *striatus*.

Habitat. The Cassius Blue is most common in residential areas, but may appear almost anywhere in southern Florida.

Life History. *Broods*. The Cassius Blue flies all year in Florida, but the number of generations has not been recorded.

Early Stages. Females oviposit between flower buds, and the caterpillars feed on the buds and flowers. There are four larval instars, and the caterpillars are tended by ants that feed on the sugary secretions from the abdominal glands. The ants' presence may deter predation and parasitism. The early stages have been described in great detail by Downey and Allyn (1979). In general, the caterpillar is green with a russet red overtone (Haskins 1933).

Adult Nectar Sources. In Florida the Cassius Blue is especially fond of shepherd's needle and lippia, but will also nectar at a variety of flowers, including those of trees and shrubs.

Caterpillar Host Plants. In Florida, ornamental leadwort (*Plumbago capensis*) is used most often, but more usual hosts elsewhere are members of the pea family, including hairy milk pea (*Galactia volubilis*), lima bean (*Phaseolus limensis*), and rattlebox (*Crotalaria incana*).

85. *Leptotes cassius theonus*

Marine Blue
Leptotes marina (Reakirt), 1868

PLATE 17 · FIGURE 99 · MAP 86

Synopsis.   The Marine Blue may be distinguished from the Cassius Blue (*Leptotes cassius*) by the dorsal purplish color of the males and the lack of white areas on the female. It is not resident in our area, but occasionally strays from the Southwest to states bordering the Mississippi, and is found as far north as Illinois, Nebraska, and Colorado with some frequency. It is resident from southern Texas and the Southwest through Mexico to Central America. The larvae feed on leadwort (*Plumbago*) and a variety of legumes.

Miami Blue
Hemiargus thomasi bethunebakeri Comstock & Huntington, 1943

PLATE 17 · FIGURE 100 · MAP 87

Etymology.   The species is named after W. Donald Thomas of Miami, Florida, a friend of the late Harry K. Clench, the species' describer. George T. Bethune-Baker (1856–1944) was a businessman from Birmingham, England, who was active at the British Museum of National History and studied the classification of blues.

Synopsis.   The Miami Blue is a distinctive Antillean endemic with differing seasonal forms. The butterfly has declined in recent years and is now restricted to only a few sites on the Florida Keys.

Butterfly.   Male forewing: $\overline{X}$ = 1.2 cm, range 1.0–1.3 cm; female forewing: $\overline{X}$ = 1.2 cm, range 1.0–1.3 cm. The very broad white ventral submarginal band, the dorsal turquoise color of both sexes, and the orange-capped marginal eyespot on the hind wings should serve to distinguish the Miami Blue from the Ceraunus Blue (*Hemiargus ceraunus*) and Cassius Blue (*Leptotes cassius*), which are found in the same habitats. The winter (dry-season) form is a much lighter blue than the summer (wet-season) form and has narrow black borders. The winter-form adults may live 3 or 4 months, an inordinately long time for such a small butterfly.

Range.   The Miami Blue, as its name suggests, formerly occurred from the mainland of peninsular Florida south through the Keys. Urbanization has caused its extirpation on the Floridian mainland, and the butterfly is now limited to a few locations on the Keys. Other subspecies occur on the Antilles from the Bahamas south and east to St. Kitts.

Habitat.   This butterfly is found along trails and in openings of Tropical hardwood hammocks or along their edges.

Life History.   *Broods.* The Miami Blue is found in all

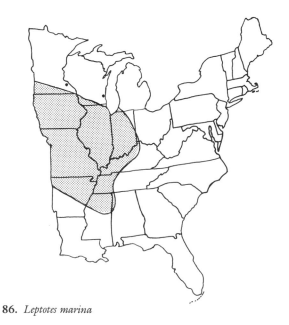

86. *Leptotes marina*

87. *Hemiargus thomasi bethunebakeri*

months. There is one long winter generation in December–April, during most of which time the adults are probably in reproductive diapause. From May through November there is probably a succession of shorter generations, the exact number of which is unknown.

Early Stages. Females deposit single eggs just above lateral buds, and the caterpillars feed on host flowers and pods. The stages have not been described.

Adult Nectar Sources. The adults nectar at shepherd's needle and probably a variety of other flowers.

Caterpillar Host Plants. Previous reports may refer to another Antillean butterfly, Lucas' Blue (*Hemiargus ammon*). The primary host is balloon vine (*Cardiospermum*), although oviposition has also been seen on snowberry (*Chiococca alba*).

Ceraunus Blue
Hemiargus ceraunus antibubastus (Hübner), 1818
PLATE 17 · FIGURE 101 · MAP 88

Etymology. The genus name is derived from the Greek term *hemi*, "half," and *Argus*, a Greek proper name from several myths. The species is derived from the Greek *keraunos*, "thunder and lightning."

Synopsis. The Ceraunus Blue is probably the most common, widespread blue in the Americas. It is known to undertake huge emigrations at times, and is well-adapted to disturbed habitats.

Butterfly. Male forewing: $\overline{X} = 1.1$ cm, range 1.0–1.2 cm; female forewing: $\overline{X} = 1.1$ cm, range 0.9–1.3 cm. The Ceraunus Blue is characterized by its *lack* of distinguishing features. It lacks the tails of the Eastern Tailed Blue (*Everes comyntas*), the broad ventral submarginal white bands found on the Miami Blue (*Hemiargus thomasi*), and the row of enlarged ventral postmedian spots shown on the forewing of the Reakirt's Blue (*Hemiargus isola*).

Range. The southeastern subspecies is limited to Florida, Georgia, the Carolinas, and the Gulf states, although it may be a permanent resident only in Florida. Other geographic forms range from southern Texas and the Southwest south through the Antilles, Mexico, and Central America to Argentina.

Habitat. The Ceraunus Blue is found in a variety of open situations, from lowland to moderate elevations in both dry and wet Tropical Life Zones. Specific habitats include turkey oak or open pine woods, beach dunes, vacant lots, pasture, and second-growth scrub.

Life History. *Behavior.* Males patrol erratically through the habitat during most warm daylight hours. In Central America, large numbers may emigrate from the lowlands to higher, cooler elevations at the beginning of the dry season. *Broods.* Broods have not been well documented. Most references state there are at least three generations. It is possible that this species, like the Miami Blue, has a long-lived winter form and several generations of a shorter-lived summer form. It is found all year in Florida, but mainly in late summer in other southeastern states.

Early Stages. The egg is a delicate light green and is turban-shaped. The caterpillars feed on flower buds and may be either green with or without a mauve dorsal stripe, red with greenish color below, or even yellow. The body is covered with short silver-white hairs. The pupa is translucent green (Comstock and Dammers 1934).

Adult Nectar Sources. This blue nectars at shepherd's needle and a variety of other plants. It seems to favor composites.

Caterpillar Host Plants. The Ceraunus Blue is known to feed on a variety of legumes, including all three families, as well as herbs, shrubs, and trees. Hosts reported include rosary pea (*Abrus precatorius*), *Rhynchosia minima*, and poison pea (*Macroptilium lathyroides*) in the pea family (Fabaceae); blackbead (*Pithecellobium unguiscati*), acacia, and mimosa in the mimosa family; and partridge pea (*Cassia brachiata*) and hairy partridge pea (*Cassia aspera*) in the Cesalpinia family.

88. *Hemiargus ceraunus antibubastus*

Reakirt's Blue
Hemiargus isola (Reakirt), 1866
PLATE 17 · FIGURE 102 · MAP 89

Synopsis. Reakirt's Blue is chiefly at home in arid and semiarid habitats and is more tolerant of cold winters than its close relatives. Periodically, it colonizes habitats far to the north of its zone of permanent residence and may even appear in the Arctic-Alpine Life Zone of western mountains.

Butterfly. Male forewing: $\overline{X} = 1.1$ cm, range 0.9–1.2 cm; female forewing: $\overline{X} = 1.3$ cm, range 1.2–1.3 cm. The submarginal row of enlarged round black spots on the ventral forewing serves to distinguish Reakirt's Blue.

Range. The Reakirt's Blue is resident from southern Texas and the arid Southwest south through the middle elevations of Mexico and Central America to Costa Rica. Annually, it emigrates northward through the mountains and plains, appearing by early summer. In our area it regularly occurs in the states adjacent to the Mississippi River, and on rare occasions has appeared in Indiana, Ohio, and Michigan (Remington 1942).

Habitat. In our area this butterfly appears in grasslands or fields where suitable host plants are present.

Life History. *Behavior.* The males patrol erratically during most daylight hours. Oviposition takes place at midday. Like many blues, the butterflies sleep on weedy vegetation with the head downward.

Broods. There appear to be three broods in our area: April, mid-June to mid-July, and September to early October.

Early Stages. The caterpillars feed on buds and flowers, but have not been described (Remington 1952).

Adult Nectar Sources. The adults visit a variety of herbaceous flowers, including white sweet clover and spearmint.

Caterpillar Host Plants. Many plants of the pea and mimosa families are fed upon. These include mesquite (*Prosopis juliflora*), acacias, mimosa tree (*Albizzia julibrissin*), indigo bush (*Dalea*), indigo (*Indigofera*), and yellow sweet clover (*Melilotus indicus*).

89. *Hemiargus isola*

Eastern Tailed Blue
Everes comyntas (Godart), 1824
PLATE 18 · FIGURES 103 & 104 · MAP 90

Etymology. The genus may have been named after Evere, a Brabantine city in Belgium. The species may have been named after John Comyn, Lord de Babenoch, who died in 1306.

Synopsis. In many parts of the eastern United States, the Eastern Tailed Blue is the most common butterfly, although its low flight and small size make it somewhat inconspicuous.

Butterfly. Male forewing: $\overline{X} = 1.21$ cm, range 0.99–1.37 cm; female forewing: $\overline{X} = 1.14$ cm, range 0.97–1.37 cm. The only other blue in the East with tails is the Western Tailed Blue (*Everes amyntula valeriae*). The latter species may be distinguished by its larger size, chalky white ground color, and reduced ventral pattern. The Eastern Tailed Blue is seasonally variable. Early spring butterflies are smaller, with the ventral markings more clear-cut. The spring female has much blue dorsally, whereas summer females are uniformly black-brown.

Range. The Eastern Tailed Blue ranges from New England and southern Quebec west to central Colorado and south through our area, becoming rare in northern Florida. It ranges farther south through Mexico and Central America to at least Costa Rica. Isolated populations occur in the valleys of California and Oregon.

Habitat. This ubiquitous blue is found in almost any open, sunny situation where suitable host legumes are present. These include such man-made habitats as roadsides, landfills, and reclaimed land.

Life History. *Behavior.* Males patrol during most daylight hours in the vicinity of the host plants. Mated pairs may be seen from midday to midafternoon (1100–1500, rarely 1030 or 1830 hr). In Virginia, the male was the carrier in most instances. Oviposition usually occurs during a brief period in midafternoon (1315–1350 hr), although occasional females are observed as early as 1030 or as late as 1500. The eggs are inserted between tightly packed young flower buds. When at rest, the butterflies actively rub the hind wings together in the manner of many hairstreaks. They also may perch with their wings open at about a 45° angle, particularly in early spring. This is presumed to be a form of dorsal basking for heat gain. Most Lycaenids never perch with open wings. Newly emerged males may be found in large numbers taking moisture on wet soil along streams or dirt roads. The butterflies roost for the night in sites receiving late-afternoon sun.

Broods. There are probably no less than three generations in the northern parts of this blue's range, and there are at least four broods and a partial fifth in Virginia (early April–mid-November). Farther south, there may be six or seven broods in such areas as Mississippi where the species is found from early February through November.

Early Stages. The caterpillars usually feed on flower buds and flowers. Development is direct, and there are four larval instars, as in most blues. The caterpillar is dark green with a dark-brown middorsal stripe and faint brownish oblique lateral stripes. The body is covered with tubercles, each bearing a short hair. The chrysalis is whitish green with a darker head and thorax. A very complete description of all stages is given by Lawrence and Downey (1966). The mature larva overwinters, and it pupates in the spring.

Adult Nectar Sources. The Eastern Tailed Blue, with its short proboscis and low flight, is precluded from most nectar sources. A seasonal sequence of short-tubed or open flowers of plants within a few inches of the ground is preferred, including cinquefoils, wild strawberry, winter cress, white clover, dogbane, butterflyweed, fleabane, beggar's tick, white sweet clover, shepherd's needle and asters.

Caterpillar Host Plants. A wide variety of pea family legumes are utilized, including vetches (*Vicia*), clovers (*Trifolium*), beggar's tick (*Desmodium*), bush clover (*Lespedeza*), alfalfa (*Medicago sativa*), and yellow sweet clover (*Melilotus indicus*). At any one locality, subsequent generations of the blue switch from one host to another as the flowering sequence advances.

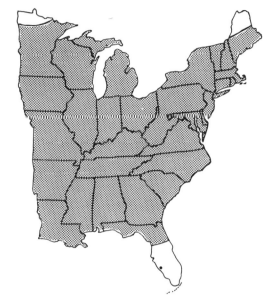

90. *Everes comyntas*

Western Tailed Blue
Everes amyntula valeriae Clench, 1944

MAP 91

Synopsis. The Western Tailed Blue is a more northern and western relative of the common Eastern Tailed Blue (*Everes comyntas*). It barely enters our area, occurring only in the northern Great Lakes states.

Butterfly. This butterfly is extremely similar to the Eastern Tailed Blue, but is generally larger and has more rounded forewings and a smaller orange spot at the anal angle. It is more variable than its eastern counterpart.

Range. This blue ranges from Alaska south to southern New Mexico, Arizona, and northern Baja California, Mexico. It extends eastward to northern Minnesota, northern Michigan, and the Gaspé Peninsula.

Habitat. The Western Tailed Blue occurs in open areas with scattered shrubbery. It tends to occur in native habitats, whereas its more eastern and southern relative is a more frequent denizen of disturbed areas. Although it may occupy Lower Austral situations in southern California, it is most often found in Canadian and Hudsonian Life Zones.

Life History. *Behavior*. Males perch and patrol during most daylight hours. The butterflies perch and fly higher above the ground than do Eastern Tailed Blues. A mated pair was seen in late morning.

Broods. The Western Tailed Blue is univoltine throughout most of its range, although is has a partial second brood in the southern Rocky Mountain area and is multiple-brooded in southern California.

Early Stages. The eggs are laid singly on flowers or young seed pods. The larvae feed mainly on young seeds, and the nearly mature caterpillars overwinter and eventually pupate inside the pods. The egg is pale green, while the larva is variable, ranging from solid green to yellow-green or straw with pink and maroon markings. The pupa is pale tan or gray-green with a dark brown dorsal band and blotches.

Caterpillar Host Plants. The larvae feed on the flowers and young seeds of various legumes, usually those with inflated pods, such as milkvetch (*Astragalus*), locoweed (*Oxytropis*), vetches (*Lathyrus* and *Vicia*), and false lupine (*Thermopsis*). Clover (*Trifolium*) has been reported, but seems doubtful.

91. *Everes amyntula valeriae*

Spring Azure
Celastrina ladon (Cramer), 1780
PLATE 18 · FIGURE 105 · MAP 92

Etymology. The genus is named for the plant bittersweet (*Celastrus*). The species is named after Ladon, a dragon with 100 heads, who guarded the gardens of the Hesperides and was killed by Hercules (Greek myth).

Synopsis. The Spring Azure is the most obvious common blue in our area, although the Eastern Tailed Blue outnumbers it in most areas. In recent years, it has been realized that, instead of a single species, several sibling species are involved. Seasonal and geographic variation confuses matters.

Butterfly. Male forewing: $\bar{X}$ = 1.42 cm, range 1.37–1.54 cm; female forewing: $\bar{X}$ = 1.42 cm, range 1.37–1.52 cm. The Spring Azure is extremely variable both seasonally and geographically, and the sexes are dimorphic. Dorsally, the males are blue with a narrow black border, while the females have black costal and outer portions of the forewing and have a row of black submarginal spots on the outer margin of the hind wing. The spring brood may have individuals of three different forms. All have different patterns on the ventral wing surfaces. Form "lucia" has smeared marginal brown bands on both wings and a large brown patch in the center of the hind wing, form "marginata" has the brown marginal bands only, and form "violacea" lacks the bands altogether and may have dorsal wings more iridescent blue. Forms "lucia" and "marginata" become rarer as one goes southward. The second brood and later-emerging individuals are usually

form "neglecta," which has large areas of white scaling on the upper wings of males and females, a more lustrous blue above, smaller black spots below, and more extensive black on the female forewings above. The factors responsible for these forms are not well understood.

The two other members of the group considered separate species by us are the Dusky Blue (*Celastrina ebenina*) and the Appalachian Blue (*Celastrina neglectamajor*). Within its range the Dusky Blue is most similar to the "violacea" form of the Spring Azure. Its males are sooty black dorsally, while the females are a dull gray-blue. Ventrally, the submarginal row of lunules and the black spots are always clear-cut in the Dusky Blue, not blurred or absent as in the comparable form of the Spring Azure. The Appalachian Blue is most similar to the late-spring form of the Spring Azure, but is significantly larger. Dorsally, its males lack any infusion of white scaling, and females have a more clear-cut pattern. Ventrally, Appalachian Blues are almost pure white, with the spot pattern strongly reduced but never blurred. The two or three submarginal black spots near the hind-wing anal angle are usually present and distinct, even if all others are absent. There will probably be several other siblings recognized in the future.

Range. The Spring Azure ranges from Labrador and the Maritime Provinces west and north to the Pacific Coast and Alaska. It ranges south throughout the United States, where suitable hosts and habitats permit, and occurs south in the mountains of Mexico and Central America to Panama. It is absent from most of peninsular Florida.

Habitat. The Spring Azure ranges from Lower Austral through Canadian Life Zones wherever suitable woody hosts are present. These may include everything from well-advanced old fields to mature forests and their edges. The butterfly also occurs in wooded freshwater marshes and swamps.

Life History. *Behavior.* Males patrol and occasionally perch during most daylight hours, but are most active in their search for mates from midafternoon to almost dusk, when most male–male interactions and attempted courtships are seen. One mated pair was reported at 1430 hr. Females oviposit from late morning until dusk (1030–1900 hr). Newly emerged males may be found in large numbers at moist soil along roads and streams. These blues usually fly at least 1 m above ground, and may fly regularly at 5 m or more. Spring Azures are short-lived, surviving no more than 4 days. On average, females emerge and mate on their first day, lay eggs on the second, and rarely survive to live a third or fourth day.

Broods. In the northern tier of states, there is one generation and a partial second, while farther south there are two full broods, and a continuing, but rapidly diminishing number of emergences until early fall. In Maine flight dates extend from May 1 to August 18. In Illinois they range from March 28 to September 5, and in Mississippi they begin as early as late January and last until early October.

Early Stages. The pale-green turban-shaped eggs are laid between flower buds. The larvae feed on buds and flowers and are usually tended by ants. Development is direct to the pupa, and at least some pupae of each generation enter diapause until the following spring. The mature larva is velvety, and varies in color from yellow-green to pinkish, and has a dark dorsal stripe and oblique, lateral greenish stripes. The pupa is light yellow-brown marked variously with black.

Adult Nectar Sources. Individuals of the first brood seldom visit flowers, but may be found at spicebush, wild plum, and winter cress. Individuals of the second generation may be found at flowers of dogbane, New Jersey tea, blackberry, privet, common milkweed, and a variety of others.

Caterpillar Host Plants. Each generation selects a different combination of larval hosts because of the staggering of flowering periods. Hosts selected are usually woody shrubs or trees with fairly compact bunches of small flowers. For example, at one locality in northern Virginia, first-brood females oviposit between flower buds of dogwood (*Cornus florida*) and wild cherry (*Prunus serotina*), while those of the second brood select New Jersey tea (*Ceanothus americana*), osier dogwood (*Cornus stolonifera*), viburnum (*Viburnum*), and staghorn sumac (*Rhus typhina*). Emerging adults shift to areas of suitable host abundance as the season progresses. Additional hosts fed upon in the East are meadowsweet (*Spiraea salicifolia*), black snakeroot (*Cimicifuga racemosa*), blueberry (*Vaccinium*), wingstem (*Actinomeris alternifolia*), and others.

Females, especially of the second or later generations, oviposit on plants upon which their larvae cannot complete development (C. Oliver, personal communication), including elderberry (*Sambucus*) and viburnum.

92. *Celastrina ladon*

Appalachian Blue
Celastrina neglectamajor (Tutt), 1908
PLATE 18 · FIGURE 106 · MAP 93

Etymology. The species name is a combination of the Latin *neglectus* (neglected) and *maior* (major or greater).

Synopsis. A sibling, the Appalachian Blue is restricted to the Appalachians and flies between the first and second generation of the Spring Azure.

Butterfly. Male forewing: $\overline{X} = 1.64$ cm, range 1.57–1.73 cm; female forewing: $\overline{X} = 1.67$ cm, range 1.57–1.77 cm. This butterfly is extremely similar to the "neglecta" form of the Spring Azure (*Celastrina ladon*).

Range. This butterfly is restricted to the Appalachians, from southern New York southwest at least to southeastern Virginia.

Habitat. The Appalachian Blue is found in rich, deciduous forest habitats, especially near streams. It is limited to the Transition Life Zone.

Life History. *Broods.* The Appalachian Blue is univoltine and flies from mid-May to early June. It occurs between the first and second broods of the Spring Azure, and is temporally isolated as a result.

Caterpillar Host Plant. Maple-leaf viburnum (*Viburnum acerifolium*) is utilized in the Delaware Valley.

Dusky Blue
Celastrina ebenina Clench, 1972
PLATE 19 · FIGURE 109 · MAP 94

Etymology. The species name is derived from ebony, meaning black.

Synopsis. The Dusky Blue was long thought to be a brown form of the Spring Azure, but recently was realized to be a separate species. Its caterpillars feed on young leaves of goat's beard, a plant of rich, moist deciduous forests.

Butterfly. Male forewing: $\bar{X} = 1.4$ cm, range 1.3–1.5 cm; female forewing: $\bar{X} = 1.3$ cm, range 1.0–1.5 cm. The sooty black of the male is distinctive, while the female is a dull gray-blue above.

Range. The Dusky Blue ranges from southwest Pennsylvania and central Maryland south in the Appalachians to North Carolina. It also occurs in the upper Ohio River drainage, eastern Missouri, and northwestern Arkansas (Clench 1972).

Habitat. The butterflies are found in shaded, moist, rich deciduous woodland in rolling or mountainous terrain. In addition to their host, the following woody plants are typical of Dusky Blue colonies: maples, birches, dogwood, beech, spicebush, tulip poplar, and basswood.

Life History. *Behavior.* Males fly at 1–3 m in patrolling flights. Males, and rarely females, visit moist spots on roads or near streams. Females oviposit on the underside of young, unfurled host leaflets from 1030 to at least 1200 hr.

Broods. With a few exceptions, the Dusky Blue has a single spring generation that flies from early April to early May.

Early Stages. The caterpillars feed on young host leaves, and development is direct, but most pupae overwinter in nature. The eggs are a pale gray blue and are turban-shaped. The mature caterpillars are whitish blue-green with three longi-

93. *Celastrina neglectamajor*

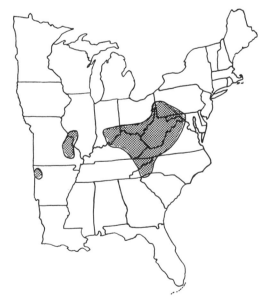

94. *Celastrina ebenina*

tudinal yellow-white stripes, one dorsal and two lateral. There are yellow-white spots and dashes scattered over the body, and minute star-shaped tubercles, each bearing a single hair, give the caterpillar a velvety appearance (Wagner and Mellichamp 1978).

Adult Nectar Sources. The only nectar flower reported for the Dusky Blue is wild geranium. Most individuals feed from the underside of the flowers, inserting their proboscises between the sepals and petals.

Caterpillar Host Plant. The only known host is Goat's beard (*Aruncus dioicus*) of the rose family.

Silvery Blue
Glaucopsyche lygdamus (Doubleday), 1841
PLATE 18 · FIGURES 107 & 108 · MAP 95

Etymology. The genus name is a combination of the Latin *glaucus* (gray-blue) and the Greek *psyche* (soul). The species is named for Lygdamus, a slave of Cynthia, the mistress of Propertius.

Synopsis. The Silvery Blue is a widespread, variable butterfly, but it is usually very local. There is no mistaking the male's iridescent silvery blue in flight.

Butterfly. Male forewing: $\overline{X}$ = 1.3 cm, range 1.1–1.4 cm; female forewing: $\overline{X}$ = 1.3 cm, range 1.0–1.5 cm. Dorsally, the males are iridescent silvery blue, with a narrow black border, while the females are darker blue with fairly wide black outer margins. Ventrally, the wings are brown to gray, with an irregular postmedian row of round black spots. The typical subspecies *lygdamus* occurs over most of our area, while in states adjacent to Canada, the subspecies *couperi* occurs.

Range. The Silvery Blue occurs from Nova Scotia west across Canada to British Columbia and the Pacific Coast. It ranges north to central Alaska and south to northern Baja California, Mexico. In the East, it occurs from the northernmost states south to Georgia, northern Alabama, and Arkansas.

Habitat. The Silvery Blue is usually found in open areas within woodlands, often in regions with rocky or sandy soils. Its specific habitats in the East range from damp-wooded valleys and rocky, moist woods to pine woods and brushy fields. It is usually not found below the Transition Life Zone.

Life History. *Behavior.* Males patrol in the general area of the caterpillar host.

Broods. The Silvery Blue is univoltine. The flights are progressively earlier southward. In Maine, the butterflies are found from June 14 to 27, in Illinois they fly from April 13 to May 30, and in Georgia from March 26 to April 19.

Early Stages. The caterpillars feed on flowers and developing pods. They are assiduously tended by ants that "milk" the caterpillars for the sugary secretions produced by special glands in the middle of the tenth abdominal segment. The caterpillars are variably colored, being green to purplish with a dark dorsal stripe and oblique white dashes. The pupa, which overwinters, is attached to debris near the base of the host.

Adult Nectar Sources. Silvery Blues are particularly fond of composites.

Caterpillar Host Plants. In the East, several vetches are the favored hosts. Carolina vetch (*Vicia carolina*) is the only host reported for the typical subspecies, while veiny peavine (*Lathyrus venosus*), and white sweet clover (*Melilotus alba*) have been reported in Maine, Michigan, and southern Ontario.

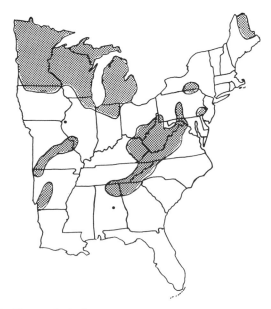

95. *Glaucopsyche lygdamus*

Nabokov's Blue
Lycaeides argyrognomon nabokovi Masters, 1972

MAP 96

Etymology. The species name is derived from the Greek *argyros* (silver) and *gnomon,* an astronomical instrument or the raised stylus of a sundial that casts the shadow. The subspecies was named in honor of Vladimir Nabokov, Russian émigré, Cornell professor, author, and student of blues.

Synopsis. Nabokov's Blue is a more boreal relative of the Melissa Blue. It has many subspecies in western North America and Canada.

Butterfly. Male forewing: $\bar{X}$ = 1.6 cm, range 1.4–1.7 cm; female forewing: $\bar{X}$ = 1.6 cm, range 1.5–1.7 cm. One feature distinguishing the Nabokov's Blue from the Melissa Blue is mentioned under the latter species. There are other, subtle differences.

Range. The Nabokov's Blue is just one race of a species that extends across the boreal portions of both North America and Eurasia. In North America it occurs from Newfoundland and Nova Scotia north and west to Alaska, British Columbia, and central California. It also ranges south to central Colorado. In our area, Nabokov's Blue occurs in northern Michigan, Minnesota, and Wisconsin (Masters 1972b).

Habitat. In the East this butterfly occurs in openings within mixed evergreen forest in the Canadian Life Zone. In northern Michigan, such forest consists of aspen, oak, and jack pine.

Life History. *Behavior.* Males patrol close to their hosts, searching for receptive females, during most warm daylight hours. Mated pairs have been observed in early afternoon (1310–1340 hr), with males as the active carriers. Females oviposit from mid- to late afternoon (1500–1730 hr). The eggs are laid singly on host stems or debris beneath it. *Broods.* Unlike the Melissa Blue and its relatives, the Nabokov's Blue and other related subspecies have only a single generation each year. Flight dates in Minnesota and Wisconsin range from July 1 to August 8.

Adult Nectar Sources. Adults visit dogbane; white, alsike, and hop clover; orange hawkweed; and yarrow for nectar.

Caterpillar Host Plants. Caterpillars of western populations feed on legumes, while in eastern Canada adults oviposit on several plants in the heath family, including crowberry (*Empetrum nigrum*), laurel (*Kalmia polifolia*), and Labrador tea (*Ledum palustre*). Nabokov's Blue utilizes dwarf bilberry (*Vaccinium caespitosum*) in Wisconsin (Nielsen and Ferge 1982).

96. *Lycaeides argyrognomon*

Melissa Blue, Karner Blue
Lycaeides melissa (Edwards), 1873

MAP 97

Etymology. The genus name is probably another derivation of Lycaon (see Lycaena). Melissa means honey bee in Greek. The subspecies *samuelis* (Karner Blue) was named in honor of Samuel Hubbard Scudder (1837–1911), who was a librarian at Harvard and an outstanding student of insect fossils, Orthoptera, and Lepidoptera.

Synopsis. Although the Melissa Blue is a widely adaptable denizen of many western habitats, including alfalfa fields, the eastern populations are limited to sandy pine barrens and beach dunes. The butterfly is declining in several states and has been extirpated from several others. The butterfly is listed as endangered under New York state law.

Butterfly. Male forewing: $\bar{X}$ = 1.4 cm, range 1.2–1.4 cm; female forewing: $\bar{X}$ = 1.4 cm, range 1.4–1.6 cm. This species is exceedingly difficult to distinguish from Nabokov's Blue. The Melissa Blue has a continuous black terminal line along the ventral hind-wing margin, while in Nabokov's Blue the line is broken into small black triangles at the vein endings.

Range. The Karner Blue, subspecies *samuelis,* occurs in small isolated colonies from New Hampshire and New York west across southern Ontario, northern Indiana, and Michigan to southern Wisconsin. The typical subspecies occurs from southwestern Canada south through the northern plains, and western states to northern Baja California, Mexico. It enters our area in western Iowa and Minnesota.

Habitat. Subspecies *samuelis* inhabits sandy pine barrens, lakeshore dunes, and sandy pine prairies. In our western area, typical Melissa Blues may be found on dry ridges within the tall-grass prairie biome.

Life History. *Behavior.* Males patrol near the host plants during most daylight hours. Mated pairs have been seen in the afternoon (1400–1700 hr).

Broods. The butterfly is bivoltine in the East. The first flight extends from late May through June and the second from late July through August.

Early Stages. Females may lay eggs on various parts of the host plant or nearby sticks and pebbles. The caterpillars feed on young host leaves and are assiduously tended by ants that feed on sugary secretions from the caterpillar's abdominal glands. There are five larval instars. The larva's body is pale green, with short, pale-brown hairs, while the pupa is pea green with a yellow-tinged abdomen.

Caterpillar Host Plants. In the East, the subspecies *samuelis'* only recorded host is lupine (*Lupinus perennis*), which requires fire or other periodic disturbance to aid in germination and competition with woody plants. Caterpillars of the Melissa Blue's western population feed on legumes, such as various lupines, alfalfa, and wild licorice (*Glycyrrhiza*).

Saepiolus Blue
Plebejus saepiolus amica (Edwards), 1863

MAP 98

Etymology. The genus name is derived from the Latin *plebs* (common people), while the species name is probably derived from the Latin *saepio* (to limit, to protect, or to delineate). The subspecies name is also derived from the Latin *amica*, "friendly, peaceful."

Synopsis. The Saepiolus Blue moved across southern Canada into Maine and the Maritime Provinces during the 1920s. It is a common western species and may continue to expand its range southward.

Butterfly. Male forewing: $\overline{X}$ = 1.3 cm, range 1.2–1.5 cm. Where found, the Saepiolus Blue might be confused with the Spring Azure (*Celastrina ladon*) or Silvery Blue (*Glaucopsyche lygadmus*). Saepiolus has gray-white ventral wing surfaces, with greenish scaling at their bases, and several rows of small oblong spots. Where the Spring Azure occurs with the Saepiolus it has heavy dark margins, and the Silvery Blue has larger, round black spots ventrally.

Range. The Saepiolus Blue occurs from the Maritime Provinces and northern Maine westward across southern Canada to the northern parts of the Great Lakes states. From there it ranges north and west to Alaska, the Rocky Mountains, and Pacific Coast states. In our area it inhabits the

97. *Lycaeides melissa*

98. *Plebejus saepiolus amica*

northern portions of Minnesota, Wisconsin, Michigan, and Maine. There is one record from New Hampshire.
Habitat. In the East, this species is found in bogs, moist, open fields, and along roadsides within the Canadian Life Zone.
Life History. *Behavior.* Males patrol close to the ground in the area of the host plants. Mating occurs at midday (1130 hr). Males may congregate at moisture in open fields and streams.

Broods. The Saepiolus Blue is univoltine, with flight dates in the East extending from May 18 to July 26.
Early Stages. Eggs are laid in clover flower heads and the caterpillars feed on the developing flowers. Larvae hibernate half-grown and complete their feeding and development the following year. Larvae have both green and red color phases.
Caterpillar Host Plant. In the East, alsike clover (*Trifolium hybridum*) is the only known host.

Acmon Blue
Icaricia acmon (Westwood & Hewitson), 1852

Synopsis. The Acmon Blue is a rare waif east of the Great Plains, but is the commonest blue in many portions of the western states. It occasionally appears in western Minnesota, and a colony, probably the result of an accidental introduction, was once found near Camden, New Jersey. Both sexes have a bright, submarginal, dorsal hind-wing band, salmon pink in the male, orange-red in the female.

Subfamily Riodininae: Metalmarks

Metalmarks are distinguished by the extension of the male's prothoracic coxa beyond its articulation with the front leg. The family is primarily tropical, and only three species are found in our area. This group is so varied and so little studied that no biological generalizations seem warranted. Our species all belong to the cryptic genus *Calephelis,* which, as caterpillars, feed on leaves of several composites.

Little Metalmark
Calephelis virginiensis (Guérin-Méneville), 1831
PLATE 19 · FIGURE 110 · MAP 99

Etymology. The species name is derived from the state of Virginia.
Synopsis. The Little Metalmark is the only metalmark found on the Atlantic and Gulf Coastal plains in the southeastern states. The species of *Calephelis* are very difficult to separate.
Butterfly. Male forewing: $\overline{X}$ = 1.1 cm, range 1.0–1.2 cm; female forewing: $\overline{X}$ = 1.1 cm, range 1.0–1.1 cm. This species is best distinguished by its locality, its small size, its uniform rusty orange dorsal wing surfaces, and the lack of checkered fringes. There are several closely related species in Texas that might also occur in Louisiana.
Range. The Little Metalmark is found in coastal plain habitats from southeastern Virginia south to Florida and west to eastern Texas.

Habitat. The only reported habitat for the Little Metalmark is grassy areas in flat, open, pine woods with sandy soil.
Life History. *Behavior.* Male *Calephelis* perch on low vegetation during most sunny daylight hours. This species usually selects broad leaves or flower heads as perching sites.
Broods. In southeastern Virginia there are three generations: late April–early May, July, and late September–early October. In Mississippi, where the Little Metalmark is found from March 12 to October 20, there may be four or five generations.
Early Stages. The larva spends the daylight hours hiding under a host leaf, where it blends effectively with the arachnoid tissue (R. Cavanaugh, personal communication). At night and on cloudy days it feeds on top of the leaf. The

caterpillar is pale green with dorsal and lateral rows of long, white, hairlike setae. There is a rust-brown spot to either side of the dorsal row of hair tufts from abdominal segments 5 to 10. The chrysalid, which is attached to the lower leaf surface by a silk button and girdle, has a pale-green abdomen and cream or pale-yellow wing cases.

Adult Nectar Sources. Adult nectar plants are all various short-flowered composites: yarrow, blue mist flower, lance-leaved coreopsis, and fine-leaved sneezeweed.

Caterpillar Host Plants. Larvae use yellow thistle (*Cirsium horridulum*) in Texas (Kendall 1976) and North Carolina (R. Cavanaugh, personal communication).

Northern Metalmark
Calephelis borealis (Grote & Robinson), 1866
PLATE 19 · FIGURES 111 & 112 · MAP 100

Etymology. The species name is derived from the Latin *borealis,* "northern" or "of the northern wind."

Synopsis. The Northern Metalmark actually has a more southern distribution than the Swamp Metalmark (*Calephelis muticum*). It has very localized colonies in rocky terrain, usually in the vicinity of shale or serpentine barrens.

Butterfly. Male forewing: $\overline{X}$ = 1.3 cm, range 1.3–1.4 cm; female forewing: $\overline{X}$ = 1.4 cm, range 1.3–1.5 cm. The Northern Metalmark is often confused with the Swamp Metalmark, where their ranges narrowly overlap in the upper Midwest. The Northern Metalmark often has a black smudgy area across both wings above, and more crescent-shaped and connected metallic marks ventrally.

Range. The Northern Metalmark has three distinct population aggregates, one extending from extreme southeastern New York south in the Appalachians to southern Virginia, another in the upper Midwest, and a third in central Kentucky.

Habitat. This metalmark is most often found near streams in open woodlands, adjacent to barren habitats on shale, serpentine, or limestone.

Life History. *Behavior.* The Northern Metalmark perches on broad leaves with its wings open at about 150°. In overcast weather or when disturbed, the butterflies perch upside down beneath broad leaves.

Broods. This species is univoltine everywhere in its range. Flight dates range from July 1–20 in New Jersey to June 16–July 13 in Virginia.

Early Stages. Single eggs are laid under leaves of young host plants and the larvae develop to the sixth instar by late summer. The larvae overwinter in leaf litter and complete their development the following spring (Randle 1953). There are eight instars in all. The egg is a flattened sphere with large polygonal cells. The larva is green with black dots, and the dorsal and lateral surfaces are covered with long white setae (dos Passos 1936).

99. *Calephelis virginiensis*

100. *Calephelis borealis*

Adult Nectar Sources. The Northern Metalmark avidly visits flowers of butterflyweed as well as those of several composites, such as yarrow, sneezeweed, ox-eye daisy, and goldenrod. They will also nectar at white sweet clover.

Caterpillar Host Plants. Roundleaf ragwort (*Senecio ob-* *ovatus*) is the only documented larval host, but golden rag-wort (*Senecio aureus*) and common fleabane (*Erigeron phil-adelphicus*) have been suggested as hosts since adults are often found in association with them.

Swamp Metalmark
Calephelis muticum McAlpine, 1937
MAP 101

Etymology. The species is named for the specific epithet of its first known host, swamp thistle (*Cirsium muticum*). *Mu-ticum* means shortened or curtailed in Latin.

Synopsis. The Swamp Metalmark is adapted to bogs and marshes in the vicinity of the Great Lakes. The species has long been confounded with the Northern Metalmark (*Cal-ephelis borealis*).

Butterfly. Male forewing: $\overline{X}$ = 1.2 cm, range 1.1–1.4 cm; female forewing: $\overline{X}$ = 1.2 cm, range 1.1–1.2 cm. This species averages slightly smaller than the Northern Metalmark, and has uniform red-brown dorsal wing surfaces with no blackish areas. Its metallic lines are narrower than those of the North-ern Metalmark.

Range. The Swamp Metalmark occurs from northern In-diana and southern Michigan west to southeastern Wis-consin. Another group occurs in Missouri and northern Arkansas. There are isolated records from other states (see map). The Swamp and Northern Metalmarks occupy adja-cent habitats near some northeastern Indiana bogs.

Habitat. This small, mothlike butterfly is found only in wet meadows, marshes, and bogs. In Michigan, its habitat is openings in tamarack–poison sumac bogs.

Life History. *Broods.* The Swamp Metalmark is univoltine in the Great Lakes states (June 30 to August 12), but is bivoltine in Arkansas and Missouri (May 30 to June 19 and August 14 to September 6).

Early Stages. Early stages have been described by McAlpine (1938) and are virtually identical to those of the Northern Metalmark. There are eight or nine instars.

Adult Nectar Sources. Black-eyed susans are utilized in southern Michigan and no doubt other composites are visited as well.

Caterpillar Host Plants. The only known hosts are swamp thistle (*Cirsium muticum*) and roadside thistle (*Cirsium altissimum*).

101. *Calephelis muticum*

Family Libytheidae: Snouts

The snouts are the smallest butterfly family. The medium-sized adults have tremendously elongated labial palpi and reduced front legs in the male, while the forewing apex is falcate and squared off. The family occurs worldwide.

Little is known of the reproductive behavior of snouts except that the males seem to patrol around the host, hackberry, during the afternoon. The adults almost always perch with closed wings, and when inactive and perching on dead twigs they are excellent dead leaf mimics. The butterflies have fairly short proboscises and will feed at small (usually white or yellow) flowers, but they also feed on bird droppings. Freshly emerged males often take moisture at mud. Periodically, the Eastern Snout undertakes vast emigrations that in some instances may obscure the sun.

The only definite caterpillar hosts are various hackberry species (*Celtis*). The larvae have small heads and are slender with swollen thoracic segments. They are green and have only very short, fine setae covering their bodies. The butterflies usually have several flights each year, and the species are reputed to overwinter as larvae, although it is more likely that they overwinter as adults in reproductive diapause.

Eastern Snout Butterfly
Libytheana bachmanii (Kirtland), 1852
PLATE 20 · FIGURES 115–117 · MAP 102

Etymology. The species is named in honor of Reverend John Bachman (1790–1874). He was an ardent naturalist and close friend of John James Audubon, who married one of his daughters. He lived near Charleston, South Carolina. The nearly extinct Bachman's Warbler was named in his honor by Audubon.

Synopsis. The Snout is named for its extremely long, forward-projecting labial palpi, which give the butterfly the appearance of having a long beaklike nose. When resting on a twig, the butterfly may pass for a dead leaf. Different species of *Libytheana* occur on most of the world's continents, and all are similar in size and appearance.

Butterfly. Male forewing: $\overline{X}$ = 2.1 cm, range 1.9–2.3 cm; female forewing: $\overline{X}$ = 2.2 cm, range 2.1–2.4 cm. This is the only eastern United States member of the snout family. Dorsally, the Snout is typified by a black-brown ground color, several orange patches, and a few small white spots near the apical portion of the falcate forewing. There are two forms, each with different ventral hind-wing maculation. In one form the wings are uniformly gray, while in the other form they are a two-toned gray and black.

Range. The Eastern Snout Butterfly occurs from central New England and southern Ontario west to the edge of the Rocky Mountains and then south to Florida, Texas, Arizona, southern California, and Mexico. The butterfly does not survive the winter in the northern portion of its range.

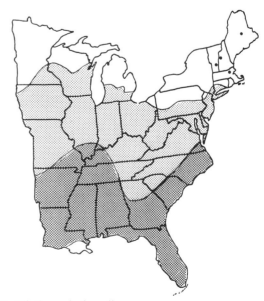

102. *Libytheana bachmanii*

125

Habitat.　The Eastern Snout Butterfly is never found far from woodlands or brushy areas, particularly those of river bottoms. It is a species of woodland edges and adjacent clearings, brushy fields, and roadsides.

Life History.　*Behavior.*　Little is known of this butterfly's reproductive behavior except that mated pairs have been seen only at night (2205–2345 hr). The butterfly is renowned for its mass emigrations (Williams 1937). During one movement, an estimated 1.25 million Snouts passed over a 250-m front each minute. A 1966 migration through Tucson, Arizona, obscured the sun and caused the street lights to be turned on at midday. These migrations are irregular and are probably the result of population explosions in the Southwest. In the East, such movements are rare but are occasionally observed along Florida's Gulf Coast. Freshly emerged males visit moist spots along streams or trails.

Broods.　Although Klots (1951) reports three or four broods, and Clark (1932) states that the species may pass from egg to adult in 15 to 17 days, it is likely the butterfly is only bivoltine. In Georgia, overwintered adults are seen from January to the end of March. There, the first generation flies from May 17 to June 24 and the second flight emerges as early as August 4. The species overwinters as an adult in the South.

Early Stages.　The caterpillar is dark green with dorsal and lateral yellow stripes. The first two thoracic segments are swollen, and the last abdominal segment is tapered abruptly, giving the caterpillar a humpbacked appearance. There are two black tubercles on the thoracic dorsum. The conical chrysalis is green and is abruptly tapered at either end.

Adult Nectar Sources.　The adults visit a wide array of flowers, including dogwood, dogbane, sweet pepperbush, goldenrod, and aster.

Caterpillar Host Plants.　The Eastern Snout Butterfly selects only hackberry as its host. In the East, *Celtis occidentalis* is used widely, while in Texas *C. laevigata, C. pallida,* and *C. reticulata* are fed upon. A report of wolfberry (*Symphoricarpos occidentalis*) is no doubt erroneous.

Family Nymphalidae

The nymphalids are the largest, most diverse family of true butterflies. We follow Ehrlich and Ehrlich (1961) in considering the satyrs, milkweed butterflies, and brushfoots a single family.

These butterflies all have strongly reduced forelegs, which are used as chemoreceptors for smell instead of for walking, and pendant pupae that hang from a silk button. The eggs are usually longer than they are broad.

Subfamily Nymphalinae: Brushfoots

The brushfoots range from small to large, and are an extremely diverse lot, with almost no generalization covering all species. Some lepidopterists would break this group into at least three separate families. Males of most species perch in wait of receptive females, although many species (e.g., fritillaries and crescents) employ the patrolling technique for mate finding. Mating and oviposition are usually afternoon activities. In general, the adults have relatively short proboscises and feed at sap flows, dung, or fermenting fruit, although most species at least occasionally visit flowers. The Zebra (*Heliconius charitonius tuckeri*) and its tropical kin are unique in that the adults collect pollen, which can be broken down and ingested through their proboscises.

There is a wide array of larval hosts. The barrel-shaped eggs are laid singly or in groups, usually under host leaves or on twigs, but in several instances eggs are laid on non-host leaves or other objects. Larvae of most species have several longitudinal rows of tubercles bearing long spinelike hairs, but some, such as the *Limenitis,* are relatively naked. Still others have dense, short hair and forked terminal segments. Most species bear a pair of long spiny tubercles, or "horns," atop their head. The caterpillars live in webbed or folded leaf shelters and feed predominantly at night.

Many species (e.g., tortoise shells, checkerspots, crescents, and emperors) feed colonially during the first several larval stages. Almost all species feed on host leaves. Most species overwinter as partially grown larvae, but some overwinter as hibernating or reproductively inactive adults.

Many nymphalines have two annual generations or are homodynamic, with as many generations as host availability and temperature conditions permit. Some species, including the *Speyeria* fritillaries, Baltimore, several crescents, and the Mourning Cloak, have only a single annual generation. The pupae are always pendant.

Gulf Fritillary
Agraulis vanillae (Linnaeus), 1758
PLATE 19 · FIGURES 113 & 114 · MAP 103

Etymology. The genus name is derived from *agraulos,* which means dwelling in the fields. The species name is derived from *Vanilla,* an orchid genus.

Synopsis. The Gulf Fritillary is the hardiest of the heliconiines, sometimes surviving the winter as far north as Missouri and Illinois. It also undergoes periodic emigrations, ranging farther north than any of its passion vine host plants.

Butterfly. Male forewing: $\bar{X}$ = 3.7 cm, range 3.1–4.2 cm; female forewing: $\bar{X}$ = 4.0 cm, range 3.8–4.5 cm. The combination of the Gulf Fritillary's pointed, bright orange, black-marked wings, the three small black-ringed white spots on the dorsal forewing, and the extensive bright silver markings on the ventral hind wing serves to distinguish it from other fritillaries.

Range. This species occurs from the southernmost portions of the United States south through the West Indies and Central America to Argentina. In the eastern United States it is resident in much of Florida, adjacent Georgia, and southern portions of the Gulf states. At rare intervals emigrants are found as far to the north as Minnesota, Wisconsin, New York, and Pennsylvania. Rarely, temporary breeding populations occur in late summer as far north as does *Passiflora*, e.g., Illinois, Missouri, and Virginia.

Habitat. In contrast to most heliconiines, this butterfly occurs in open fields, pastures, and suburban gardens. Basically, the Gulf Fritillary is resident in the Tropical Life Zone.

Life History. *Behavior.* The Gulf Fritillary is neither as long-lived as the Zebra nor as complex in its behavior patterns. Although the males patrol in search of suitable mates with a fairly rapid flight, they have not been shown to trap-line (for definition, see Zebra, "Behavior," and Glossary). Gulf Fritillaries sometimes join in huge migrations with other butterflies along the Gulf Coast of Florida. These are homologous to the up-slope emigrations seen in tropical countries at the onset of the dry season.

Broods. The number of annual generations of Gulf Fritillaries has not been documented. The butterfly has been found in Florida during all months, but less extensively in other states. In Georgia, the dates range from January 6 to October 31, while in Louisiana adults have been found only from March 22 to November 7. The species is usually most common from late August through October, when most observations of emigrants and migratory flights are reported.

Early Stages. The eggs are laid singly on host leaves, and complete development requires 1 month. The mature caterpillar is glossy black, with two dorsal and two lateral red-orange stripes. In addition, it has four longitudinal rows of long, setose, black spines. The chrysalid is strongly depressed at the intersection of thorax and abdomen and is tan, mottled with green or pink.

Adult Nectar Sources. In Florida, lantana and shepherd's needle are often visited. Throughout its range, the Gulf Fritillary favors lantana but also visits other plants, particularly composites, cordias, and others.

Caterpillar Host Plants. In our area, maypops (*Passiflora incarnata*) has been reported as a host in several locations (Louisiana, Missouri, and Virginia). In other portions of the range, many other *Passiflora* serve as hosts, although running pop (*Passiflora foetida*), another weedy passion vine, has been reported most frequently.

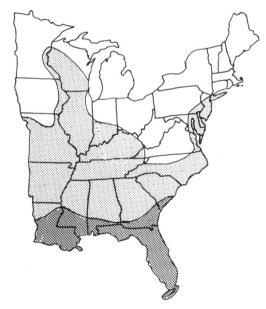

103. *Agraulis vanillae*

Julia
Dryas iulia largo Clench, 1975
PLATE 24 · FIGURE 139 · MAP 104

Etymology. The genus name is derived from Dryas, father of Lykurgos, King of Thrace (Greek mythology). The species is derived from Julia, feminine form of Julius. The subspecies name is derived from one of the Florida Keys.

Synopsis. Julia is found in our area only in the very southern part of Florida and the Keys. Unlike the Zebra and Gulf Fritillary, it seldom wanders to the north.

Butterfly. Male forewing: $\bar{X}$ = 4.1 cm, range 4.0–4.4 cm; female forewing: $\bar{X}$ = 4.2 cm, range 3.9–4.4 cm. Dorsally, the male is a bright orange and is almost unmarked except for a narrow black border on the hind wing and a black mark near the forewing costa. The forewing apex projects out beyond the outer margin. The female is dull brown-orange above with somewhat more extensive black markings than the male.

Ventrally, the wings of both sexes are plain yellow-orange with faint mottling.

Range. Julia ranges from southern Florida and southern Texas south through the West Indies and Central America to Brazil and Bolivia. Strays from Texas occasionally appear in Missouri, Arkansas, and Louisiana.

Habitat. Julia is never found far from hardwood hammocks or other subtropical woodland, although it will venture into open fields to visit appropriate nectar plants. To the south of us, it occurs in a wider range of tropical and subtropical situations.

Life History. *Behavior.* Julia is intermediate between the Zebra and Gulf Fritillary in its general habits, but its flight is the most rapid. It trap-lines nectar sources (see under Zebra, "Behavior"), not feeding on pollen, and the males patrol in search of mates. It will enter forested areas more readily than the Gulf Fritillary but is not so much of a forest denizen as the Zebra. Courtships have been observed in the morning (0900–0930 hr).

Broods. The number of broods has not been defined, but Julias have been seen during every month in southern Florida.

Early Stages. The mature caterpillar has a white head with three black triangular spots on the face and two black spines on the vertex (Rickard 1968). The body is dark brown, marked transversely with fine black lines and spots. There is a broad white stripe along each side that is broken by red diagonal spots. The black spines are long and branched. The pupa is a mottled light gray-brown with prominent finlike projections on the dorsum of each abdominal segment (Muyshondt 1973). There are silver spots in the dorsal projection.

Adult Nectar Sources. In Florida, adults favor shepherd's needle and lantana.

Caterpillar Host Plants. Julia caterpillars feed on various passion vines. Hosts have not been recorded in Florida, but *Passiflora lutea* var. *glabrifolia* is selected in Texas, and *Passiflora tuberosa* is reported in the West Indies.

104. *Dryas iulia largo*

Zebra
Heliconius charitonius tuckeri Comstock & Brown, 1950
PLATE 20 · FIGURE 118 · MAP 105

Etymology. The genus name is derived from Helicon, a mountain in Boeotia devoted to Appolon and the Muses. The species name is derived from Charites, the three graces or goddesses of beauty (Greek mythology). The subspecies was named in honor of Reverend Royal Tucker of Winter Park, Florida, a friend of Dr. A. B. Klots.

Synopsis. The Zebra is unique among the butterflies of our area, being both unusual in its habits and our only representative of a tropical genus known for its complex mimetic relationships.

Butterfly. Male forewing: $\overline{X}$ = 3.9 cm, range 3.4–4.8 cm; female forewing: $\overline{X}$ = 3.9 cm, range 3.4–4.8 cm. The Zebra has long, narrow black wings and several narrow horizontal yellow stripes. Ventrally, there are several small red spots at the base of the hind wings. Zebras are poisonous and distasteful to birds and other vertebrates.

Range. The Zebra occurs from the extreme Southeast and southern Texas south through the West Indies and Central America to Venezuela and Peru. In the eastern United States, it is resident in peninsular Florida and the Keys. Occasionally, emigrants will establish breeding populations in coastal South Carolina, southern Georgia, and along the Gulf, but these are usually eliminated by freezing winter weather. Vagrants have been found in Illinois, Missouri, and Arkansas.

Habitat. These butterflies favor moist subtropical and tropical forests that have sunlit openings. They also occur in young second-growth forest. They venture out from forest edges into nearby fields to visit favored flowers. In Florida, they may be found in hardwood hammocks or mixed pine-oak woods.

Life History. *Behavior.* Zebras have a slow, direct flight characterized by very shallow, rapid wing beats. These butterflies may live as long as 6 months, and they have complex behavior patterns. Each individual follows a set foraging route each day to best take advantage of local flower sources. This behavior is known as "trap-lining" and is also used by some tropical bees, hummingbirds, and bats. Instead of feeding on nectar, Zebras collect pollen on their modified proboscises. The pollen is broken down by enzymes, and the amino acids and other nutrients within are absorbed through the walls of the proboscis. Females lay only a few eggs each day over a period of several months. Each evening, Zebras gather to form communal sleeping roosts of 25–30 individuals. Zebras may shift from one roost to another periodically (Cook et al. 1976). Thus, Zebras learn and remember features of their local environment very well.

Broods. The Zebra is found throughout the year in southern Florida. There are probably no synchronized broods per se.

Early Stages. Female Zebras lay groups of 5–15 eggs on young host leaves at branch tips. The caterpillar is creamy white, with three bands circling each segment and six rows of delicately branched black spines. The head is white, marked with black. The pupa is elongate and is mottled tan and dark brown. It has a twisted frontal projection and a large keellike formation on the abdominal dorsum. It also has a median series of dorsal spines and a row of smaller spines lining the costal margin of the wing cases (Beebe et al. 1960).

Adult Food Sources. A favorite source of pollen in Florida

and elsewhere in the Zebra's range is lantana. Shepherd's needle is also visited in Florida.

Caterpillar Host Plants. Caterpillars of the Zebra feed on young leaves of several species of passion vine (*Passiflora*).

105. *Heliconius charitonius tuckeri*

Variegated Fritillary
Euptoieta claudia (Cramer), 1775

PLATE 20 · FIGURES 119 & 120 · MAP 106

Etymology. The genus name is derived from the Greek *euptoietos*, "easily scared." The species name is derived from Claudius, the name of several Roman emperors.

Synopsis. The Variegated Fritillary is a vagrant in much of its range. Although none of its stages can survive freezing weather, the butterfly ranges far northward almost every summer, only to die out each winter.

Butterfly. Male forewing: $\overline{X}$ =2.7 cm, range 2.0–3.5 cm; female forewing: $\overline{X}$ = 3.2 cm, range 2.8–3.8 cm. Although the Variegated Fritillary varies little geographically, its size ranges widely (Mather 1974). It is not really like any other eastern butterfly, although it might be confused with the Great Spangled Fritillary (*Speyeria cybele*). The present species has a more pronounced forewing apex and a slightly scalloped and angled hind-wing margin. More importantly, its ventral wing surfaces lack the silver spots of the Great Spangled Fritillary.

Range. The species occurs from the eastern and southwestern United States sporadically south through the Greater Antilles and Central America to Argentina. In the eastern United States, it is resident in the Southeast, but individuals

have been recorded from every state except Vermont and Rhode Island.

Habitat. The Variegated Fritillary favors open pioneer habitats, such as meadows, pastures, abandoned crop fields, orchards, roadsides, and landfills.

Life History. *Behavior.* Males patrol back and forth short distances in dry, open, flat areas in search of receptive females. Their flight is low and darting. Mating usually occurs in late morning (1030–1120 hr) and occasionally in late afternoon (1545). Either sex may be the carrying partner. The butterflies move north, usually sporadically as individuals, but on occasion they may be seen moving in greater numbers, as we observed in western Iowa during June 1980.

Broods. The Variegated Fritillary is potentially a continuous breeder in the absence of freezing weather. In Florida, Georgia, and Mississippi, adults are found from February or March to November or December. In these states there are probably four broods. In the central states, such as Pennsylvania, Illinois, and Virginia, the species is usually seen in late summer or fall, but may arrive in April (Illinois) or as late as July (Michigan). In mild winters, the species may overwinter as far north as the Delaware Valley and Virginia, in which case three broods may be produced.

Early Stages. The mature caterpillar is red-orange with longitudinal rows of black spines and has a dorsal stripe of alternating white and black patches. There are also two white stripes along each side that are interrupted periodically with black. The chrysalis is an iridescent pearly white, marked with dark-brown patches and points.

Adult Nectar Sources. The Variegated Fritillary appears to be catholic in its flower visitation habits, selecting dogbane, common milkweed, swamp milkweed, butterflyweed, tickseed sunflower, peppermint, and red clover.

Caterpillar Host Plants. Larvae of the Variegated Fritillary feed on a number of unrelated plants. Maypops (*Passiflora incarnata*) is utilized in Mississippi, North Carolina, and Virginia, while Shapiro (1974) found larvae feeding on common blue violet (*Viola papilionacea*) and northern downy violet (*V. fimbriatula*) in New York. Field pansy (*Viola rafinesquii*) is eaten by the first brood in Georgia and Virginia. Other hosts recorded, all in different families, are: may apple (*Podophyllum peltata*), stonecrop (*Sedum*), moonseed (*Menispermum*), purslane (*Portulaca*), and beggar's ticks (*Desmodium*)

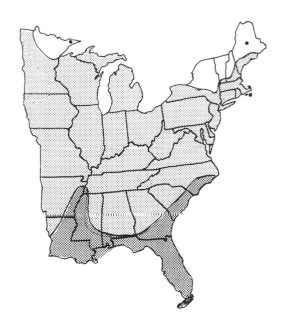

106. *Euptoieta claudia*

Diana
Speyeria diana (Cramer), 1775
PLATE 21 · FIGURES 121–124 · MAP 107

Etymology. The genus was named in honor of Adolph Speyer (1812–1892), a German lepidopterist who specialized in butterfly study. Diana is the Roman goddess of light and life, the virgin goddess of the moon and hunting.

Synopsis. Diana is one of two spectacular examples of sexual dimorphism among our eastern butterflies. (The other is the Tiger Swallowtail.) Its striking blue-black female and orange and black male are both magnificent but would likely be thought of as separate species by the uninitiated. The female is assumed to be one of several butterflies that mimic the distasteful Pipe Vine Swallowtail (*Battus philenor*).

Butterfly. Male forewing: $\overline{X}$ = 4.3 cm, range 4.2–4.5 cm; female forewing: $\overline{X}$ = 5.0 cm, range 4.6–5.4 cm. Both sexes are distinctive and superficially unlike other fritillaries. Dorsally, the male's wings are predominantly black basally, with the outer two-fifths orange. A row of black dots lies just external to the black area, with each dot situated between two black-lined veins. Ventrally, the hind wing is nearly immaculate, with a submarginal row of narrow metallic silver spots. Dorsally, the female's wings are blue-black, with the outer portion of the hind wing strongly blue. Both wings have an iridescent sheen in newly emerged individuals. In addition, there are three rows of white spots on the outer portion of the

forewing, and a marginal row of whitish lines on the hind wing. Ventrally, the wings are basically charcoal black. The underside of the forewing has the basic *Speyeria* pattern in pure black, but the hind wing is nearly immaculate, with the basal two-thirds dark brown or rusty. A rare form of the female occurs that is green instead of blue. There is an extensive description of Diana's variation in the Clarks' (1951) publication on Virginia butterflies.

Mimicry. The Diana female is one of five butterflies that mimic the distasteful Pipe Vine Swallowtail (*Battus philenor*). (Refer to the latter species' account for a full discussion.) The Diana mimicry is unusual in that the male is clearly *not* mimetic, as is the case with the Tiger Swallowtail. In insect mimicry where only one sex is mimetic, it is almost always the female, probably because it is more important evolutionarily for the females, with their cargo of eggs, to survive longer.

Range. The Diana once occurred in four population groupings, although two have now disappeared. One group probably occurred in the Ohio River drainage from extreme western Pennsylvania to Illinois, but disappeared in the 1800s. Another group was known from the tidewater area of southeastern Virginia, including the species' type locality (Jamestown, Virginia), but has not been found there since the 1950s. The largest population occurs in the southern Appalachians from central Virginia and West Virginia southwestward in the mountains to northern Georgia and Alabama. A fourth population persists in the Ozark Mountains of Arkansas and Missouri.

Habitat. Diana's favored haunts are forested valleys and mountainsides with moist rich soil. The Clarks (1951) point out that thick undergrowth, usually with alders and rhododendrons, is usually present. The habitat in coastal Virginia was in or near old pine woods with running streams or extensive cold seepage. These forests are now mostly destroyed by logging.

Life History. *Behavior.* Male Dianas patrol their deep woodland habitats, flying up to 10 m above ground when soaring over a patch of unsuitable habitat. Mating has not been reported, but oviposition behavior was reported by Harris (1972). The female walks on the forest floor, laying single eggs on dead leaves and twigs in a seemingly haphazard manner. Periodically she flies a short distance and repeats the process. Egg-laying is probably delayed as late in the summer as possible, since the eggs hatch in the late fall, with the first-instar larvae overwintering.

Broods. All *Speyeria*, including Diana, are univoltine. The males begin flying at least a week earlier than the females. Typical flight extremes in the Virginia mountains are June 18 to August 16 (males) and July 10 to September 4 (females). Diana has been found as early as June 6 and as late as September 21 in Georgia.

Early Stages. The mature caterpillar is velvety black, with rows of fleshy black setose spines that are deep orange at their base. The dorsal mesothoracic spines are longer and bend forward over the head. Between each pair of dorsal spines there are two white dots. The pupa is brown and cylindrical, with a depression and several elevations on the anterior dorsum.

Adult Nectar Sources. Milkweeds are the preferred nectar source of Diana. Butterflyweed is the favorite, but common milkweed and swamp milkweed are also visited frequently. Other plants, such as butterflybush, ironweed, and red clover, are also visited on occasion. Male Dianas visit flowers throughout the day, but females do not appear at milkweed patches until midafternoon.

Caterpillar Host Plants. Caterpillars in captivity readily feed on violets, but the host plants used in nature are unknown.

107. *Speyeria diana*

Great Spangled Fritillary
Speyeria cybele (Fabricius), 1775
PLATE 21 · FIGURES 125 & 126 · MAP 108

Etymology. The species is named after Cybele, the Greek and Roman great mother of gods.

Synopsis. This is the most common large orange butterfly throughout much of the eastern United States in midsummer. It is not unusual to see one or more nectaring Great Spangled Fritillaries on every milkweed plant.

Butterfly. Male forewing: $\bar{X}$ = 3.8 cm, range 3.5–4.0 cm; female forewing: $\bar{X}$ = 4.5 cm, range 4.2–4.7 cm. The Great Spangled Fritillary is sometimes confused with Aphrodite, a closely related fritillary, but may be separated from it by its

less reddish color on the dorsal wing surfaces, its broader light submarginal band on the ventral hind wing, and its lack of a black spot below the cell on the dorsal forewing.

Range. This species ranges from coast to coast and from the southern United States north to the southern portions of Canada. In the eastern United States, the Great Spangled Fritillary occurs regularly south to the northern portions of Georgia, Alabama, Mississippi, and Arkansas.

Habitat. Almost any open area, particularly a moist flat situation, is the preferred habitat. Naturally, suitable host violets must be present. Since many individuals wander widely, the species may be found in a variety of habitats where they would not breed.

Life History. *Behavior.* This fritillary is a patrolling species, and males have been observed to fly a somewhat circular beat around open fields or meadows. An area of less than one hectare is patrolled at a height of about 1.5 m. Individuals disperse widely and are often seen flying a rapid, direct flight 3–5 m above ground. The female acts as the carrier for mated pairs, which have been seen from 1415 to 1525 hr.

Broods. The Great Spangled Fritillary is univoltine, with males emerging about a month before the females. Some females survive to the end of the summer. In Georgia, individuals emerge as early as June 3. In most of the East, the first males appear in mid-June and fly to mid-August; females emerge in mid-July, while some individuals fly until late in September or early October.

Early Stages. The eggs are laid in late summer and are deposited on or near their violet hosts. The larvae hatch in the fall and overwinter without feeding until the following spring. In the spring they feed on young violet foliage at night and rest off the host by day. The mature caterpillar is velvety black, with the ventral portion chocolate brown. The black spines are red-yellow at their base. There is a pair of gray transverse dots between each dorsal pair of spines. The pupa is variably colored, being dark-brown mottled with red-orange or light brown.

Adult Nectar Sources. Milkweeds, especially common milkweed and butterflyweed, composites, including thistles, joe-pye weed, purple coneflower, and ironweed, as well as dogbane, are favored. Other plants, such as vetch, bergamot, red clover, mountain laurel, and verbena, are utilized on occasion.

Caterpillar Host Plants. Various species of violets are known to be larval hosts. Shapiro (1974) reported round-leaved yellow violet (*Viola rotundifolia*) to be eaten on Staten Island, New York, while common blue violet (*Viola papilionacea*) is utilized in Pennsylvania.

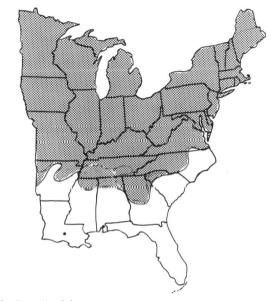

108. *Speyeria cybele*

Aphrodite Fritillary
Speyeria aphrodite (Fabricius), 1787
PLATE 22 · FIGURES 127–129 · MAP 109

Etymology. The species is named after Aphrodite, the ancient Greek goddess of love and beauty.

Synopsis. The Aphrodite Fritillary occurs with the Great Spangled Fritillary in many areas but tends to occur at slightly higher elevations and more northern latitudes.

Butterfly. Male forewing: $\bar{X}$ = 3.3 cm, range 3.1–3.4 cm; female forewing: $\bar{X}$ = 3.8 cm, range 3.7–4.0 cm. Aphrodite is more like the Mountain Silver-spot (*Speyeria atlantis*) than the Great Spangled Fritillary (*Speyeria cybele*). Aphrodite lacks the black borders on the outer margins of the dorsal forewing, and has a more reddish orange ground color and narrower pale marginal band on the ventral hind wing. In the East the species has more extensive geographic variation than the Great Spangled Fritillary does.

Range. Aphrodite ranges from the Maritime Provinces and New England west across southern Canada, the northern Midwest, and the Great Plains to the Rocky Mountains. In the East, it ranges south to northeastern Georgia and Tennessee in the mountains.

Habitat. Aphrodite is more habitat-restricted than the Great Spangled Fritillary, and is found in habitats with more acidic soils. Upland brushland, dry fields, open oak woods, openings in shale barrens, bogs, glades, and high mountain pastures have been mentioned. The species is found in the Transition Life Zone in the East.

Life History. *Behavior.* Males patrol during the warmer hours of daylight in search of receptive females. One mated pair was observed at midday (1230 hr), with the female acting as the carrier.

Broods. Aphrodite has a single annual generation, and individuals of this species appear later in the season and disappear earlier than those of the Great Spangled Fritillary. Flight extremes for Maine are July 13 to September 14; for Illinois, June 15 to September 2; and for Georgia, June 29 to August 25.

Early Stages. Larvae are extremely similar to those of the Great Spangled Fritillary. The pupa is red-brown or gray and is mottled and striped with black.

Adult Nectar Sources. Common milkweed is visited in many areas throughout Aphrodite's range in the East. Viper's bugloss is utilized avidly in the mountains of West Virginia. A variety of other plants may also be visited.

Caterpillar Host Plants. On Staten Island, New York, Shapiro (1974) found Aphrodite to utilize northern downy violet (*Viola fimbriatula*), lance-leaved violet (*V. lanceolata*), and primrose-leaved violet (*V. primulifolia* var. *acuta*).

Regal Fritillary
Speyeria idalia (Drury), 1773
PLATE 22 · FIGURES 130–132 · MAP 110

Etymology. The species is named after Idalium, a town of ancient Cyprus.

Synopsis. This large, spectacular butterfly is declining in much of its range, but it is still common in tall-grass prairie remnants. This species is a relict that has probably existed in eastern North America since long before the Ice Age (Hovanitz 1963).

Butterfly. Male forewing: $\bar{X}$ = 4.0 cm, range 3.6–4.3 cm; female forewing: $\bar{X}$ = 4.7 cm, range 4.2–5.0 cm. This is the only *Speyeria* in our area with a largely reddish orange forewing and a largely black hind wing. The ventral hind wing is blackish gray with the normal *Speyeria* spot pattern, except that the spots are flat white, not metallic silver. The wings have greenish iridescent highlights when the adults are newly emerged.

Habitat. The Regal Fritillary inhabits tall-grass prairie in its Midwest metropolis, but it is found in other open grassy situations elsewhere. Damp meadows or pastures with boggy or marshy areas are frequently utilized in the East, but dry mountain pastures are also selected in some areas. The butterfly is restricted to the Upper Austral and Transition Life Zones.

Range. The Regal Fritillary formerly occurred from the Maritime Provinces south in the Piedmont and Appalachians to North Carolina, and westward across the northern half of the United States to eastern Colorado and Montana. Now the species no longer occurs in the Maritimes or much of New England. Elsewhere, the species seems to have declined precipitously and is common only in the few remaining untilled areas found sporadically in the prairie states.

Life History. *Behavior.* The Regal Fritillary is a patrolling species, with males searching with a low but steady flight in late morning and early afternoon. Occasionally, males will dip close to the ground in clumps of vegetation. Courtship and mating were reported by Clark (1932). A flushed female will perch atop a plant in an open area while the male follows

and flutters all around her, changing position with great rapidity. In mated pairs, the female is the carrier, and, if newly emerged, will visit flowers while carrying the quiescent male.

Female Regal Fritillaries become more active and wander more extensively in late summer, when they search for oviposition areas. Clark observed females flying about 1.5 m above the ground and dropping into the grass every 30 m or so. On each such occasion the female will walk through the vegetation for 10 to 15 min, depositing her eggs on various plants regardless of the presence of violets, the only acceptable larval host.

The butterflies warm up in the morning by perching with wings open at right angles to the sun. In these instances the forewing covers the hind wing.

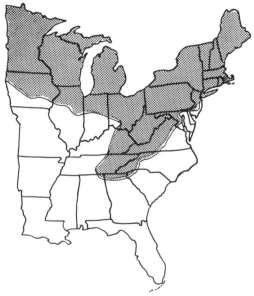

109. *Speyeria aphrodite*

Broods. The species is univoltine. Males emerge a week or more earlier than females. Some ranges of flight dates are: Maine, July 6–August 3; Michigan, June 28–September 7; Virginia, late May–mid-October, with the main male flight from mid-June to mid-July and female flight from early July to mid-August.

Early Stages. The mature larva is velvet black with ochre-yellow to dull-orange mottlings. The dorsal spine rows are silver white, tipped with black, while the lateral rows have yellow or orange bases. The chrysalis has a brown and yellow abdomen, with pink-brown wing cases. The entire chrysalis has scattered dark-brown patches.

Adult Nectar Sources. Milkweeds and thistles are favored, although red clover is visited in pastures. Regal Fritillaries were seen in large numbers nectaring at Sullivant's milkweed at midday in western Iowa, while Clark reports swamp milkweed to be favored near Washington, D.C. Other plants, such as common milkweed and mountain mint, are occasionally sipped at.

Caterpillar Host Plants. Larvae feed on bird's foot violet (*Viola pedata*) and probably other violets.

Mountain Silver-spot
Speyeria atlantis (Edwards), 1862
PLATE 23 · FIGURES 133 & 134 · MAP 111

Etymology. The species is named after Atlantis, the mythical lost continent of the Atlantic Ocean.

Synopsis. Most butterfly aficionados in the East must travel in order to find this most boreal of the eastern *Speyeria*, which is restricted to cool Canadian Life Zone habitats.

Butterfly. Male forewing: $\overline{X}$ = 3.0 cm, range 2.7–3.3 cm; female forewing: $\overline{X}$ = 3.1 cm, range 2.8–3.3 cm. This northern silver-spot is the smallest and most highly silvered of our eastern *Speyeria*. Its distinctive characters are its black marginal borders on the dorsal forewings and the purplish brown ground color on the ventral hind wings.

Range. The species occurs from the Maritime Provinces west across the northern states and southern Canada to Manitoba and Minnesota. The species also occurs south in the Appalachians to Virginia and West Virginia. Another group of populations occurs in the mountains of the western United States and the Rocky Mountains of Alberta and British Columbia. The latter group, including many subspecies, was treated as a separate species, *Speyeria electa* by Howe (1974).

Habitat. In the East, the species is essentially restricted to the Canadian Life Zone, and is found in cool, open woods, along woodland streams, upland pastures, and clearings.

Life History. *Behavior*. Little is known of this species' behavior. J. A. Scott (1975b) found that males of the Colorado representative patrol all day in search of females.

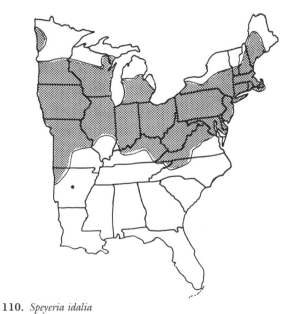

110. *Speyeria idalia*

111. *Speyeria atlantis*

Broods. The Northern Silver-spot has a single generation each year. The species flies over a narrower period than other *Speyeria*. Some sample flight extremes are: Maine, June 23–August 20, and Michigan, June 30–September 10.

Adult Nectar Sources. The Mountain Silver-spot visits a variety of flowers. Adults seem to be particularly fond of common milkweed, mountain laurel, and mints. Some other plants visited include virgin's bower, crown vetch, spiraea, burdock, ox-eye daisy, and boneset.

Caterpillar Host Plants. Violets are the only known larval hosts. Northern blue violet (*Viola septentrionalis*) has been reported by Shapiro (1974) to be a host in New York.

Mormon Fritillary
Speyeria mormonia eurynome (Edwards), 1872

This small fritillary has been found a few times in western Minnesota, although we do not have the specific records. It may be distinguished from other eastern *Speyeria* by its smaller size, paler ground color, and finer black markings above. It is found west to British Columbia, north to southern Alaska, and south in the mountains to central New Mexico and central California. As do other *Speyeria*, it has a single flight of adults each summer and its caterpillars feed on violets. It feeds on a wide variety of flowers. We observed it visiting goldenrod in eastern North Dakota.

Bog Fritillary
Boloria eunomia dawsoni (Barnes & McDunnough), 1916
PLATE 23 · FIGURES 135 & 136 · MAP 112

Etymology. The genus is named after Mt. Bolor, a mountain in Asia. The species is named after Eunomia, daughter of Zeus.

Synopsis. The Bog Fritillary is a butterfly of acid bogs, known in our area only from a few locales in Maine and Michigan.

Butterfly. Male forewing: $\bar{X}$ = 2.0 cm, range 1.8–2.1 cm; female forewing: $\bar{X}$ = 2.0 cm. The Bog Fritillary is similar to the Silver-bordered Fritillary dorsally, but is darker orange, with more extensive black markings. The species is most distinct ventrally, with a pattern similar to the Silver-bordered, except that the pale marks on the hind wing are dull white, not silvered, and the submarginal row consists of circular white spots, not black as in other *Boloria*.

Range. The Bog Fritillary is holarctic, occurring across Eurasia and North America. In the New World it extends from nothern Maine and Labrador northeastward across Canada to Alaska. It also occurs southward in the Rocky Mountains to Colorado. In the East it is known only in northern Maine, in the eastern portion of Michigan's upper peninsula, and in northern Wisconsin and Minnesota.

Habitat. This butterfly frequents acid sphagnum bogs with black spruce, tamarack, leatherleaf, labrador tea, and other bog plants. The insect seems to favor open areas around the bog's perimeter.

Life History. *Behavior.* The males patrol all day, with a low, fast, flickering flight. Nielsen (1964) observed that males, upon encountering each other, engage in "aerial ma-neuvers, sometimes rising above the tallest spruce trees before disengaging."

Broods. The Bog Fritillary is univoltine. In northern Michigan the species flies for a week or so in mid-June. On Mt. Katahdin, records range from June 6 to July 9, with the peak

112. *Boloria eunomia dawsoni*

in late June. Grey (1965) reported the species to fly for only a single day each year in New Brunswick, but this is probably erroneous.

Adult Nectar Sources. Adults have been observed visiting Labrador tea in Michigan and Wisconsin. Goldenrod is relied upon in Alberta.

Caterpillar Host Plants. Willow is a preferred larval host, but alpine smartweed (*Polygonum viviparum*) has been reported as a host in North America. *Polygonum* and *Viola* have been reported in Europe.

Silver-bordered Fritillary
Boloria selene (Denis & Schiffermüller), 1775
PLATE 23 · FIGURES 137 & 138 · MAP 113

Etymology. The species is named after Selene, Greek goddess of the moon.

Synopsis. The Silver-bordered Fritillary is the little jewel of our eastern *Bororias*. It is extremely variable geographically (Kohler 1977) and is usually limited to very local populations.

Butterfly. Male forewing: $\bar{X}$ = 2.0 cm, range 1.9–2.2 cm; female forewing: $\bar{X}$ = 2.2 cm, range 2.0–2.5 cm. This is the only eastern *Boloria* with rows of metallic silver spots on the ventral surface of the hind wings.

Range. This small, holarctic fritillary ranges across northern and central Eurasia, the central lowlands of Alaska and southern Canada, and the northern United States (Kohler 1977). In the East, the Silver-bordered Fritillary ranges from Labrador and the Maritime Provinces west to Alberta and south to Illinois, the mountains of Virginia, and the Atlantic coastal plain of Maryland.

Habitat. This fritillary occurs in a variety of wet, open habitats, some of which may be altered by man's activities. Wet meadows, marshes, or bogs, often with willows and usually with taller vegetation than the Meadow Fritillary's haunts, are normal for the Silver-bordered Fritillary.

Life History. *Behavior.* According to J. A. Scott (1975a), males patrol fairly level land in wet meadows during most daylight hours. In West Virginia a single male was observed in a boggy marsh for several hours. During this period he repeatedly patrolled a fairly regular oval route covering about 2 hectares. His route coverage varied little, and he flew with a rapid flight about 1–1.5 m above the ground, pausing occasionally to visit flowers.

Broods. The Silver-bordered Fritillary is univoltine in the northern part of its range, but in the East the butterfly has two generations in the northern tier of states, and three generations farther south. For example, near New York's Finger Lakes, the dates for the two broods are June 3–July 7 and July 29–September 8. Similarly, in Michigan's upper peninsula the first brood appears as early as June 6 and the second generation flies until August 30. Areas where three generations have been documented are the lower Delaware Valley (May, late June–July, and late August–September) and mountain regions of Virginia (May–early June, mid-July–mid-August, and late August through September).

Early Stages. The mature larva is dark gray, variably mottled with dark patches. The body spines are pale yellow, but the dorsal mesothoracic pair are black and extremely long, projecting forward over the head. The pupa is dark to light brown and has dorsal rows of sharp conical tubercles on the abdomen.

Adult Nectar Sources. Favored nectar sources are composites, including black-eyed susans and goldenrods.

Caterpillar Host Plants. Violets are probably the chief larval hosts, but other plants are suspected. It is possible that females may deposit eggs on plants other than the true host, as do other nymphalines. This is perhaps why such unlikely host plants as willow (New York), strawberry, and blueberry (Europe) have been reported.

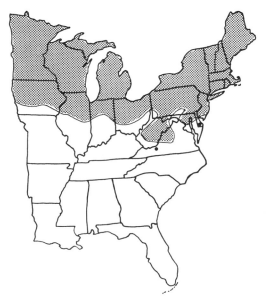

113. *Boloria selene*

Meadow Fritillary
Boloria bellona (Fabricus), 1775
PLATE 24 · FIGURES 140 & 141 · MAP 114

Etymology. The species is named after Bellona, goddess of war and sister of Mars (Roman myth).

Synopsis. The Meadow Fritillary, already the most widespread *Boloria* in the East, has been expanding its range southward and eastward during recent years. It is the only eastern representative that has become adapted to man-made habitats, such as hayfields.

Butterfly. Male forewing: $\bar{X}$ = 2.1 cm, range 1.8–2.3 cm; female forewing: $\bar{X}$ = 2.3 cm, range 2.1–2.4 cm. This is the only eastern *Boloria* that has a squared-off (truncate) forewing apex. In addition, the Meadow Fritillary may be distinguished from the Silver-bordered Fritillary by its lack of metallic silver spots on the ventral hind wing. All other eastern *Bolorias* are very rare and are characterized under their respective accounts.

Range. The Meadow Fritillary ranges from southern Quebec east across southern Canada and the northern and central states to eastern Colorado and the Dakotas. The species also ranges northward through the Canadian Rockies to northern British Columbia and Northwest Territories. In the East, the species ranges south in the Appalachian Mountains to North Carolina and Tennessee. In recent years the species has spread from the mountains into lowland areas where it was previously unknown. Areas of Kentucky, Ohio, Virginia, and Maryland have been occupied as a result.

Habitat. Relatively flat meadows, pastures, and hayfields are the usual habitats in the East. These situations are usually wet, and occasionally mountain bogs or marshes are selected. Until recent years, only Canadian and Transition Life Zone habitats were known, but the species has now spread into Upper Austral Life Zone habitats.

Life History. *Behavior.* Males patrol appropriate habitats during warm daylight hours using low (.5 m) flight. Mated pairs, with the female as carrier, have been found perched on tall vegetation during early afternoon (1310–1545 hr).
Broods. The Meadow Fritillary is bivoltine in the northern and central portions of the East. In Maine flight dates are May 10 to June 20 and July 3 to September 12, while there is a partial third generation as well in more southerly areas, e.g., Delaware Valley (late-April–mid-May, mid-June–mid-July, late August–mid-October).

Early Stages. The mature caterpillar is shiny green with a velvet band along each side. The tubercles and spines are yellow-brown. The chrysalis is yellow-brown. The third- or fourth-instar caterpillars overwinter.

Adult Nectar Sources. Meadow Fritillaries favor composites, particularly yellow-flowered species, for their nectar supply. Black-eyed susans are frequent choices for the mid-summer generations, but others, such as dandelion and ox-eyed daisy are also selected. Less frequently, they visit plants of other families, e.g., dogbane and verbena.

Caterpillar Host Plants. Larvae utilize violets, including woolly blue violet (*Viola sororia*) and northern white violet (*V. pallens),* in New York.

114. *Boloria bellona*

Frigga Fritillary
Boloria frigga saga (Staudinger), 1861
MAP 115

Etymology. The species is named after Frigga, principal wife of Odin and goddess of the clouds, sky, and conjugal love (Teutonic myth).

Synopsis. The Frigga Fritillary is a northern species that occurs in our area in only a few sphagnum bogs in northern Michigan, Wisconsin, and Minnesota.

Butterfly. The adults are very similar to those of the Meadow Fritillary but may be distinguished by their rounded forewing apex, the dorsal dark basal areas, and a squarish white patch along the inner margin of the ventral hind wing.
Range. The Frigga Fritillary is holarctic, ranging extensively in subarctic taiga habitats across northern Eurasia and North America. In North America, the species occurs through much of Alaska and Canada, ranging south in the Rocky Mountains to Colorado. In the East, it occurs only in northern Michigan, Minnesota, and Wisconsin.
Habitat. In the East, the preferred habitat seems to be wet sedgy bogs with willow, dwarf birch, and leatherleaf (Hubbell 1957). Tamarack may also be present.
Life History. *Behavior.* According to Scott (1975) males patrol all day in low spots of willow bogs.
Broods. The Frigga Fritillary is univoltine. Ferge and Kuehn (1976) found that it flew from May 24 to 31 in northern Wisconsin.
Caterpillar Host Plants. Larval hosts are willow (*Salix*) in Alberta, dwarf birch (*Betula*) in Michigan (M. C. Nielsen, personal communication), and cloudberry (*Rubus chamaemorus*) in Europe.

Freija Fritillary
Boloria freija (Thunberg), 1791
MAP 116

Etymology. The species is named after Freya, who was the goddess of love and fertility in old Norse mythology and later a wife of Odin in Teutonic legend.
Synopsis. A somewhat more boreal species than the Purple Lesser Fritillary, in our area the Freija inhabits black spruce bogs (in the northwestern Great Lakes region).
Butterfly. Male forewing: $\overline{X}$ = 1.8 cm, range 1.7–1.8 cm. The distinguishing field marks are discussed under the Purple Lesser Fritillary, which may be distinguished from Freija only with difficulty.
Range. The Freija Fritillary is holarctic and occurs across northern Eurasia and North America. On our continent, it is found from Labrador and Quebec west and north across the Canadian taiga and Arctic to Alaska. It ranges south to Washington and down the Rocky Mountains to northern New Mexico. In our area it has been found only in northern Minnesota, Wisconsin, and Michigan's upper peninsula.
Habitat. The Freija occurs in black spruce–sphagnum bogs with Labrador tea, cranberry, and cottongrass, often at the margins of the bogs.
Life History. *Behavior.* Male Freija Fritillaries patrol in search of receptive females during the warmer daylight hours in open areas, usually at the edges of their bog habitat. When clouds obscure the sun, the butterflies immediately cease flight, dropping into the vegetation and remaining quiescent.
Broods. This species is univoltine, with its flight occuring during June and early July. It has been found in Wisconsin during early June (Masters 1968b).

115. *Boloria frigga saga*

116. *Boloria freija*

Caterpillar Host Plants. Larval hosts appear to be primarily members of the heath family (Ericaceae). Dwarf bilberry (*Vaccinium caespitosum*) has been reported in Washington, while rhododendron, bearberry (*Arctostaphylos uva-* *ursi*), and crowberry (*Empetrum*) have been documented in Eurasia. An unrelated plant, *Rubus*, has also been reported for the species in Europe.

Purple Lesser Fritillary
Boloria titania (Esper), 1793
PLATE 24 · FIGURE 142 · MAP 117

Etymology. The species is named after the Titans, sons and daughters of the god Uranus (Greek myth).

Synopsis. An extremely localized butterfly in the East, the Purple Lesser Fritillary is restricted to bogs or meadows in only a few northern locations.

Butterfly. Male forewing: $\overline{X}$ = 2.0 cm, range 1.9–2.2 cm; female forewing: $\overline{X}$ = 2.0 cm, range 1.9–2.2 cm. This species is very similar to the Freija Fritillary, but may be distinguished by the markings on the ventral hind wing. The Freija has distinct white marks on the submedian costal and discal areas, which the Purple Lesser Fritillary lacks, and there is a median black line that is wavy and somewhat broken in the Purple Lesser, but distinctly zig-zag or sawlike on the Freija.

Range. A holarctic species, the Purple Lesser Fritillary occurs in northern boreal areas of both Eurasia and North America. In North America it ranges from New Hampshire and the Maritime Provinces west and north through Canada to Alaska. It occurs south in the Rocky Mountains to New Mexico and is also found in the Cascade and Olympic Mountains of Washington. In the East it occurs only in the White Mountains of New Hampshire (subspecies *montinus*), in northern Maine, and in northern Minnesota.

Habitat. The Purple Lesser Fritillary occurs in black-spruce–sphagnum bogs (Minnesota) or in damp areas adjacent to ice-cold mountain springs (New Hampshire).

Life History. *Behavior.* The males patrol in search of females during the warmer daylight hours.

Broods. Like most northern *Boloria*, this butterfly is univoltine. It flies from mid-July to early August in New Hampshire.

Early Stages. Eggs are laid in midsummer, and the newly hatched first-instar caterpillars hibernate before completing development during the following spring and summer.

Adult Nectar Sources. Adult food sources are chiefly goldenrods (*Solidago graminifolia*, *S. rugosa*, and *S. squarrosa*). Aster has also been reported.

Caterpillar Host Plants. In eastern Canada, the Purple Lesser Fritillary has been found feeding on dwarf willows (*Salix arctica* and *S. herbacea*) as well as alpine smartweed (*Polygonum viviparum*). In Washington it feeds on false bistort (*Polygonum bistortoides*).

Patch Butterfly
Chlosyne lacinia adjutrix Scudder, 1875

Synopsis. The Patch has been found as a stray, sparingly, in western Missouri (Barry and Jackson Counties). It normally ranges west to southern California and thence south through Mexico to Argentina. It is unlike its close checkerspot relatives in being predominantly black above and below, with an orange-yellow median hind-wing band and a few other white and yellow spots.

Farther south and west, the Patch has a number of other color forms as well as numerous intermediates. The butterfly is multiple brooded and utilizes various sunflowers (*Helianthus*) and their relatives as its caterpillar food plants.

Gorgone Checkerspot
Chlosyne gorgone (Hübner), 1810
PLATE 24 · FIGURES 143 & 144 · MAP 118

Etymology. The genus name is an anagram of *synchloe*, which was a derivation of the Greek *syn*, "with," and *khloe*, "light green of spring." The species is named after the Gorgons, daughters of Forkys, king of the sea.

Synopsis. The Gorgone Checkerspot is intermediate in size between the other checkerspots and crescents. An old painting by Abbot, upon which Boisduval and LeConte based their species *ismeria,* long thought to be a rare, missing species, actually represents a pale southeastern form of the Gorgone Checkerspot.

Butterfly. Male forewing: $\overline{X}$ = 1.6 cm, range 1.5–1.7 cm; female forewing: $\overline{X}$ = 1.9 cm, range 1.7–2.0 cm. The Gorgone Checkerspot is distinctive but somewhat similar to the Silvery Checkerspot and the Pearl Crescent. Dorsally, it may be distinguished from the former by the hind wing's submarginal spot row, which consists of solid black spots, and from the latter by its more distinctly checkered fringe. On the ventral hind wing there is a pattern of strongly contrasting white and brownish marks, and a marginal row of strongly scalloped white marks.

Range. The Gorgone Checkerspot occurs from Michigan and Minnesota west to Manitoba, thence south through the Great Plains, Mississippi Valley, and the east slope of the Rocky Mountains to the Gulf Coast and northern Mexico. Isolated populations occur at several points in the Appalachians from New York to Kentucky. The large pale form (*ismeria*) occurs in Georgia and the more mountainous portions of adjacent Alabama and South Carolina.

Habitat. This checkerspot occurs in open areas. In the main part of its range it occurs on prairie slopes and ridges, as well as in grassy areas near streams. In New York it is found on a dry, grassy burn scar with blueberry and ground pine (*Lycopodium*), while in Georgia it is found in open hardwood forests.

Life History. *Behavior.* Scott (1975) observed courtship and mating to occur during most daylight hours in Colorado. He observed perching behavior among males on ridge tops and patrolling behavior on slopes and in valleys. In Iowa we observed perching males only on ridge tops, especially in late afternoon (1740–1830 hr). Perching males sat basking in the late afternoon sun with their wings spread at about a 90° angle.

Broods. The Gorgone Checkerspot is univoltine in the northern portion of its range, e.g., Michigan's upper peninsula (July 24–August 7), while two generations are probably the rule in the central portion of its range, e.g., Illinois (May 8–September 13). In Georgia, three broods occur (April 17–May 18, June 3–July 16, August 18–September 1).

Early Stages. Caterpillars feed communally. The larvae have yellow bodies with dorsal and lateral black stripes, and their heads are black. There are the usual six rows of black barbed spines on the body. The chrysalis is a mottled light gray.

Adult Nectar Sources. Composites, especially yellow-flowered species, are favored, e.g., lance-leaved coreopsis (in Louisiana) and stiff coreopsis (in Iowa). A variety of other plants are visited for nectar, e.g., white sweet clover, common milkweed, shrub houstonia, and fleabane.

Caterpillar Host Plants. Host plants are several Asteraceae. Sunflower (*Helianthus*) and crosswort (*Lysimachia*) are two that have been recorded.

117. *Boloria titania*

118. *Chlosyne gorgone*

Silvery Checkerspot
Chlosyne nycteis (Doubleday & Hewitson), 1847
PLATE 25 · FIGURES 145 & 146 · MAP 119

Etymology. The species is named after Nykteus, King of Thebes (Greek legend).

Synopsis. The Silvery Checkerspot is a widespread, common eastern checkerspot adapted to disturbed areas. For example, it may be found in city parks, as in the Bronx, New York.

Butterfly. Male forewing: $\overline{X} = 1.9$ cm, range 1.8–2.1 cm; female forewing: $\overline{X} = 2.3$ cm, range 2.2–2.4 cm. This checkerspot closely resembles Harris' Checkerspot dorsally, but looks more like a crescent ventrally. The Silvery Checkerspot may be distinguished from Harris' by the presence of one or more white, centered submarginal spots on the dorsal hind wing, and, on the ventral hind wing, by the presence of a "crescent" mark on the outer margin and the lack of red-orange markings.

Range. The Silvery Checkerspot ranges from the Maritime Provinces west to Manitoba and south to the Florida panhandle, Mississippi, Texas, and Arizona.

Habitat. This species is usually found in moist habitats near streams or rivers; usually, the habitats are somewhat open, with second-growth scrub, and the soil is usually sandy or gravelly and acidic. In Louisiana the species is found in an unusual habitat, a live oak–sweet gum forest.

Life History. *Behavior.* The species has a low, slow flight with alternate flaps and glides. Males perch and patrol open areas, often utilizing dirt roads or stream beds. Mated pairs have been seen from 1505 to 1530 hr in Indiana.

Broods. In the northern portion of its range, the Silvery Checkerspot is univoltine. For example, in northern Michigan the species flies from June 7 to July 25. The species is bivoltine farther south, e.g., in New York (June 5–July 12 and August 10–September 15) and in Georgia (May 5–26 and June 28–August 25). In Mississippi the species flies from early April to early September, leaving the possibility that three broods might occur there.

Early Stages. The females lay a cluster of about 100 green-white eggs under a host leaf. The resultant caterpillars feed communally by skeletonizing the leaf tissue, periodically moving from leaf to leaf as a group (Ehle 1957). Winter is passed by partially grown hibernating larvae. The mature caterpillar varies from black-brown to black, with rows of long, shiny, black spines, the lowest row arising from yellow tubercles. It has broken, yellow, lateral stripes and its back and sides may be dotted with white. The chrysalis has five rows of conical tubercles on the abdomen and varies in color from yellow-green to pink-brown or gray-brown.

Adult Nectar Sources. Flowers favored by adults include dogbane, common milkweed, and red clover.

Caterpillar Host Plants. A wide variety of composites are selected by the Silvery Checkerspot throughout its range. In the East, wingstem (*Actinomeris alternifolia*) is utilized most often (Pennsylvania, Maryland), but sunflowers (*Helianthus decapetalus, H. tuberosus,* and *H. strumosus*) and purple-stemmed aster (*Aster puniceus*) are also fed upon. Farther west, other composites, such as black-eyed susan (*Rudbeckia*) and crown-beard (*Verbesina*), are eaten.

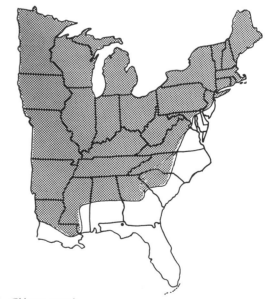

119. *Chlosyne nycteis*

Harris' Checkerspot
Chlosyne harrisii (Scudder), 1864
PLATE 25 · FIGURES 147 & 148 · MAP 120

Etymology. The species is named after Thaddeus W. Harris (1796–1856), a Harvard graduate in medicine, who later switched to entomology and became librarian at Harvard. He published on spiders and beetles and maintained a large insect collection.

Synopsis. Harris' Checkerspot is a localized butterfly of wet meadows and is always found near its caterpillar host, *Aster umbellatus*.

Butterfly. Male forewing: $\bar{X}$ = 1.8 cm, range 1.6–2.0 cm; female forewing: $\bar{X}$ = 2.2 cm, range 2.1–2.4 cm. This butterfly may be distinguished from the Silvery Checkerspot (*Chlosyne nycteis*), which is extremely similar dorsally, by its distinctive ventral hind-wing pattern which consists of a marginal red-orange row and red-orange and cream-white "checks" over the remaining portion. There is extensive geographic variation. Populations in Pennsylvania and southward (subspecies *liggettii*) are composed primarily of individuals that are primarily blackish dorsally, while to the north and west the butterflies are largely orange (typical *harrissii* and subspecies *albimontana*).

Range. The Harris' Checkerspot has a predominantly east-west distrubtion, with the species being found from the Maritime Provinces south to New Jersey thence westward to Manitoba, Illinois, and Wisconsin. Isolated populations occur even farther south in the Appalachians, since there are isolated West Virginia colonies and a single authenticated specimen has been collected in the mountains of Georgia.

Habitat. The species is highly colonial in brushy, damp meadows, pastures, bog edges, and marshes that have good stands of the caterpillar host. Klots (1951) states that the presence of blue flag (*Iris*) is also a good indicator.

Life History. *Broods.* Harris' Checkerspot is univoltine, with its flight in early summer, usually in June but sometimes extending into July. In Maine flight dates are June 8–July 21; in Michigan's upper peninsula, June 27–July 30; and in New York's Finger Lakes, June 10–July 1.

Early Stages. The pale-yellow eggs are laid in groups of about 25 under host leaves, and the resultant young caterpillars feed communally, spinning a slight web as they feed. The partially grown caterpillars overwinter at the base of the host plant. The mature caterpillar is a deep red-orange, with a dorsal longitudinal black stripe and two transverse stripes on each body segment. The last two segments are black. The spines are long and black, with divergent black setae. The chrysalis is white, spotted with black, brown, and orange, and has several rows of subconical spines on its abdomen.

Caterpillar Host Plant. The only known larval host is flat-topped white aster (*Aster umbellatus*). Caterpillars have been found on this plant in many states, including Massachusetts, Maine, New Jersey, New York, and Pennsylvania.

120. *Chlosyne harrisii*

Elf
Microtia elva H. Bates, 1864

Synopsis. A stray of this small tropical checkerspot was once found in Jefferson County, Missouri. It is a regular resident of the seasonally dry lowlands of Mexico and Central America.

Seminole Crescent
Phyciodes texana seminole (Skinner), 1911
MAP 121

Etymology. The species name is derived from Texas, while the subspecies name is a Muskhogean Indian tribe of Florida, originally composed of immigrants from the Lower Creek villages of the Chattahoochee River.

Synopsis. The Seminole is the only crescent in our area that belongs to the largely tropical subgenus *Anthanassa*. Its range is extremely limited.

Butterfly. Male forewing: $\bar{X}$ = 1.7 cm, range 1.6–1.8 cm; female forewing: $\bar{X}$ = 1.9 cm, range 1.9–2.1 cm. Unlike our other eastern crescents, the Seminole is largely black above, with a few white spots and bands, as well as reddish brown on the basal portion of both wings. It has been suggested that the southeastern populations may represent a separate species.

Range. The Texan Crescent, of which the Seminole is a subspecies, ranges from central Florida north to coastal South Carolina, thence west through coastal Louisiana, Texas, rarely Kansas and Nebraska, and southern Colorado to southern California. From the southern United States the species ranges south through Mexico to Guatemala. The Seminole Crescent includes all of the range in the Southeast. Individuals of typical *texana* have appeared as rare strays in Illinois, Missouri, and Arkansas.

Habitat. In the Southeast the species is found close to the banks of streams and rivers. In Colorado, Scott (1975) reports the Texan Crescent is found in gulches and dry streambeds.

Life History. *Behavior.* In contrast to other crescents, male Seminole Crescents perch and occasionally patrol during most daylight hours in their search for females.
Broods. The Seminole Crescent is bivoltine in Georgia (May 22–June 3, August 15–September 24), and may have three flights in Florida, since there are records from March, May, and September.

Caterpillar Host Plants. Larval hosts have been recorded by Kendall for the Texas Crescent, and all belong to the family Acanthaceae—*Dicliptera brachiata*, Brazilian plume (*Jacobinia carnea*), and several others.

Cuban Crescent
Phyciodes frisia (Poey), 1832
PLATE 26 · FIGURE 151 · MAP 122

Etymology. The species name is derived from Friesland, a historical European region. The Frisian people were of Teutonic stock and occupied the lowlands of what is today coastal Germany and the Netherlands.

Synopsis. The Cuban Crescent is alleged to have five subspecies, occurring primarily in tropical America. The typical subspecies occurs in the Caribbean islands.

Butterfly. Male forewing: $\bar{X}$ = 1.5 cm, range 1.4–1.5 cm; female forewing: $\bar{X}$ = 1.8 cm, range 1.6–2.0 cm. The Cuban Crescent is somewhat similar to the Pearl Crescent but may be distinguished by the concave outer margin of the forewing

and the alternating orange and black transverse bands on the dorsal hind wing.

Range. Typical *frisia* ranges through many of the Caribbean islands and extreme southern Florida. Other subspecies of *frisia* range from Missouri (once), Texas, and southern Arizona (rarely) south through Central America and South America to Argentina.

Habitat. The Cuban Crescent occurs in open fields and other second-growth situations.

Life History. *Broods.* The number of generations is unrecorded. The Cuban Crescent flies during most of the year in

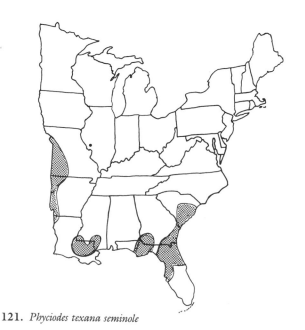

121. *Phyciodes texana seminole*

122. *Phyciodes frisia*

southern Florida. Subspecies *tulcis* flies during every month in Costa Rica.

Early Stages. The eggs are laid in clusters under host leaves and the larvae feed communally at night. The caterpillar is mottled yellow and gray. The subspiracular spines are yellow, while the remainder are black (Chermock and Chermock 1947).

Caterpillar Host Plant. In southern Florida, the Cuban Crescent feeds on shrimp flower (*Beloperone guttata*) of the Acanthaceae family, which is primarily tropical.

Phaon Crescent
Phyciodes phaon (Edwards), 1864
PLATE 25 · FIGURES 149 & 150 · MAP 123

Synopsis. The Phaon is a southern crescent superficially similar to the Pearl Crescent and often flying with it in the Southeast. Its host plant *Lippia*, of the verbena family, is unusual for crescents.

Butterfly. Male forewing: $\overline{X}$ = 1.5 cm, range 1.4–1.6 cm; female forewing: $\overline{X}$ = 1.6 cm, range 1.5–1.7 cm. This butterfly is a member of the tharos group, to which it is superficially similar. It may be distinguished by its more checkered fringe, yellow median spot band on the dorsal forewing, greater amount of black, more reddish ground color on the ventral forewing, and more whitish or cream color on the ventral hind wing (summer form only). The Phaon Crescent also has a short-photoperiod seasonal form, "hiemalis," which has the same general characteristics as the homologous form for the Pearl Crescent, i.e., a distinctly dark-grey or brown ventral hind wing with a more checkered appearance above, including a pale line near the outer margin.

Range. The Phaon Crescent occurs from the Outer Banks of North Carolina to the Gulf States, Missouri, Kansas, and the Southwest south through some of the Caribbean islands and Mexico to Guatemala.

Life History. *Behavior.* The Phaon Crescent is a patrolling species, with males spending most of the warmer daylight hours in search of receptive mates, usually in the vicinity of the host-plant colony. Mating occurs from late morning to midday (1000–1230 hr) and the female is the carrier. Eggs are deposited in small clusters on the underside of host-plant leaves.

Broods. The Phaon Crescent breeds almost continuously, warm weather and host-plant conditions allowing. There are probably at least four broods and possibly as many as six or seven in some areas. Flight extremes are March 2 to December 3 in Georgia and March 7 to October 16 in Louisiana. The Phaon flies through most of the year in peninsular Florida.

Early Stages. The eggs are laid in small clusters on host leaves. The caterpillar is olive green to olive brown, with dark subdorsal bands and scalloped white subspiracular bands. The head is cream with a large brown spot. The chrysalis is brown and is speckled with black and white.

Adult Nectar Sources. The adults often visit flowers of the larval host, *Lippia*, as well as a variety of composites, including shepherd's needle.

Caterpillar Host Plants. Larval hosts are known to be several species of herbaceous, prostrate *Lippia*, fogfruit (*Lippia lanceolata*), and mat grass (*Lippia nodiflora*), which is known to serve as a host in at least California, Texas, and Florida. Riley (1975) associated the butterfly with creeping daisy (*Wedelia trilobata*).

Pearl Crescent
Phyciodes tharos (Drury), 1773
PLATE 26 · FIGURES 152 & 153 · MAP 124

Synopsis. The Pearl Crescent is one of the most common, widespread butterflies in much of the eastern United States, but it has been confounded with the similar but distinct Northern Pearl Crescent (*Phyciodes pascoensis*) for more than 200 years. Furthermore, many aspects of its biology are poorly understood.

Butterfly. Male forewing: $\overline{X}$ = 1.6 cm, range 1.4–1.7 cm; female forewing: $\overline{X}$ = 1.8 cm, range 1.7–1.9 cm. This butterfly demonstrates considerable individual variation, usually within populations, while the Northern Pearl Crescent demonstrates much more between-population variation.

The differences between the Pearl Crescent and the Phaon and Tawny Crescents are discussed under the latter species, while the differences between it and the Northern Pearl Crescent are discussed below. The males are easiest to distinguish. In the male Pearl Crescent, the antennal clubs are usually black, not orange; dorsally, the forewing postmedial and the hind-wing submarginal orange areas are not as open;

the yellow scaling on the forewing is scattered, not concentrated in patches; the patch enclosing the crescent on the ventral hind-wing outer margin tends to be black; and the thin linear markings on the ventral hind wing are blackish, not orange. The females of the two species are nearly identical, and are best told apart by location, by associated males, and the by Pearl Crescent's smaller size.

There are a spring and a fall form of the Pearl Crescent that result from caterpillars that develop under the influence of short day-length. In these forms the ventral hind wing is most distinctive, being gray or gray-brown, with an irregular median pale stripe. Dorsally, the black areas are reduced and there is a submarginal pale line on the hind wing.

The Pearl Crescent and Northern Pearl Crescent occur together in limited areas of New York, Pennsylvania, Virginia, and West Virginia. In one West Virginia locality where both species are common, differences in behavior, flight periods, genetics, and compatibility have been documented. Oliver (1981) has published results of crossing experiments between the two species, which he refers to as Type A and Type B, respectively. Electrophoretic tests of genetic variability in a number of eastern populations from New York to Florida revealed almost no between-population differences (Vawter and Brussard 1975).

Range. The species occurs from southern New England, New York, southern Michigan, Minnesota, North Dakota, Colorado south to the tip of Florida, the Gulf of Mexico, and central Mexico. A different form (subspecies *distincta*) occurs in the Southwest and adjacent Mexico. The Pearl Crescent's range has probably expanded greatly with the cutting of eastern deciduous forests and the expansion of agriculture.

Habitat. The Pearl Crescent occurs in almost any large open area where asters are present, including roadsides, vacant city lots, pastures, fields, and open pine woods.

Life History. *Behavior.* Male Pearl Crescents spend much of the day patrolling open areas in search of receptive females. Newly emerged males take moisture on damp sandy or muddy areas along dirt roads and small streams. Males bask with their wings open. Males exhibiting courtship behavior toward females were seen from 1015 to 1715 hr. Males land behind females and flutter their wings with low amplitude. Actual coupling was not observed, but mated pairs were seen from 1100 to 1745 hr. Most pairs are seen between 1200 to 1530 hr, a very restricted period.

Broods. The Pearl Crescent is a homodynamic species, having continuous broods as long as temperature and food-plant conditions permit. There are only two or three broods in the northern part of its range, e.g., in Illinois, where the species flies from April 20 to October 19 and there are probably three broods. In northern Virginia, where the species flies from April 22 to November 11, there are four broods and a partial fifth. In the Deep South (Florida, Louisiana, Mississippi), where the species has been found in all months, there might be as many as six generations.

Early Stages. Female Pearl Crescents lay their eggs in small groups on the underside of aster leaves between 1420 and 1500 hr. Females lay up to 700 eggs during their life. During the summer, Pearl Crescents develop directly, and about 35 days are required from egg to adult emergence. The larvae feed communally on the host, but, unlike those of the Tawny

123. *Phyciodes phaon*

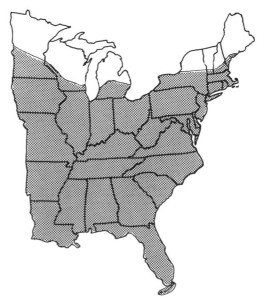

124. *Phyciodes tharos*

Crescent, do not spin webbing. In late summer, under the influence of short day-lengths, third-instar larvae enter diapause for the winter. Feeding resumes in the spring. The mature larva is a dark chocolate brown with a faint, pale, broken dorsal stripe. The head has extensive dorsal and lateral white patches. The tubercles are chocolate brown and the caterpillar has a crisp gray and white appearance.

Adult Nectar Sources. We have observed Pearl Crescents visiting more than thirty kinds of flowers in many families. Preferred nectar sources are those where the butterfly may perch on top of an inflorescence and have a view of surrounding terrain and air space. Especially favored are dog-bane, butterflyweed, swamp milkweed, black-eyed susans (which it visits in much greater frequency than other butterflies), tickseed sunflower, shepherd's-needle, asters (its larval host), peppermint, white clover, and winter cress.

Each generation at the same locality will have a different preferred nectar source, since the Pearl Crescent's flight season spans all blooming periods except those of earliest spring.

Caterpillar Host Plants. Larvae of the Pearl Crescent prefer smooth-leaved true asters of a wide variety, including *Aster pilosus, A. ericoides, A. laevis, A. solidagineus, A. texanus,* and *A. prealtas.* The species seems closely associated with *Aster simplex* in many central Atlantic Coast locations.

Northern Pearl Crescent
Phyciodes pascoensis Wright, 1905, NEW STATUS

PLATE 27 · FIGURES 157 & 158 · MAP 125

Etymology. The species name is based upon the city of Pasco, Washington, near the species' type locality. Pasco in Latin means "I feed" (pastor, shepherd).

Synopsis. The Northern Pearl Crescent, a larger version of the common eastern Pearl Crescent, occurs primarily in the mountainous West and Canada. It had not been recognized as a separate species until a few years ago (Oliver 1980).

Butterfly. Male forewing: $\overline{X}$ = 1.9 cm, range 1.7–2.0 cm; female forewing: $\overline{X}$ = 2.0 cm, range 1.9–2.2 cm. The Northern Pearl Crescent may be distinguished from the more common Pearl Crescent by several features. Refer to the discussion under the latter. Occasional hybrids may be found where the species occur together.

Range. The Northern Pearl Crescent is a generally more northern and western species than the Pearl Crescent. It ranges from Newfoundland, the Maritime Provinces, and the northern New England states west across the northern tier states and central Canada to British Columbia. It occurs south in the western mountains to Utah, the White Mountains of Arizona, and northern New Mexico. In the East it ranges south in the Appalachians to Pennsylvania, Virginia, and West Virginia.

Habitat. This species is generally found in moist open areas near streams within rocky regions. In Virginia and West Virginia, where we have observed the species, it is closely associated with shale-barren habitats.

Life History. *Behavior.* The Northern Pearl Crescent, like other crescents, is a patrolling species. In a West Virginia habitat where this species and the Pearl Crescent are found, several differences between the species were noted in the male's flight and patrolling characteristics. Both sexes fly with a less rapid wing beat and have some of the flap-and-glide features of the Silvery Checkerspot. Significantly, both sexes will readily fly through fairly dense, shaded woodland to move from one clearing to another. The flight of this species is usually higher than that of the Pearl Crescent, and individuals frequently rise to 3 to 5 m to clear a small tree or shrub. Males often patrol several hundred meters up the middle of a dry streambed or dirt road without stopping.

Broods. This species differs strikingly from the Pearl Crescent in that it is univoltine, although a partial second brood may occur at some localities. Some sample flight extremes are June 11 to July 18 in Wisconsin and June 29 to July 18 in West Virginia.

125. *Phyciodes pascoensis*

Early Stages. The early stages are similar to those of the Pearl Crescent, but the larval tubercles are light pinkish gray and the caterpillars have a less contrasted, slightly pinkish appearance (Oliver 1980).

Adult Nectar Sources. Northern Pearl Crescents have been observed at flowers of dogbane, white clover, and fleabane.

Caterpillar Host Plants. The natural larval host plants are unknown, but females will readily oviposit upon and the caterpillars will feed on panicled aster (*Aster simplex*) in captivity.

Tawny Crescent
Phyciodes batesii (Reakirt), 1865
MAP 126

Etymology. The genus is named after phycis, a fish that lives in seaweed and changes its color at different seasons (Plinius). The species is named in honor of Henry Walter Bates (1820–1892), who was Assistant Secretary to the Royal Geographical Society, an explorer of the Amazon basin, and a butterfly specialist.

Synopsis. Always an extremely localized species, the Tawny Crescent has apparently disappeared from much of its range in the East due to unknown reasons. It is often confused with the Pearly Crescent.

Butterfly. Male forewing: $\bar{X}$ = 1.7 cm, range 1.5–1.8 cm; female forewing: $\bar{X}$ = 1.9 cm, range 1.8–2.0 cm. Dorsally, the Tawny Crescent has more extensive black than the Pearl Crescent and the median spot row on the forewing tends to be a paler orange than the other light areas. Ventrally, the hind wing is nearly immaculate, and is yellow-cream colored.

Oliver (1979) carried out hybridization experiments with the Tawny and Pearl Crescents. The results demonstrated differences in developmental rates between the species, as well as heavy reduction in hybrid viability and fertility. In summary, these two species are distinct and reproductively isolated. Tawny Crescent larvae raised under artificial long-day conditions produced adults with heavier dark patterns above and below, but these were not like the spring or fall short-day phenotypes of the Pearl Crescent.

Range. The Tawny Crescent ranges from southern Quebec and Ontario west to Nebraska and Colorado and southward to Pennsylvania and Michigan. It has been found farther south in the Appalachians in Virginia, Kentucky, North Carolina, and Georgia.

Habitat. In the northern part of its range, the Tawny Crescent is found in low-lying moist meadows or pastures, while farther south it is found on top of dry, rocky bluffs above rivers or on dry hillsides or rocky upland pastures, usually with much *Andropogon* grass.

Life History. *Broods.* The Tawny Crescent, unlike the Pearl Crescent, has a single annual generation. Emergence occurs progressively earlier as one proceeds south. Sample flight extremes are: New York (Onondaga County), June 20 to July 21; Michigan's upper peninsula, June 10 to July 7; Virginia, June 1 to July 1; and Georgia, May 26. A partial second brood sometimes occurs in Michigan. The Tawny Crescent is purported to fly between the first and second broods of the Pearl Crescent, but this is not always the case.

Early Stages. Eggs are laid in batches under leaves of the host and hatch in 7 to 8 days. The first- and second-instar larvae live communally in webs on the host. Larvae enter diapause in the third instar and complete development during the following spring. The mature larvae is brown with a pinkish tinge and a broad, pale dorsal stripe. Compared to that of the Pearl Crescent, the pupa is more rounded, with dorsal projections almost absent (McDunnough 1920).

Caterpillar Host Plants. Wavy-leaved aster (*Aster undulatus*) and possibly other true asters are the known larval hosts for this species. Panicled aster (*Aster simplex*) was selected by released females in Pennsylvania (Oliver 1979).

126. *Phyciodes batesii*

Baltimore
Euphydryas phaeton (Drury), 1773
PLATE 27 · FIGURES 159 & 160 · MAP 127

Etymology. The genus name is derived from the Greek *euphyes,* "shapely, comely," and *dryas,* "dryad." The species is named after Phaeton, the son of Helios, god of the sun in Greek mythology.

Synopsis. The Baltimore is unique in appearance among eastern butterlies, but is closely related to several widespread checkerspots found in the West, such as Edith's Checkerspot (*Euphydryas editha*), which has been intensively studied by Paul Ehrlich and numerous students. Both the adult and immature stages of the Baltimore are distasteful and emetic to vertebrate predators.

Butterfly. Male forewing: $\overline{X}$ = 2.5 cm, range 2.0–2.8 cm; female forewing: $\overline{X}$ = 2.8 cm, range 2.6–3.3 cm. The Baltimore is a black butterfly with a comlete row of red-orange spots on all wing margins and several rows of small pale-yellow spots over the remainder of the wings; these markings are most pronounced ventrally. This coloration is typical of unpalatable butterflies, and the black and red markings warn would be predators. Experiments with laboratory-trained blue jays demonstrated that Baltimores are indeed unpalatable: butterflies that had fed as larvae on the usual host turtlehead (*Chelone glabra*) were unpalatable and emetic, although adults that had been fed English plantain (*Plantago lanceolata*) were not emetic and were at least partially palatable (Bowers 1980). The larvae and pupae of the Baltimore were also shown to be unpalatable and emetic to the jays. Several subspecies of uncertain merit have been proposed. The most valid of these seem to be *E. phaeton phaeton* and *E. phaeton ozarkae* Masters. These two subspecies may be distinguished on the basis of adult coloration, habitat, and food plant (see below).

Range. The Baltimore ranges from Maine west to Wisconsin (and portions of southern Canada) and thence south to Georgia, Alabama, Mississippi, and Kansas. Subspecies *ozarkae* seems to occur mostly in Missouri, Arkansas, and Kansas (Masters 1968a).

Habitat. The Baltimore occurs in two basically different habitats: *E. p. phaeton* in open bogs, marshes, or wet meadows, and *E. p. ozarkae* in dry, open, or wooded hillsides.

Life History. *Behavior.* The adult male Baltimore perches close to the ground while awaiting passing females. A mated pair was seen at 1220, with the female as carrier. Eggs are laid during the afternoon in batches of 100 to 700 under leaves of the primary host plant.

Broods. The Baltimore is univoltine. The range of flight dates becomes progressively earlier southward. For example, in New York and northern Michigan, emergence does not occur until at least mid-June (in New York, June 24 to August 18; in northern Michigan, June 19 to July 26), while flight extremes are earlier in Maryland (May 17 to June 28). Irwin and Downey (1973) reported that *E. p. ozarkae* flew earlier (May 30 to June 13) than true *phaeton* (June 6 to July 4) in Illinois, but these differences may be due primarily to the amount of insolation in the two respective habitats.

Early Stages. Female Baltimores lay large masses of 100–700 lemon yellow eggs under leaves of the host. The eggs hatch in about 20 days, and the young caterpillars soon move to the tip of the plant and construct webbing, within which they feed gregariously. As they devour the host's leaves, the web is extended downward. The caterpillars develop to the third instar in about 3 weeks. At the end of the summer they build a thickened prehibernation web within which they molt to the fourth instar. In late October the caterpillars leave the plant and spend the winter in rolled leaves in the ground litter nearby (M. D. Bowers, 1978). In the spring they return to the host to complete feeding. If host material is completely exhausted, they seek alternate hosts (see below)

The mature caterpillar is bright red-orange, except for the first two and last two segments, and is transversely striped with black. There are seven rows of long, black, branched spines, each emanating from a black tubercle (chalaza). The pupa is off-white, with black and orange mottling and spots. There are several longitudinal rows of low-lying orange tubercles on the abdomen.

Adult Nectar Sources. The Baltimore visits flowers infrequently, but it has been observed nectaring at common milkweed, wild rose, and viburnum.

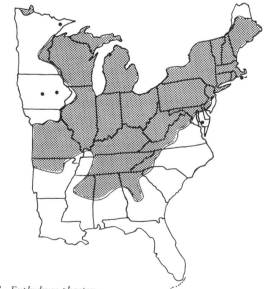

127. *Euphydryas phaeton*

Caterpillar Host Plants. The Baltimore's host plants may be divided into two groups: those upon which the females lay their eggs and the larvae feed prior to winter, and those that the larger larvae feed on in the spring, when they are more catholic in their choice of food. The classic oviposition plant for *E. p. phaeton* is turtlehead (*Chelone glabra*), but others are utilized on occasion, including hairy beardtongue (*Penstemon hirsutus*) and the exotic English plantain (*Plantago lanceolata*) (Stamp 1979). In *E. p. ozarkae* populations, females oviposit on false foxglove (*Aureolaria*: Scrophulariaceae). After overwintering, Baltimore larvae may continue to feed on their oviposition plant, but they sometimes wander and complete development on a wide range of unrelated plants, including white ash (*Fraxinus americana*), arrowwood (*Viburnum recognitum*), common lousewort (*Pedicularis canadensis*), and Japanese honeysuckle (*Lonicera japonica*). Reports of wisteria are probably incorrect and require confirmation.

Question Mark
Polygonia interrogationis (Fabricius), 1798
PLATE 26 · FIGURES 154–156 · MAP 128

Etymology. The generic name is derived from the Greek *polygonos,* "many angled." The species name is derived from the Latin *interrogatio,* "question."

Synopsis. The angle wings (genus *Polygonia*) are all woodland species that feed on sap flows and rotting fruit. They rarely visit flowers. The Question Mark is named for the white mark on the lower hind wing that closely resembles the interrogatory form of punctuation.

Butterfly. Male forewing: $\overline{X}$ = 2.8 cm, range 2.6–3.1 cm; female forewing: $\overline{X}$ = 3.2 cm, range 2.7–3.5 cm. The Question Mark may be distinguished from the Comma (*Polygonia comma*), its close relative, by the mark on the ventral hind wing, its longer taillike projection from the hind wing, and its larger size. Like the Comma, and to a lesser extent the Satyr, the Question Mark has two distinct seasonal forms. The summer form is a largely black hind wing above. In addition, the outer margins of both wings have a ventral series of small, iridescent blue linear markings. The winter form has a largely red-orange dorsal hind wing, with the outer margin of both wings above outlined with violet. The winter form has longer tails. Ventrally, there are also two hind-wing coloration types that are independent of both sex and season. In one form the ventral hind wing is a uniform slaty gray-brown, while in the other form the hind wing is strongly mottled or striate.

Range. The Question Mark occurs from the Maritime Provinces west to the eastern foothills of the Rocky Mountains and thence south to the Gulf Coast, but is rare in peninsular Florida. In the northernmost part of its range, the adults cannot survive the winter, and these areas must be re-invaded each year. In the extreme South the species is present primarily as overwintering individuals.

Habitat. The Question Mark is found in a variety of situations, which are usually characterized by the presence of trees and at least some open areas. Proximity to rivers, streams, or other wet areas is frequent but not necessary.

Life History. *Behavior.* The species spends much of the year in either hibernation or estivation, appearing briefly at certain times. The males perch and seem territorial and aggressive. Their mate-seeking behavior usually takes place in late afternoon (after 1600 hr), as is also the case for other angle wings. They perch on the trunks of trees, flying out and chasing other large insects or even birds (Hendricks 1974). After each foray, the male returns to his perch or to another one nearby. At other times of day, the butterflies feed, perch at moisture, or engage in other activities. The females lay their eggs only for a brief period at midday (1205–1330 hr). Although they are reported to lay their eggs in small stacks under host leaves, our field observations indicate they may lay single eggs in some instances. An ovipositing female flies with a bouncing flight, briefly touching down on a number of plants. After having touched a proper host, she bounces off, lands on another plant nearby, and lays an egg on its stem or under a leaf. These plants are usually not a host and may include grasses, ferns, or virtually any other plant.

These butterflies warm up in the morning or late afternoon by using dorsal basking while perching on the ground or low vegetation.

The Question Mark overwinters as an adult, with some individuals remaining in the North and many migrating to the South in the winter. There have been reports of large numbers of angle wings overwintering in such areas as South Carolina's Congaree Swamp. Each spring we have seen numbers of these butterflies flying north through northern Virginia, and migratory movements have also been seen in Ohio.

Broods. There are probably only two generations each year throughout the butterfly's range. Overwintered adults are found in the spring, and then the summer form emerges in June and flies through much of the summer, although individuals may estivate for a portion of this time. In late summer, the summer-brood individuals lay eggs that develop into the winter form, which first appears in the fall. Overwintering individuals may occasionally be seen on warm winter days, and they fly until the end of May. The summer form emerges as early as mid-May (Mississippi) or as late as June 3 (New York). Summer-form individuals fly as late as mid-September. Since the winter form emerges in late August or early September, both forms may be seen together for a few weeks in the fall.

Early Stages. The mature caterpillar is extremely variable in body and spine color. The body varies from black to yellow and often has yellow or red irregular longitudinal lines. The relatively large tubercles vary from yellow to red, and the greatly branched spines vary from yellow to red-orange and black. The pupa varies from yellow to dark brown. Its head is deeply notched frontally and there is a prominent narrow thoracic keel. The dorsum has eight metallic silver spots.

Adult Food. The normal food of Question Marks is tree sap, rotting fruit, dung, or dead animals. On rare occasions, usually during dry periods when sap flows or rotting fruit may be unavailable, they visit flowers for nectar. At such times, Question Marks have been seen at flowers of common milkweed, sweet pepperbush, and aster.

Caterpillar Host Plants. Question Mark larvae feed on plants in a small number of disparate families, including nettles (*Urtica*), false nettle (*Boehmeria cylindrica*), Japanese hop (*Humulus japonicus*), American elm (*Ulmus americanus*), red elm (*U. rubra*), and hackberry (*Celtis*). Cross vine (*Bignonia capreolata*) and basswood (*Tilia*) are questionable hosts.

Comma or Hop Merchant
Polygonia comma (Harris), 1842
PLATE 28 · FIGURES 163–165 · MAP 129

Etymology. The species name is derived from "comma," a punctuation mark.

Synopsis. The Comma is a close relative of the Eurasian *Polygonia c-album.* It is very similar in appearance to, and shares many traits with, the Question Mark.

Butterfly. Male forewing: $\overline{X}$ = 2.5 cm, range 2.3–2.7 cm; female forewing: $\overline{X}$ = 2.7 cm, range 2.4–2.8 cm. The Comma may be distinguished from the Question Mark (*Polygonia interrogationis*) by its silvery "comma" mark in the center of the ventral hind wing, its smaller size, more irregular outer wing margins, and shorter tails. It has a dark hind wing summer form and an orange hind wing winter form that differ from each other in much the same way as do the Question Mark's seasonal forms.

Range. This butterfly ranges from Canada's Maritime Provinces west to the central plains and south to Florida and the Gulf states.

Habitat. The Comma is only slightly more limited to woodland areas than the Question Mark. It is found most often in moist bottomland and riparian forests, but may also be found in almost any deciduous woodland. It occurs in many disturbed open areas as long as appropriate larval hosts are present.

Life History. *Behavior.* The behavior of the Comma is

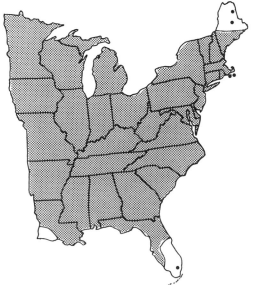

128. *Polygonia interrogationis*

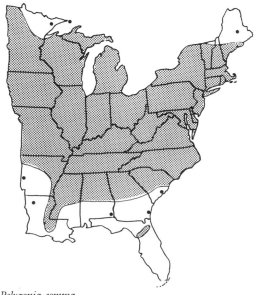

129. *Polygonia comma*

similar to that of the Question Mark in almost all regards. Commas hibernate over the winter and probably migrate southward, but to a lesser degree than the Question Mark. The males perch on sunlit tree trunks or foliage in late afternoon to await receptive females, and fly out to interact with other male Commas or other large insects. Oviposition by females takes place at midday (1220–1335 hr) in the same manner as described for the Question Mark.

Broods. The Comma is probably bivoltine. The black hind wing summer form, resulting from eggs laid by overwintered females, emerges in the summer and is not found until late fall. This generation appears as early as May 12 in Mississippi, but not until June 24 in southeastern Pennsylvania. Individuals of this generation may spend part of the summer in estivation. The orange hind wing winter forms emerge in September or October, but soon seek shelter in hollow trees or a similar refuge, where they spend the winter. Many individuals must move far south, but there are observations of only a diffuse northward movement in the spring. The winter form may be found in the spring until the end of April.

Early Stages. The eggs are laid singly or in stacks and are at first pale green. In the last instars, the caterpillars feed at night and shelter under host leaves that have had their margins drawn in with silk. The caterpillars are extremely variable, ranging from black to green-brown and white. The spines also vary accordingly from black to white. In the darker-colored larvae, the spines emanate from yellow tubercles. The chrysalis may be gray or brown, with two rows of gold or silver projections from the abdominal dorsum.

Caterpillar Host Plants. The larvae will feed on all members of the nettle and elm families. Common hosts are nettle (*Urtica*), wood nettle (*Laportea canadensis*), false nettle (*Boehmeria cylindrica*), American elm (*Ulmus americana*), and hops (*Humulus*).

Satyr Angle Wing
Polygonia satyrus (Edwards), 1869
PLATE 28 • FIGURE 166 • MAP 130

Etymology. The species name is derived from the Latin *satyrus,* a riotous, lascivious woodland deity and attendant of Bacchus that is commonly represented as part human and part goat.

Synopsis. The Satyr Angle Wing is a western species found rarely in our area.

Butterfly. Male forewing: $\overline{X} = 2.6$ cm, range 2.2–2.8 cm; female forewing: $\overline{X} = 2.7$ cm, range 2.6–2.8 cm. The Satyr is similar to the Comma (*Polygonia comma*), but differs from the Comma's winter form, which it most closely resembles, by its lack of a dark border on the dorsal hind wings and its complete lack of violet margins. The Satyr has a number of forms, some of which may be seasonal, but the species is certainly not as strikingly dimorphic as either the Question Mark or the Comma.

Range. The Satyr Angle Wing is a western species found at low elevations from coastal British Columbia south to central California and through suitable areas of most western states. In the East it ranges from Newfoundland and Maine west through extreme southern Canada and northern portions of the Great Lakes states.

Habitat. The Satyr Angle Wing is found primarily near streams, but it may wander into openings of adjacent woodlands.

Life History. *Behavior.* Male Satyr Angle Wings perch on low vegetation and tree trunks in sunlit openings in woods, where they await the passage of receptive females. The male–male encounters of this butterfy may be prolonged, and the apparently dueling insects may rise as much as 50 m in the air. Perching activity normally takes place in late afternoon and early evening.

Broods. The number of generations in the East is not documented, but at least two and possibly three broods occur at

130. *Polygonia satyrus*

low elevations in the Pacific Coast states. In Michigan the species is known only from July 6 to September 18, and only a single flight may occur there. Adults overwinter.

Early Stages. The female butterfly deposits pale-green eggs on the lower surface of nettle leaves. The young caterpillar makes a single-leaf shelter by drawing the leaf edges together with silk. The mature caterpillar is black with a broad green-white dorsal band and a similarly colored stigmatal line. The dorsal band has a fine V-shaped black mark on each segment.

The dorsal and subdorsal spine rows are green-white, while the lateral rows are black. The chrysalis is pale brown with a few darker marks, especially on the wing covers.

Adult Food. Presumably, adults feed on sap flows and rotting fruits, but there are no reports for our area. In the West, Satyr Angle Wings are found at flowers with some frequency. Records include almond and blackberry.

Caterpillar Host Plants. Nettles (*Urtica*) are the only known larval hosts.

Green Comma
Polygonia faunus Edwards, 1862
MAP 131

Etymology. The species is named after Faunus, an ancient Roman woodland deity, later identified with Pan.

Synopsis. The Green Comma is mostly a northern boreal species. A large, brightly colored subspecies (*smythi*) is found in the southern Appalachians.

Butterfly. Male forewing: $\overline{X}$ = 2.6 cm, range 2.2–2.7 cm; female forewing: $\overline{X}$ = 2.7 cm, range 2.6–2.8 cm. The Green Comma may be distinguished from other angle wings by its combination of more jagged wing margins and the submarginal series of distinct green markings on the ventral hind wings.

Range. The Green Comma ranges from Newfoundland and the Maritime Provinces west through northern New England, New York, the northern Great Lakes states, and Canada to the Pacific Coast. It occurs north to the Northwest Territories. In the East, a distinct, large, dark subspecies (*P. f. smythi*) occurs in the Appalachians from central Virginia and West Virginia south to northern Georgia.

Habitat. In New York, the Green Comma is found in spruce-fir forests and may be found occasionally in beech–maple–birch–hemlock association (Shapiro 1974). In the Virginia mountains this angle wing is found in spruce or lashorn forests or in deciduous forests with replicating spruce (Clark and Clark 1951). Wherever found, this is a species of the Canadian Life Zone.

Life History. *Behavior.* Scott (1975) reported that males perch on shrubs or rocks in gullies during late afternoon, from 1300 hr on.

Broods. The Green Comma is univoltine. Adults emerge in mid- to late summer and may be found well into the fall, e.g., mid-October in New York. It is possible that adults, especially of the southern Appalachian subspecies, may spend part of the summer in estivation. The winter is passed by hibernating adults, which then emerge in spring to mate and lay eggs. Adult longevity is probably 9 or 10 months.

Early Stages. Single green eggs are laid on the upper surface of host leaves. After hatching, the young caterpillars move to the undersides of leaves, where they then feed solitarily. The mature caterpillar has complex markings. The head is black with a white W mark on the front. The body varies from yellow-brown to brick red, the abdominal dorsum is predominantly white, and there is a broken double orange band laterally. In addition, there are transverse black and yellow bands or spots. The spines, including those on the head, are white, except for the subspiraculars, which are brown. The chrysalis is gray-brown, sometimes marked with green or dark brown, and its head is divided into a forked projection anteriorly. The thoracic dorsum has an elevated keellike protuberance with tubercles on each side. The dorsum also has numerous pale or silver-tipped, low, pointed projections.

Caterpillar Host Plants. The species had been reared in the East from caterpillars found on small pussy willow (*Salix humilis*), black birch (*Betula lenta*), and alder (*Alnus*). The Appalachian form has been reared from gooseberry (*Ribes*).

131. *Polygonia faunus*

Hoary Comma
Polygonia gracilis (Grote & Robinson), 1867
MAP 132

Etymology.   The species name is derived from the Latin *gracilis,* "slim, gracefully slender."

Synopsis.   The Hoary Comma, although found throughout much of Canada, barely enters the United States and is little studied.

Butterfly.   Male forewing: $\overline{X}$ = 2.3 cm, range 2.2–2.4 cm; female forewing: $\overline{X}$ = 2.4 cm, range 2.3–2.5 cm. The Hoary Comma may be distinguished from the Gray Comma, which it most closely resembles, by its lack of striations and the distinctly pale-gray outer portions of the ventral wing surfaces. Dorsally, the marginal area of the hind wings is black. It is probably conspecific with the western Zephyr (*Polygonia zephyrus*).

Range.   The Hoary Comma extends from nothern New England west and north across the forested boreal portions of Canada to Great Slave Lake and southern Alaska. It barely enters the United States in northern New England, New York's Adirondacks, and the northern portions of Michigan and Minnesota.

Habitat.   In New York, the Hoary Comma is found at altitudes above 1000 m, often perched on bare granite at or near mountain summits (Shapiro 1974).

Life History.   *Broods.* This angle wing is univoltine. In New York, the Hoary Comma has been found only in July, while in Maine flight dates range from June 9 to September 6. The species is not found every year in our area, and individuals in the eastern United States may represent vagrants from further north. Adults probably hibernate.

Early Stages. The caterpillar of the eastern Hoary Comma is probably much like that of the Zephyr, which is pre-dominantly black, with the dorsal portions of segments 7–12, including their spines, marked with white. The spines and their bases on segments 3–6 are red-brown. The chrysalis is similar to that of the Green Comma (*Polygonia faunus*).

Adult Nectar Source.   Klots (1951) reports that adults often take nectar from everlasting flowers.

Caterpillar Host Plants.   The larval hosts are probably species of gooseberry or currant (*Ribes*), since these plants are eaten by the Zephyr (*Polygonia zephyrus*).

132. *Polygonia gracilis*

Gray Comma
Polygonia progne (Cramer), 1776
PLATE 27 · FIGURES 161 & 162 · MAP 133

Etymology.   The species name is derived from Prokne, daughter of Pandion, king of Athens (Greek myth).

Synopsis.   The Gray Comma is found in more southern or lower altitudes than the Green Comma and is most similar in appearance to the Hoary Comma.

Butterfly.   Male forewing: $\overline{X}$ = 2.6 cm, range 2.4–2.8 cm; female forewing: $\overline{X}$ = 2.7 cm, range 2.4–2.8 cm. The Gray Comma may be distinguished dorsally by the combination of the black outer portions of the hind wings, within which there is a series of three or four small orange dots, and ventrally by the gray to gray-brown striations. Like the Question Mark and Comma, the Gray Comma has two seasonal forms. In the summer form, the outer half of the hind wing above is black, while in the winter form the hind wing above is covered with black only on the outer fourth.

Range.   The Gray Comma has the most restricted range of any angle wing that occurs in the East. It is found from the Maritime Provinces and New England west across southern Canada and the northern United States to the beginning of the Great Plains. It is also found southward in the Appalachians to North Carolina and south just east of the Mississippi River to northern Arkansas.

Habitat. This is a butterfly of rich deciduous woodlands that seems to be limited to the Transition Life Zone. Adults are found within clearings and along dirt roads.

Life History. *Behavior.* Like other angle wings, the Gray Comma is a perching species, with mate-locating activities beginning in midafternoon (1500 hr). Males perch on the foliage of trees or shrubs at edges of clearings. The flight is less rapid than that of other angle wings (except Hoary Comma). Freshly emerged adults may be found at moisture in midsummer.

Broods. The Gray Comma is bivoltine throughout its range. The summer form appears as early as mid-June and may be found until mid-August, and the winter form may be found by early October. The adults overwinter and are found the following spring in April and May, at which time they court, mate, and lay the eggs that eventually lead to the shorter-lived summer generation.

Early Stages. The pale-green eggs are laid singly on host leaves and the caterpillars feed underneath. The mature larva has an orange-brown head and a yellow-brown body with dark olive-brown blotches and lines, but there is much variation. The head and body spines are black or yellow and branched, those on the head being particularly long. The pupa may be dull green, brown, or buff salmon-pink streaked with dark brown or black.

Adult Food. Adults are found most often at sap flows and seldom visit flowers.

Caterpillar Host Plants. The Gray Comma has usually been reared from wild gooseberry (*Ribes rotundifolium*) and was reported to feed on wild azalea (*Rhododendron nudiflorum*) by Shapiro (1966). Elm (*Ulmus*) is utilized rarely

Compton Tortoise Shell
Nymphalis vau-album (Denis & Schiffermüller), 1775

MAP 134

Etymology. The generic name is derived from the Greek *nymphé,* an inferior divinity of nature. The species name is derived from *vau,* Greek for 'V,' and the Latin *album,* "white." It refers to the V mark on the ventral hind wing.

Synopsis. The presence and abundance of the Compton Tortoise Shell are unpredictable in any particular area.

Butterfly. Male forewing: $\bar{X}$ = 3.4 cm, range 3.3–3.6 cm; female forewing: $\bar{X}$ = 3.4 cm, range 3.3–3.6 cm. The adult is unique in its appearance. Like other *Nymphalis* species, the Compton has a general wing outline and a ventral coloring that are reminiscent of large broad-winged *Polygonia*, but it is the only species of either genus with brown, yellow, white, and black wings dorsally.

Range. The Compton Tortoise Shell is holarctic, occurring in northern Eurasia and North America. On our continent it is found from eastern Canada and northern New England west and north through much of the Canadian Life Zone of Canada to the northern Rocky Mountains of Wyoming and Montana, British Columbia, and Alaska. It occurs with some regularity in New York, Pennsylvania, Michigan, and the

133. *Polygonia progne*

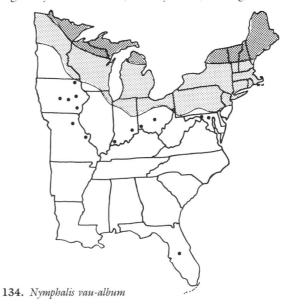

134. *Nymphalis vau-album*

northern Great Lakes states. On rare occasions single individuals are found in the central and southern Appalachians as well as in the Ozarks of Missouri. A single individual was found once in Florida!

Habitat. The Compton Tortoise Shell is a species of upland forests, either deciduous or coniferous. Its principal habitat is within the Canadian Life Zone.

Life History. *Behavior.* Shapiro (1966) reports that this species is sometimes migratory. Many of the southward incursions of Compton Tortoise Shells into the northeastern United States may result from population explosions farther north. These movements might best be termed emigrations. Like other tortoise shells, the Compton is often found on moist soil, often along forest roads.

Broods. The Compton Tortoise Shell has a single annual generation. Adults emerge in early July and are found until October or early November. During the fall, the butterflies enter hollow trees and vacant outbuildings, wherein they hibernate until the following spring (Proctor 1976). Overwintered adults are found from March to early June. It is at this time of year that mating and egg-laying occur. The adults live at least 9 months.

Early Stages. Eggs are laid in batches on the host during spring, and the caterpillars feed gregariously. The mature caterpillar is light green with the normal black, bristly setae characteristic of most nymphalids. The pupa is green and delicately reticulate, while the dorsal tubercles are golden.

Adult Food. Klots (1951) reports the adults are fond of sap and rotting fruit. The overwintered adults may visit willow flowers.

Caterpillar Host Plants. Host plants include aspen (*Populus*), gray birch (*Betula populifolia*), paper birch (*B. papyrifera*), and willows (*Salix*).

California Tortoise Shell
Nymphalis californica (Boisduval), 1852
MAP 135

Synopsis. The California Tortoise Shell, a western butterfly, has periodic population explosions that result in vast outward emigrations. During the "up" years, individuals have appeared in Vermont, 2000 miles east of the species' usual range. It has appeared in at least eight eastern states. Its native range is from British Columbia south to southern California and south through the Rocky Mountains to New Mexico. The California Tortoise Shell has one generation each year, emerging in midsummer, but the adults emigrate and overwinter. Its only confirmed larval hosts are various species of wild lilac (*Ceanothus*).

Mourning Cloak
Nymphalis antiopa (Linnaeus), 1758
PLATE 28 · FIGURES 167 & 168 · MAP 136

Etymology. The species is named after Antiopē, leader of the Amazons (Greek myth).

Synopsis. The adult of this species is probably the longest-lived butterfly in North America, often surviving in excess of 10 months. It also is one of the more conspicuous widespread large butterflies in North America and is familiar to most outdoor enthusiasts.

Butterfly. Male forewing: $\overline{X}$ = 3.9 cm, range 3.6–4.1 cm; female forewing: $\overline{X}$ = 4.3 cm, range 3.9–4.6 cm. With its dorsal purplish black ground color, broad yellow margins (white in overwintered individuals), and subtending row of violet spots, the appropriately named Mourning Cloak is like no other butterfly. The sexes are alike in appearance. It varies little geographically.

Range. The Mourning Cloak occurs almost everywhere throughout temperate and subarctic North America and ranges south to central Mexico in the highlands. In the East it is found everywhere, but it is rare in the South. In Florida and the Gulf states, the majority of records represent overwintering individuals. The Mourning Cloak also occurs in temperate Eurasia, where it is known as the Camberwell Beauty.

Habitat. The butterfly is found in wooded, open, and suburban areas everywhere that suitable hosts occur. Riparian situations are favored. Since the butterflies roam widely and undergo migrations (Teale 1955), there are probably few spots that have not been graced by a Mourning Cloak.

Life History. *Behavior.* Overwintered male Mourning Cloaks are found during spring in sunny woodland clearings, where they sit on the ground or on low branches while awaiting receptive females. This perching activity takes place from midday to late afternoon (1130–1700 hr), and mated pairs have been seen from 1400 to 1500 hr. At most times of year, adult Mourning Cloaks, which may live in excess of 10 months, engage only in resting and feeding activities. After their emergence in early summer, some feeding occurs for a few days, but the adults quickly seek out hollow trees or similar situations wherein they estivate (Young 1980). In the fall they range abroad again, as they feed in order to store energy for winter hibernation. At this time an unknown portion of each population undergoes a diffuse southward migration. As is true for most angle wings and tortoise shells, Mourning Cloaks warm themselves as necessary through dorsal basking, usually on the ground. During hot periods, the butterflies usually perch on tree trunks with their wings closed. In this position the butterflies are cryptic and escape the eyes of most would-be predators.

Broods. Although some authors report two or three generations, it is very likely that the Mourning Cloak is normally univoltine. Emergence takes place during June to early July, depending on latitude and elevation. Adults may be seen as late as November, and they may emerge from hibernation on warm, sunny winter days. The adults are probably in a state of reproductive arrest (diapause) until midwinter.

Early Stages. The female lays her eggs in a batch on a host branch or twigs, and the young larvae live in a communal web and feed gregariously on young foliage. The mature cater-pillar is black, with white marks and a row of dull red spots up the back, and contrasting red-orange prolegs. There are several rows of bronze-black spines. The chrysalis is dark yellow-brown with black-tipped dorsal tubercles and is over-laid with a pale blue-gray bloom.

Adult Food. Mourning Cloaks favor sap flows of trees, especially oaks. Trees with fresh sap flows kept active by sapsuckers may be visited daily, usually in the morning (Young 1980). The Mourning Cloaks alight above the flow and walk down the trunk to it. They feed head downward. They may also feed on fermenting fruit. Use of floral nectar is limited, but they may occasionally be found at common milkweed, dogbane, mountain andromeda, New Jersey tea, and others.

Caterpillar Host Plants. The usual hosts are willows—including black willow (*Salix nigra*), weeping willow (*S. babylonica*), and silky willow (*S. sericea*)—American elm (*Ulmus americana*), paper birch (*Betula papyrifera*), aspen (*Populus tremuloides*), cottonwood (*Populus deltoides*), and hackberry (*Celtis occidentalis*). The mature larvae wander and may be found on plants that are never fed upon.

135. *Nymphalis californica*

136. *Nymphalis antiopa*

Milbert's Tortoise Shell
Nymphalis milberti (Godart), 1819
PLATE 29 · FIGURE 169 · MAP 137

Etymology. The species is named after a Mr. Milbert, a friend of Godart, who collected North American butterflies in 1826. His collection is now in the Museum of Natural History in Paris.

Synopsis. The Milbert's Tortoise Shell is a close relative of the Old World Small Tortoise Shell (*Nymphalis urticae*). Both species feed on nettles.

Butterfly. Male forewing: $\bar{X}$ = 2.3 cm, range 2.1–2.5 cm;

female forewing: $\overline{X}$ = 2.5 cm, range 2.3–2.8 cm. With its dorsal clear-cut black-blue spotted margins, broad orange and yellow postmedian band, and black basal area, Milbert's Tortoise Shell could be confused with only the California Tortoise Shell (*Nymphalis californica*), which is an extremely rare immigrant in the East. The sexes are alike.

Range. Milbert's Tortoise Shell inhabits the Canadian Life Zone from Newfoundland west to the Rocky Mountains and Sierra Nevada of California. It ranges north to British Columbia. It may take temporary residence in the Transition Life Zone, and in the East it occasionally strays south to Arkansas, North Carolina, and Georgia.

Habitat. In the East, Milbert's Tortoise Shell is found in moist pastures and fields adjacent to woodlands. Wet areas with nettles must be nearby. In the western mountains, habitats selected are usually adjacent to streams but may be treeless. Adults may range far from suitable hosts.

Life History. *Behavior.* As is true for related species, Milbert's is a perching species. Scott (1975) reports that males perch behind shrubs on hilltops or on logs next to gulley banks while awaiting receptive females. Perching usually occurs from 1230 hr until late afternoon. Adults hibernate in hollow trees and abandoned outbuildings.

Broods. Shapiro (1966) reports four broods in the Delaware Valley (May, July, late August, and late September–early October), but some of these appearances may represent emergences from summer estivation, as in the Mourning Cloak. It is most likely that two broods are the rule. Overwintered adults appear in March and newly emerged adults first appear in May or June. Late-season individuals may be found on occasion until mid-November.

Early Stages. The pale-green eggs are laid in large groups of up to 900 under host leaves, and the resultant first- and second-instar caterpillars feed gregariously within a web.

Older larvae feed solitarily and live in folded leaf shelters. The caterpillars are variable but generally are black with lateral greenish yellow and subdorsal orange stripes; the body is also scattered with spots and branched black spines. The chrysalis is either gray or pale golden green flecked with dark brown.

Adult Food. Milbert's Tortoise Shells visit flowers, e.g., lilac, thistles, and goldenrods, much more avidly than their close relatives, but they probably also feed on sap flows and rotting fruit.

Caterpillar Host Plants. This species feeds exclusively on nettles (*Urtica dioica U. gracilis*). There are reports of other hosts, but they are probably incorrect.

137. *Nymphalis milberti*

American Painted Lady
Vanessa virginiensis (Drury), 1773
PLATE 29 · FIGURES 170 & 171 · MAP 138

Etymology. The genus may be derived from a combination of the Dutch *van*, "from," and the Greek *nesos*, "island." (Also, Vanessa is a feminine name.) The species is named for Virginia.

Synopsis. Although similar in general appearance to the Painted Lady, the American Painted Lady is only occasionally migratory and is resident throughout the East.

Butterfly. Male forewing: $\overline{X}$ = 2.7 cm, range 2.4–3.0 cm; female forewing: $\overline{X}$ = 2.8 cm, range 2.4–3.1 cm. The American Painted Lady may be distinguished from the Painted Lady (*Vanessa cardui*), its close relative, by its two large black-ringed blue eyespots in the postmedian area of the ventral hind wing. As is true for the Painted Lady, there are

two seasonal forms of the American Painted Lady. The fall-winter form is generally smaller and paler, with reduced black markings, while the summer form is larger and more brightly colored.

Range. The American Painted Lady ranges from southern Canada south through the entire United States as well as Central America (mountains) to northern South America (Colombia and Galapagos Islands). It also occurs rarely on the larger Caribbean islands. It may not be capable of overwintering in the northernmost states.

Habitat. The American Painted Lady favors open areas with low vegetation, including meadows, flats near streams, vacant lots, and beach dunes.

Life History. *Behavior.* Males perch on the leeward side of hilltops or on low vegetation in flat areas when there are no hills nearby. Perching activity lasts from midday to late afternoon, while mated pairs are seen from 1520 hr to dusk. Oviposition occurs from late morning to early afternoon (1010–1300 hr). The butterflies spend considerable time on bare ground with their wings almost fully open. Freshly emerged males imbibe moisture at damp spots along dirt roads and streams.

Broods. The number of yearly emergences of the American Painted Lady seems to vary from two in the northern part of its range to three or four in the South. In the Delaware Valley, where there are three broods, flights occur in late May to mid-June, late June to July, and from mid-August to early November. Farther south, in Mississippi and Florida, adults have been seen in every month, but it is not definite whether the species is reproductive there during the winter.

Early Stages. The delicate yellow-green eggs are laid singly on upper leaf surfaces under the tomentum. The caterpillar lives and feeds solitarily within a leaf nest that is tied together with silk and bits of interwoven host material. The caterpillar has harlequin coloration—black with circumferential white bands, each thinly lined with black—and has two round white spots amid the black on each of the first six segments. There are four longitudinal rows of black, branched spines; each spine projects from a broad, deep-red base. The chrysalis is either gray-white marked with brown and overlaid with an olivaceous tinge, or bright golden green marked with purple-brown.

Adult Nectar Sources. The American Painted Lady relies almost entirely on floral nectar. In Virginia, spring individuals favor winter cress; in summer, dogbane is selected most often; and in fall several composites, notably goldenrod

and aster, are favored. A variety of other flowers are occasionally visited, for example, marigold, zinnia, buttonbush, sweet pepperbush, selfheal, common milkweed, vetch, and privet.

Caterpillar Host Plants. The larvae feed on leaves of the aster family's everlasting tribe, including sweet everlasting (*Graphalium obtusifolium*), pearly everlasting (*Anaphalis margaritacea*), and plaintain-leaved pussy toes (*Antennaria plantaginifolia*). Other members of the family, such as wormwood (*Artemisia*), burdock (*Arctium*), and ironweed (*Vernonia*), are used on occasion, but reports of feeding on plants of other families require confirmation.

138. *Vanessa virginiensis*

Painted Lady or Cosmopolitan
Vanessa cardui (Linnaeus), 1758
PLATE 29 · FIGURES 172 & 173 · MAP 139

Etymology. The species name is derived from the Latin *carduus*, "thistle."

Synopsis. This is the most widely distributed butterfly in the world (see Range). It is well known for its periodic mass migrations.

Butterfly. Male forewing: $\overline{X} = 3.0$ cm, range 2.5–3.3 cm; female forewing: $\overline{X} = 3.2$ cm, range 3.0–3.4 cm. The Painted Lady is most similar to the American Painted Lady (*Vanessa virginiensis*) but may be distinguished by its small, indistinct submarginal eyespots on the ventral hind wing. This species is reported to have two forms (Clark 1932). The summer form is larger and brighter and has blue pupils in the submarginal spots on the dorsal hind wing. The winter or

migratory form is smaller and duller and has entirely black wing spots.

Range. The Painted Lady occurs throughout Eurasia, Africa, North America, and Central America. It is not a permanent resident in the eastern United States but is a periodic colonist from the deserts of northern Mexico. It occasionally survives the winter in the North.

Habitat. This butterfly prefers open areas, including old fields, vacant lots, and gardens. Because of its migratory movements, it may be seen almost anywhere.

Migration. In western North America, in some years great northward emigrations of Painted Ladies emanate from northern Mexico during the spring (Williams 1970). On occasion, migrations are seen in the East, as in Iowa and Illinois during 1966 to 1968. The butterflies fly rapidly in a direct line 1.5–2 m above the ground, and pass directly over obstructions, such as buildings or trees, without moving around them. Diffuse southern movements have been reported in the fall on a few occasions, e.g., during 1968 and 1970 in New York and in Georgia. These movements are not the strict migrations undergone by the Monarch but usually result from population explosions in favorable years.

Life History. *Behavior.* Male Painted Ladies perch and sometimes patrol during the afternoon (1330–1930 hr) in their search for receptive females. In the West, males often fly to hilltops, where they perch on shrubs, but in much of the East the males select bare ground in open areas for their perch sites. Mated pairs have been found from 1200 to 1930 hr.
Broods. Since the Painted Lady is not a regular resident in the East, the number of broods may vary from year to year. At most there are three or four generations, but usually only one or two. The butterflies are most common in summer (July–October) but may be entirely absent in the East after harsh winters. Only in the Gulf states can the Painted Lady survive some winters.
Early Stages. The caterpillars live and feed solitarily under loose webbing on leaves of the host. Pupation usually occurs on the host. The caterpillars are variable in color, but are generally yellow-green with black mottling and a black head capsule. The body has several rows of branched spines. The chrysalis is tan or gray, with short, metallic gold projections.
Adult Nectar Sources. The Painted Lady is an avid flower visitor, being partial to composites 1–2 m high, especially thistles, but also blazing star, ironweed, joe-pye weed, aster, and cosmos. Red clover is a favorite, and milkweeds, buttonbush, and privet are visited on occasion.

Caterpillar Host Plants. The female Painted Lady is catholic in her selection of host plants: more than 100 have been recorded (Williams 1970). Plants of the families Asteraceae (especially thistles), Boraginaceae, Malvaceae [especially hollyhock (*Althaea*) and common mallow (*Malva neglecta*)], and various Fabaceae are favored.

139. *Vanessa cardui*

Red Admiral
Vanessa atalanta rubria (Fruhstorfer), 1909
PLATE 30 · FIGURES 175 & 176 · MAP 140

Etymology. The species is named after Atalantē, daughter of Schoineus, King of Boeotia (Greek myth).
Synopsis. The Red Admiral, with its swift flight and nervous habits, is difficult to approach or observe. Although the Painted Lady is considered by some to be in a different genus, hybrids between these two species are known.
Butterfly. Male forewing: $\overline{X}$ = 3.1 cm, range 2.6–3.3 cm; female forewing: $\overline{X}$ = 3.2 cm, range 2.9–3.5 cm. Dorsally, the Red Admiral, with its black wings and red bands along the hind-wing margin and across the forewing, is distinctive. Although the Red Admiral is not very variable geographically, it has distinct summer and winter forms. The summer form is larger and brighter, with an interrupted forewing band, while the winter form is smaller and duller.
Range. The Red Admiral occurs in Europe, Asia Minor, and North Africa, and in North America from northern Canada south through the United States and Mexico to Guatemala. It occurs natively on some Caribbean islands as well as Hawaii and New Zealand, where it was probably introduced. The species cannot survive harsh winters, and many portions of the northern United States and Canada must be recolonized by migrants each spring.
Habitat. The Red Admiral's typical habitat is rich, moist, bottomland woods with suitable host plants, but the butterfly may also be found in disturbed areas near towns and farms. At various times and places it occurs in all Life Zones from Subtropical to Canadian.
Life History. *Behavior.* The Red Admiral is similar to its relatives in that the males perch in the afternoon (Bitzer and Shaw 1981), usually in sunlit forest openings or on forest edges, but it differs in its predilection for shady areas. Oviposition takes place from midday to early afternoon. It has an extremely rapid, erratic flight that is very difficult to follow, particularly in and out of shaded woods. The species is

migratory, and returning adults may be seen flying north each spring as isolated individuals (late March and April in northern Virginia). Thousands were seen in Florida in April and May of 1953 and 1955, while a great migratory flight was observed in Maine during June 1957. Adults hibernate successfully as far north as New York (Shapiro 1974).

Broods. Two broods are the rule throughout most of the Red Admiral's range. Along the northernmost portions of the United States and in Canada, only one generation may be possible, while three may be possible in the southern states. In the Delaware Valley, the first brood appears in mid-June and flies to mid-July, while the second appears in August and September.

Early Stages. Young caterpillars live in a folded leaf shelter, while older ones live and feed within a nest composed of several young leaves tied together with silk. Pupae of the second generation may occasionally overwinter. The cater-pillars are variably colored blackish to yellow-green and have lateral black and yellow stripes. The body has several rows of branched spines. The pupae vary from gray-brown to red-brown, are faintly marked with black, and have golden dorsal tubercles.

Adult Food. Red Admirals feed head down at sap flows on trees, at fermenting fruit, and on bird droppings, as do angle wings and tortoise shells. They seldom visit flowers but may do so when dry conditions prevent suitable sap flows. The flowers selected include common milkweed, dogbane, red clover, alfalfa, Canada thistle, aster, and sweet pepperbush.

Caterpillar Host Plants. Larval hosts selected include members of the nettle family, such as stinging nettle (*Urtica dioica*) and tall wild nettle (*U. gracilis*), false nettle (*Boehmeria cylindrica*), wood nettle (*Laportea canadensis*), pellitory (*Parietoria pennsylvanica*), and possibly hops (*Humulus*).

Mimic
Hypolimnas misippus (Linnaeus), 1764
MAP 141

Synopsis. The Mimic is a butterfly of the Old World tropics (Asia and Africa) and is believed to have been introduced into the Caribbean during the slave trade era. The female has two forms, each of which mimics a form of an Old World Monarch relative (*Danaus chrysippus*).

Butterfly. Male forewing: X = 3.2 cm, range 2.5–3.8 cm; female forewing: X = 4.0 cm, range 3.8–4.2 cm. The Mimic is sexually dimorphic, with the male strikingly different from either of the two female forms. The male is purplish black dorsally, with a large ovoid white area centered on each wing and a white spot in the apical area. The most frequent female form is orange dorsally with a black marginal band on the hind wing and black on the costal and apical half of the forewing. A band of white spots cuts the black apical area transversely.

Range. The Mimic's principal ranges are the tropics and subtropics of Africa and Asia. In the New World it is resident on a number of the Caribbean islands and is found sporadi-

140. *Vanessa atalanta*

141. *Hypolimnas misippus*

cally in the southeastern United States (Florida, Mississippi, North Carolina), usually as vagrants, but occasionally reproducing.

Life History.　*Early Stages.*　The larvae feed gregariously. The mature caterpillar is black or dark gray with lighter gray bands. The head is reddish and bears two long branched horns. There are rows of finely branched white or pale-gray spines along the body.

Caterpillar Host Plants.　Mallows (*Malvaceae*), morning glory (*Ipomoea*), and purslane (*Portulaca*) have been recorded as larval hosts.

Buckeye
Junonia coenia Hübner, 1822
PLATE 30 · FIGURES 177 & 178 · MAP 142

Etymology.　The genus name is possibly derived from Juno, wife of Jupiter (Roman myth). The species name is possibly derived from either Jan Henri de Coene or Constantinus Fidelio Coene, both important French painters.

Synopsis.　The beautiful eyespotted Buckeye belongs to a small group of species that have beguiled and confounded biologists for several decades because of the present species' variability and close relationship with the Caribbean Buckeye (*Junonia evarete*).

Butterfly.　Male forewing: $\overline{X}$ = 2.5 cm, range 2.0–2.8 cm; female forewing: $\overline{X}$ = 2.8 cm, range 1.7–3.1 cm. This species and the Caribbean Buckeye may be distinguished from other eastern United States butterflies by the combination of the two red-orange bars in the forewing cell and the three submarginal eyespots (one on forewing, two on hind wing) on the dorsal wing surfaces. Where the two species occur together in southern Florida, the Buckeye may usually be distinguished by its significantly larger uppermost hind-wing eyespot, and because the pale area that cuts across the apical area of the forewing and partially encircles the eyespot there is uniformly white. Intermediates between the two species and inherently wide variability lend uncertainty to the identification of some individuals. A detailed behavioral and numerical study of the two species where they co-occur in coastal Texas demonstrated some behavioral and ecological differences and confirmed the general color pattern separation (Hafernik, 1982).

Adding some confusion is the fact that the Buckeye has seasonal forms. The wet-season (or summer) form is somewhat smaller and is brown or tan ventrally, while the dry-season (or winter) form is larger and brick red or purplish rose beneath. The dry-season form probably undergoes a reproductive diapause where it can survive as an adult through the winter in the South. The species is emigratory; first-brood individuals from the Deep South repopulate more northern areas in the late spring and summer. A small form occurs on the sand dunes of the Carolina coast.

Range.　The Buckeye ranges from southern Canada through most of the United States to northern Mexico. Throughout most of its range the species cannot survive the winter, and year-round populations occur only in the southern portions of its range. It can survive the winter on coastal dunes as far north as North Carolina.

Habitat.　The Buckeye requires a combination of open areas with low vegetation and at least some areas of bare ground. Examples of suitable situations include roadsides, dry fields with dirt roads, power-line cuts, open pine flatwoods, beach dunes, and railroad tracks. The species is resident in Lower Austral and Tropical Life Zones, but colonizes Upper Austral and Transition Life Zones each summer.

Life History.　*Behavior.*　Male Buckeyes perch on bare ground in fields and on trails, periodically patrolling back and forth, only to return to the general vicinity of their last perch. If a male pursues another male or some other moving object more than 30 m, he is unlikely to return. During cool weather adults perch with outspread wings, and in hot weather they sit with tightly closed wings oriented to intercept as little direct radiation as possible.

J. A. Scott (1975) describes courtship and mating behavior of the Buckeye. Males must be stimulated by a female in flight. At first, the male hovers above and just downwind from the female, then alights behind her and beats his wings somewhat more slowly than before but with greater amplitude. Then the male may nudge the female until she raises her wings. Finally, he moves up beside the female and curves his abdomen toward her until coupling occurs. The female may make several short flights with the male in pursuit before coupling actually occurs. The female indicates rejection by a posture of spread wings and raised abdomen. Mating lasts and an average of 27 minutes. Females usually mate only once in their lifetimes, but some mate twice. Rarely, a female may be mated three times. Perching by males, courtship, and mating all occur continually between 0815 and 1500 hr.

Broods.　Buckeyes normally have two or three generations in the northern and central states but may have more in the Deep South, where they are resident. In the Delaware Valley there are two emergences of the summer form (late May–June and mid- to early August), and the winter form emerges in September and early October. In Mississippi, Georgia, and Florida, the Buckeye is found all year. It is likely that winter-form adults there are in reproductive arrest from late fall to early spring. Thus the potential longevities of the two forms are probably very different.

Early Stages. The caterpillars feed solitarily on leaves of their host. The caterpillar is largely blackish, with broken lateral and dorsal yellow lines. Its head is orange and black with fleshy orange projections. The body has four rows of iridescent blue-black spines, the lateral rows originating from orange knoblike projections (chalazae).

Adult Nectar Sources. Buckeyes favor composites, including such species as chickory, knapweed, tickseed sun-flower, gumweed, and aster. Other flowers, such as dogbane and peppermint, are visited.

Caterpillar Host Plants. Members of the snapdragon family, plantain family, and acanth family are usual larval hosts. These include snapdragon (*Antirrhinum*), false foxglove (*Aureolaria*), toadflax (*Linaria*), plantain (*Plantago lanceolata* and *P. rugelii*), and ruellia (*Ruellia nodiflora*). Verbena (Verbenaceae) is an occasional host, while reports of stonecrop (*Sedum*) and false loosestrife (*Ludwigia*) must be seriously doubted.

Caribbean Buckeye
Junonia evarete genoveva (Stoll), 1782
PLATE 29 · FIGURE 174 · MAP 143

Synopsis. The Caribbean Buckeye is an extremely variable tropical species that enters the United States in southern Florida, southern Texas, New Mexico, and Arizona. It flies with the Buckeye in southern Florida.

Butterfly. Male forewing: $\overline{X}$ = 2.7 cm, range 2.4–2.8 cm; female forewing: $\overline{X}$ = 2.9 cm, range 2.6–3.2 cm. The Caribbean Buckeye usually may be distinguished from the more common Buckeye by the approximately equal, small-sized eyespots on the upper hind wing, and the pale transverse apical forewing patch, which often includes much orange. The Caribbean Buckeye has dry- and wet-season forms, but they are not as markedly different as those of the Buckeye. Riley (1975) states that the eyespots on the ventral hind wing are strongly reduced or absent. Some suggest that yet another species of *Junonia* is found near black mangroves in Florida (Baggett, personal communication).

Range. The Caribbean Buckeye ranges from the Subtropical Life Zone in the southern United States (southern Florida in our area) south through Central America and most of the Caribbean islands to South America. In the Tropical Life Zones where it occurs, it usually is confined to areas with a distinct dry season.

Life History. *Behavior.* In Costa Rica the species emigrates to higher, moister elevations during the dry season, when the adults are in reproductive diapause.

Broods. Where it occurs, the Caribbean Buckeye may be found year-round as an adult. The butterfly is primarily nonreproductive in the dry season, and probably has three or four generations in the wet season. In Florida, reproduction probably occurs from March through October. Nonreproductive winter adults presumably live much longer than those that occur in the warmer months.

142. *Junonia coenia*

143. *Junonia evarete*

Early Stages. Overwintered adults mate and lay eggs at the onset of the wet season. Riley (1975) describes the caterpillar as black, with the dorsal portion of the head often yellowish. The body is sprinkled with minute white points and has a yellow-edged dark dorsal line and two lateral rows of orange spots. There are five rows of black or yellow branched spines along the body.

Adult Nectar Sources. Adult food sources have not been recorded for the United States populations; however, in Central America a variety of composites, *Cordia* (Boraginaceae), and *Casearia* (Flacourtiaceae) are favored.

Caterpillar Host Plants. Larval hosts are primarily Verbenaceae and include lippia (*Lippia*), blue porterweed (*Stachytarpheta*), and Acanthaceae, such as ruellia (*Ruellia*).

White Peacock
Anartia jatrophae guantanamo Munroe, 1942
PLATE 31 · FIGURE 181 · MAP 144

Etymology. The genus name may be derived from *Anarta*, "sea cockle" (Pliny). The species is named after *Jatropha*, a tropical plant. The subspecies is named after a bay in Cuba.

Synopsis. The White Peacock belongs to a tropical genus, the species of which are common in weedy situations almost everywhere in tropical and subtropical habitats. Its specific name *jatrophae* is a misnomer, since *Jatropha*, a tropical plant once believed to be its host, is probably not fed on by the larvae.

Butterfly. Male forewing: $\overline{X}$ = 2.6 cm, range 2.5–2.7 cm; female forewing: $\overline{X}$ = 2.9 cm, range 2.7–3.2 cm. The White Peacock is one of two predominantly whitish nymphalids in the East, and the only one that occurs in Florida. It has six distinct, round, black submarginal spots, two on each hind-wing and one on each forewing. The hind-wing margin is slightly scalloped, and there is pale orange and brownish infuscation, especially along the margin. Two seasonal forms occur: a larger, paler, dry-season (or winter) form and a smaller, darker, wet-season (or summer) form. As in other tropical species with dry forms, it is presumed that the dry form is longer-lived and undergoes reproductive diapause. The White Peacock is sometimes emigratory in the fall.

Range. The White Peacock ranges from the coastal area of South Carolina and southern Texas south through Florida, the Caribbean islands, and the lowlands of Mexico and Central America to Venezuela. Individuals occasionally stray northward to Cape Hatteras, North Carolina.

Habitat. Rawson (1976) has characterized the haunts of the White Peacock in Florida as low ground along shallow ditches and the borders of ponds and streams. Wherever found, this butterfly prefers either open flat areas adjacent to water courses, places that are seasonally flooded, or sites with a high water table, such as sand-dune deflation plains.

Life History. *Behavior.* Male White peacocks perch close to the ground on low vegetation and await the appearance of receptive females. They react to almost any passing small object and often interact with other males with a dancing, twisty flight. Males seldom return to the perch after such interactions. Mate-location activity takes place during most daylight hours. The butterfly's normal flight is slow but erratic. Rawson (1976) reports that ovipositing females lay a single egg on a plant, and then fly some distance away before depositing another.

Broods. White Peacock adults occur year-round. The number of broods of the wet form is unknown. The dry-form adults are presumed to be in reproductive arrest from the late fall to early spring.

Early Stages. Early stages have been described in some detail by Rawson (1976). Eggs are laid singly on the underside of host leaves. The egg is pale yellow and barrel-shaped, and is 0.16 mm in height. The egg hatches in about 4 days, and the young larva consumes the eggshell. There are three larval instars, each stadium lasting about 4 days. The mature larva is 27–30 mm long and is dark brown to black, with the ventral surface dull yellow-brown. There is a row of small silver spots on the front end of each segment. The broadened bases of spines and prolegs are dull orange. There are five longitudinal rows of black, branched spines running the length of the body. The head is black, with two long dorsal projections, the

144. *Anartia jatrophae guantanamo*

ends of which are clublike. The pupa is smooth, un-ornamented, and pale green, with a thin black middorsal line and two to four light spots on each segment's dorsal surface. The adult emerges in 7 to 10 days.

Adult Nectar Sources. Shepherd's needle is a favorite nectar plant in Florida. In Central America, *Cordia* (Boraginaceae), *Casearia* (Flacourtiaceae), and several composites are selected.

Caterpillar Host Plants. It is extremely unlikely that *Jatropha manihot*, a member of the Euphorbia family, is ever fed upon by the caterpillars of this butterfly. Instead, members of the verbena family—*Lippia*—and the acanth family—water hyssop (*Bacopa monniera*) and ruellia (*Ruellia occidentalis*)—are the usual hosts.

Hübner's Anartia
Anartia chrysopelea Hübner, 1824

Synopis. Hübner's Anartia is endemic to Cuba and the Isle of Pines, but has strayed to the lower Florida Keys (Key West, Big Pine Key). With its brown wings and white medial bands, it is completely unlike any other butterfly likely to be found in Florida.

Malachite
Siproeta stelenes biplagiata (Fruhstorfer), 1907
PLATE 30 · FIGURES 179 & 180 · MAP 145

Synopsis. The large tropical Malachite is a recent colonist in southern Florida. It may be a mimic of a tropical American heliconiine, the Bamboo Page (*Philaethria dido*), which has a remarkably similar color pattern.

Butterfly. Male forewing: $\overline{X}$ = 4.1 cm, range 3.8–4.3 cm; female forewing: $\overline{X}$ = 4.4 cm, range 4.2–4.7 cm. The Malachite is the only large nymphalid in our area that is mostly pale, translucent jade green, particularly ventrally. There are two seasonal forms, the reproductive wet-season (summer) form being smaller, with blackish smudged marks between the postmedian green band and the orange marginal band below. The dry-season (winter) form is larger and has silvery marks replacing the black smudges. The green band seems more extensive in the dry form.

Range. The Malachite occurs from extreme southern Florida and southern Texas south through the Caribbean islands, Mexico, and Central America to Brazil.

Habitat. In Florida, the Malachite is found in mango, citrus, and avocado groves where its weedy host plant is common. In Central America it is most common in seasonally dry evergreen or semideciduous forests, particularly those adjacent to rivers.

Life History. *Behavior.* Males perch on shrubby vegetation in forest (orchard) openings and occasionally patrol back and forth with a slow, sailing flight. Adults normally rest with their wings open at about a 60° angle. At night they rest in loose aggregations under the leaves of low shrubs. They may spend much of the dry season (winter in Florida) in similar aggregations. In Costa Rica the newly emerged dry forms move to higher, moister elevations for the duration of the dry season.

Broods. The number of generations is not well documented, but there are probably two or three during the warmest, wet portion of the year, with the dry-season morph overwintering as an adult (October–March in Florida).

Early Stages. Early stages have been described by Bates

145. *Siproeta stelenes*

(1923). The eggs are laid singly on leaves of the small herbaceous host. The mature caterpillar is velvety black with dark-red or purple segmental divisions. The prolegs are pink, and there are two large, spined horns on the head. The segments bear four to seven warts on branched spines.

Adult Food. The Malachite favors rotting fruit but occasionally feeds on bird droppings and at flowers. Flowers of trees and lianas are preferred, but herbaceous plants, such as lantana, and composites are utilized on occasion. Malachites may visit flowers up to 12 m high and may feed from dawn to dusk.

Caterpillar Host Plants. The larval host is usually cajetin (*Blechum brownei*, family Acanthaceae), but ruellia (*Ruellia coccinea*) of the same family has also been reported.

White Admiral
Limenitis arthemis arthemis (Drury), 1773

Red-Spotted Purple
Limenitis arthemis astyanax (Fabricius), 1775

PLATE 31 · FIGURES 182–185 · MAP 146

(horizontal hatching—*arthemis;* diagonal hatching—*astyanax;* dark pattern—blend zone).

Etymology. The genus name is derived from the Latin *limen*, "threshold, entrance of house." Arthemis was a goddess, sister of Apollo, who was represented as a huntress and is associated with the moon (Greek myth), while Astyanax was the son of Hector and Andromache (Greek myth).

Synopsis. Formerly considered to be two species, *Limenitis arthemis astyanax* (Red-spotted Purple) and *Limenitis arthemis arthemis* (White Admiral) are known to hybridize freely in a belt extending from southern New England across the Great Lakes states. They are now considered a single species. The Red-spotted Purple is an excellent Batesian mimic of the Pipe Vine Swallowtail (*Battus philenor*).

Butterfly. Male forewing: $\bar{X}$ = 3.8 cm, range 3.5–4.1 cm; female forewing: $\bar{X}$ = 4.4 cm, range 3.5–4.8 cm. The two geographic forms of *arthemis* are so disparate in appearance that, despite the genetic evidence, it is still difficult to accept the fact they are the same species. Typical *astyanax* is black dorsally with purplish blue on the outer portion of the hind wings. The ventral forewing surface is largely iridescent blue; there are two orange costal forewing spots and a submarginal row of orange spots on the hind wing. The iridescent blue dorsal hind wing, and the ventral orange submarginal spot band are particularly important elements in its mimetic resemblance of the Pipe Vine Swallowtail (Platt and Brower 1968). The White Admiral has broad white postmedian bands on both wings above and below. In addition, there is a submarginal row of red spots on the dorsal hind wing and the wings are largely purplish red (except for the white band below). Frequent intermediates displaying various character combinations of the two forms may be found in or adjacent to the zone of overlap. Occasional white-banded individuals appear as far south as Virginia and Kentucky.

Range. The species occurs from the Maritime Provinces of Canada west and north across Canada to central Alaska. The butterfly is found in the United States east of the Rocky Mountains wherever forests occur and south to central Florida and central Texas. Populations also occur in southern Arizona and southern New Mexico. The White Admiral form generally inhabits Canadian Life Zone areas south to a line through north central New England, New York, Pennsylvania, Michigan, and Minnesota. The Red-Spotted Purple form occurs south of the above zone in Transition and Austral Life Zones.

Habitat. The White Admiral form is most often found in northern deciduous or mixed woods dominated by birch or aspen, while the Red-spotted Purple form is found in a variety of young to mature deciduous or mixed deciduous–evergreen woodlands. It is most common in moist upland, valley bottom, or coastal plain situations.

Life History. *Behavior.* In contrast to the Viceroy, Red-spotted Purple males normally rest 10 m or more above the ground on tree foliage, often along forest roads or in glades. Periodically, they patrol back and forth. They do not seem as faithful to particular sites as the Viceroy. Shapiro (1974) reports that mate location and mating take place from midday to midafternoon.

Females have been observed ovipositing at midday (1200–1300 h). The female flutters from branch to branch of the selected tree anywhere from 4 to 10 m above the ground. She usually selects the leaf at the end of a branch and lands just above it. Then, while backing up and probing with her abdomen, she lays an egg at the very tip.

Male *arthemis* of both forms are frequent visitors to moist soil and sand, and even wet pavement. Although this activity is prevalent among freshly emerged individuals, worn males are seen at moisture on occasion. While perched, the Red-spotted Purple males slowly raise and lower their wings. This may be an example of behavioral mimicry, since Pipe Vine Swallowtails usually flutter their wings when visiting flowers. The closely related Viceroy does not raise and lower its wings when at moisture.

Broods. The White Admiral is usually double-brooded wherever found. Adults appear as early as late May or mid-June, with a peak flight in late June and the first half of July. The second generation is usually partial, with most larvae resulting from first-brood eggs entering larval diapause. Adults of the second brood may fly as late as late August to early September. An extreme date of September 20 was reported for Michigan's upper peninsula. The Red-spotted Purple most often has two full broods and a partial third. Virginia flight dates are fairly typical (mid-May June, early July–August, late August–mid-October). Emergence occurs earlier in the South, with early dates of March 24 for Florida and April 10 for Georgia.

Early Stages. The eggs are round and very pale green when laid. The larvae and pupae are very similar to those of the Viceroy, but may be distinguished by the number of spiracular tubercles.

Adult Food Sources. Red-spotted Purples have been seen feeding at carrion, dung, rotting fruit (apples, pears), and sap flows. They visit flowers much less frequently than the Viceroy does. Females are the sex most often seen nectaring, usually at woody plants with small white flowers, such as privet, viburnum, sweet pepperbush, Hercules' club, and spiraea. Heights of 2 to 4 m are usual for such visits. White Admirals also utilize aphid honeydew and may visit flowers more frequently.

Caterpillar Host Plants. The Red-spotted Purple larva selects wild cherry (*Prunus serotina* and *P. virginiana*) most

frequently, but also feeds on poplars and aspens (*Populus alba, P. grandidentata, P. tremuloides,* and *P. deltoides*) as well as black oaks (*Quercus velutina, Q. ilicifolia,* and *Q. phellos*). There are also records for gooseberry (*Ribes*), hawthorn (*Crataegus*), and deerberry (*Vaccinium stamineum*). Reports of willow require confirmation. White Admiral larvae use black and yellow birch (*Betula lenta* and *lutea*) or aspen (*Populus*) most often, but have also been reported to select hawthorn (*Crataegus*), American hornbeam (*Carpinus caroliniana*), basswood (*Tilia*), and shadbush (*Ameliancher*).

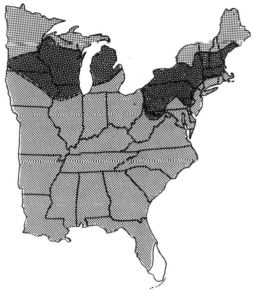

146. *Limenitis arthemis arthemis* & *Limenitis arthemis astyanax*

Viceroy
Limenitis archippus (Cramer), 1776
PLATE 32 · FIGURES 187 & 188 · MAP 147

Etymology. The species is named after Archippus, a Greek poet or comedian.

Synposis. The Viceroy's most notable feature is the close similarity of its color pattern to that of the distasteful Monarch. Thus, it is a Batesian mimic and presumably suffers less predation by birds as a result (Platt et al. 1971). In portions of its range, where the Monarch is rare, the Queen is the model, and the Viceroy's pattern has evolved to more closely match that species.

Butterfly. Male forewing: $\bar{X}$ = 3.6 cm, range 3.1–3.7 cm; female forewing: $\bar{X}$ = 3.8 cm, range 3.3–4.0 cm. Throughout most of its range in the East, both sexes of the Viceroy closely resemble the Monarch (*Danaus plexippus*). It may be dis-

tinguished by its possession of a postmedial black band on the hind wing, and by its possession of one row of white spots within its black marginal forewing and hind wing bands, instead of two as in the Monarch. In Florida and adjacent portions of Georgia, the subspecies *floridensis* has a deep mahogany brown color and more closely resembles the Queen (*Danaus gilippus*). The white spots in the hind wing marginal band of the Florida populations are almost obsolete as they are in the Queen. Sterile male hybrids resulting from crosses between the Viceroy and the Red-spotted Purple or White Admiral are rarely found in nature.

Range. The Viceroy ranges from central Canada south through many low-lying moist areas of the United States east of the Sierra Nevada and Cascade Mountains to northern Mexico. The species occurs from Tropical through Transition Life Zones.

Habitat. The Viceroy occurs in many kinds of open or shrubby areas, which are usually devoid of much topographic relief and are wet or adjacent to wet areas, such as, for example, wet meadows, marshes, margins of ponds and lakes, railroad tracks, and roadsides.

Life History. *Behavior.* Male Viceroys perch on the ground or on low vegetation no higher than 1.5 m and periodically patrol back and forth over about a 20 m distance. They usually return to the same or a similar perch. Apparently, some perch and patrol areas are ideal, since different males will occupy the same perch over successive generations and years. In northern Virginia, one marked male was found perching in almost exactly the same spot on several occasions over a 3-week period. The perch-patrol behavior takes place during most daylight hours. The Viceroy has a slow flap-and-glide flight similar to that of other *Limenitis* species. Upon encountering each other, two males will abruptly soar 50 m or more into the air. Usually one or both will return to the same area. Mating has been observed in the afternoon (1300–1700 hr), with the female as the carrier. Females lay eggs during midafternoon (1305–1520 hr), and the eggs are laid singly at the very tip of the host leaves. If the tip of a leaf has been eaten by another insect, the Viceroy female will move on before laying another egg. She usually selects low saplings and deposits only two or three eggs on each plant before flying on. Newly emerged males sit on moist sand or mud to imbibe moisture.

Broods. In the northern United States the Viceroy is bivoltine. For example, flight extremes for Maine are June 5–September 16, and for Michigan's upper peninsula, June 26–September 16. Throughout most of its range the Viceroy has three generations, or at least two and a partial third, with some larvae of the second generation entering diapause until the following spring (Clark and Platt 1969). In the lower Delaware Valley, flight extremes for the three generations are mid-May to mid-June, July to mid-August, and late August to late September. Farther south, the Viceroy flies for a greater portion of the year. In Mississippi, dates range from mid-April to early December, and in Florida the species is found all year except in cold weather.

Early Stages. The spherical, pale-green eggs are laid singly at the extreme tip of host leaves. The caterpillar eats the eggshell after its emergence. Larvae resulting from first-brood adults and some resulting from the second feed on leaves of the host at night, but rest under leaves or twigs during the day. The young larvae leave a dangling ball of leaf pieces and grass attached to a more basal portion of the leaf upon which they are feeding. This ball may reach the size of a small pea. Some first-instar larvae resulting from second-brood adults and all larvae from the partial third generation construct a curious, rolled leaf-tip shelter within which they spend the winter.

The shelter is separated from the remainder of the leaf by the bare midrib, from which the caterpillars have eaten all the adjacent leaf blade.

The Viceroy larvae are excellent imitations of fresh bird droppings. They are generally olive green with a white dorsal splotch, white sides, and a few black spots. There is a pair of slightly spiny antennalike structures arising from the thoracic dorsum. In addition, there are a number of knobby protuberances along the back. The rows of spines prevalent on other nymphalid larvae are absent. Clark (1932) states that Viceroy caterpillars found on willows are predominantly greenish (as described above), while those on cottonwood are darker and largely brownish. The pupa is black-brown, with the abdominal portion largely white. The dorsal portion of the thorax is raised, and there is a large spatulate protuberance arising from the anterodorsal portion of the abdomen.

Adult Food. Viceroys' tastes vary seasonally. Early in the season, first-brood individuals are seldom seen at flowers and instead utilize carrion, decaying fungi, aphid honeydew, and animal dung and take moisture at wet spots. Individuals of later generations, which emerge when conditions are warmer and drier, are more often seen at flowers, usually of tall, erect plants with terminal inflorescences. Composites appear to be favored, including joe-pye weed, aster, goldenrod, Canada thistle, and shepherd's needle. Flowers of other plants are also utilized on occasion.

Caterpillar Host Plants. All larval hosts belong to the family Salicaceae. They include willows (*Salix carliniana, S. cordata, S. discolor, S. nigra,* and *S. longifolia*) and poplars (*Populus alba, P. deltoides, P. nigra* var. *italica,* and *P. tremuloides*).

147. *Limenitis archippus*

Dingy Purple Wing
Eunica monima (Stoll), 1782
MAP 148

Synopsis. This tropical butterfly is not resident in our area, but on rare occasions individuals appear in southern Florida as immigrants from Cuba. The species' native range includes the West Indies and the mainland from Mexico south through Central America. It is smaller, has more rounded wings, and is dingier than the Florida Purple Wing (*Eunica tatila tatilista*). One of its caterpillar hosts is a tropical prickly ash (*Zanthoxylum pentamon*).

Florida Purple Wing
Eunica tatila tatilista Kaye, 1926
PLATE 32 · FIGURES 189 & 190   MAP 149

Etymology. The genus is named after Eunicus, a Greek comedian (fifth century B.C.).

Synopsis. The Florida Purple Wing is the only member of its group that is resident in the United States. Its habitat on the Florida Keys is seriously threatened by development. Large numbers of purple-wing species are known to occur throughout the American tropics.

Butterfly. Male forewing: X̄ = 2.6 cm, range 2.3–2.9 cm; female forewing: X̄ = 2.7 cm, range 2.5–2.9 cm. This is the only medium-sized brownish black nymphalid with a pattern of white spots on the apical portion of the forewing. There is a postmedian row of black spots on each ventral hind wing. Dorsally, the wings have a bluish-purple iridescent sheen. The only similar species is the smaller Dingy Purple Wing (*Eunica monima*), which has smoother wing margins, fewer, duller spots on the forewing, and less pronounced iridescence.

Range. The Florida Purple Wing is restricted to extreme southeastern Florida, the Florida Keys, and the larger Caribbean islands. It is most abundant on Cuba. The typical species, *Eunica tatila*, ranges from Mexico south through the Central American lowlands, especially in areas with a pronounced dry season.

Habitat. The Florida Purple Wing is a forest butterfly wherever it occurs. In Florida it is restricted to hardwood hammocks. In Central America it is found most often in evergreen forests adjacent to rivers.

Life History. *Behavior.* Little has been reported of the Florida Purple Wing except that it is usually found perching head downward on tree trunks within hardwood hammocks.

148. *Eunica monima*

149. *Eunica tatila tatilista*

It is probably migratory, as are other species in the genus. *Broods*. The number of broods is unknown. In Florida, adults have been seen in every month except June. In the dry lowland forests of Costa Rica, the species is most abundant about a month after the onset of wet-season rains (June–July).

Adult Food. Purple wings feed most often at rotting fruit, dung, and sap flows; they rarely visit flowers. In Central America, adults were seen visiting cordias, *Casearia, Baltimora* (Asteraceae), and a milkweed liana.

Texas Bagvein
Mestra amymone (Ménétriés), 1857
MAP 150

Etymology. The generic name could have been derived from Mestria, a Roman surname, while the species was named for Amymone, daughter of Danaus and grandmother of Palamedes.

Synopsis. This fragile, weak-flying butterfly occasionally strays north into the Plains states from its permanent breeding area in southern Texas. There are very few records of it occurring in the area covered by this book.

Butterfly. Male forewing: $\overline{X}$ = 2.1 cm, range 1.9–2.2 cm; female forewing: $\overline{X}$ = 2.2 cm, range 2.0–2.3 cm. Dorsally, the Texas Bagvein is gray and white with a broad marginal orange band on each hind wing. The forewings of the female are broader and have more extensive white.

Range. This species is resident from southern Texas south to southern Central America, especially in seasonally dry tropical lowlands. Rarely, it emigrates northward to Iowa, Minnesota, and South Dakota.

Habitat. Masters (1970a) reports that the Texas Bagvein is found in open areas, roadsides, and wood edges.

Life History. *Early Stages*. Early stages have not been reported.

Adult Food. The species is reported to visit flowers, but the kinds are not specified.

Caterpillar Host Plant. Texas Bagvein larvae feed on noseburn (*Tragia neptifolia*).

150. *Mestra amymone*

Ruddy Dagger Wing
Marpesia petreus (Cramer), 1776
PLATE 31 · FIGURE 186 · MAP 151

Etymology. The genus is named after Marpessa, a legendary Greek nymph, daughter of Euénos, king of Aitolia. The species is derived from the Latin *petraeus*, "rocky."

Synopsis. This bright orange, long-tailed tropical butterfly is the only breeding dagger wing in the East. Many other dagger wings are found throughout the American tropics.

Butterfly. Male forewing: $\overline{X}$ = 3.8 cm, range 3.5–3.9 cm; female forewing: $\overline{X}$ = 4.2 cm, range 4.0–4.5 cm. The Ruddy Dagger Wing is characterized by its predominantly orange wings, with three thin transverse black lines on each wing; elongate, truncate forewing tips; and, above all, a long black tail projecting posteriorly from each hind wing. The Cuban Dagger Tail (*Marpesia eleuchea*), a stray found rarely in the Florida Keys, is similar, but its forewing apex is less pronounced, and the outermost black line on the forewing is sharply angled. Julia (*Dryas iulia*) is also bright orange and is found in the same habitats. It may be distinguished from the Ruddy Dagger Wing by its long, narrow wings and lack of tails.

Range. The Ruddy Dagger Wing is resident in southern Florida and the mainland of Latin America from Mexico south to Brazil. In the Caribbean, it occurs only on the Windward and Leeward Islands as sporadic migrants.

Habitat. In Florida the Ruddy Dagger Wing is usually found in hardwood hammocks or along the sunny edges of such forests. Throughout its range it is usually a species of evergreen or semi-evergreen forests within seasonally dry tropical lowlands.

Life History. *Behavior.* Ruddy Dagger Wing males perch high (5–10 m) in trees along forest edges or in forest light gaps. The species may be migratory at the end of the wet season, presumably returning to its breeding habitat at the end of the following dry season. Dagger wings are long-lived butterflies, and Benson and Emmel (1973) have reported that related species roost upside down under large leaves, to which they return night after night for long periods.

Broods. Ruddy Dagger Wings are found during most months in southern Florida, but are most common from May through July. It is possible there are only two broods, a reproductive one emerging during the early wet season (June) and a migratory diapausing one emerging during the dry season.

Early Stages. Strohecker (1938) reports that the mature caterpillar is purplish above and white below. The abdominal segments are yellow above, with segments 5, 7, 9, and 11 each bearing a weak upright threadlike appendage. There are two hornlike projections on the head.

Adult Nectar Sources. Adults have been observed feeding at giant milkweed in Florida. In Central America, Ruddy Dagger Wings are found at flowers of trees or large shrubs, such as *Cordia* (Boraginaceae) and *Casearia* (Flacourtiaceae).

Caterpillar Host Plants. Wild banyan tree (*Ficus citrifolia*) and common fig (*F. carica*) are known larval hosts, and other dagger wings are known to feed on plants of the fig family (Moraceae). Cashew (*Anacardium occidentale*) has been reported as a host, but is unlikely.

151. *Marpesia petreus*

Cuban Dagger Wing
Marpesia eleuchea Hübner, 1818

Synopsis. The Cuban Dagger Wing is endemic to the Greater Antilles and has been found once or twice on the Florida Keys. It is very similar to the larger Ruddy Dagger Wing, buts its forewing apex is less falcate and the black median forewing line is distinctly arcuate.

Subfamily Charaxinae: Leaf Wings

The adult leaf wings are characterized by the presence of a thoracic structure termed the *parapatagium*. The adults usually have falcate forewing apices and one or more projections from the hind-wing margin.
The males perch with closed wings while awaiting receptive females. The adults prefer to feed at rotting fruit, sap flows, or dung, but will occasionally take nectar at flowers. Their flight is usually strong, irregular, and rapid.
The caterpillars feed on a variety of woodland trees or shrubs. The caterpillars have granulose bodies with forked terminal segments.

Tropical Leaf Wing
Anaea aidea (Guérin-Ménéville), 1844

Synopsis. The Tropical Leaf Wing was once found as a stray in southern Illinois. Its normal range is extreme southern Texas south through Mexico and Central America. This leaf wing is extremely similar to the Goatweed Butterfly (*Anaea andria*) but is darker dorsally and shows less sexual dimorphism.

Florida Leaf Wing
Anaea floridalis Johnson and Comstock, 1941
PLATE 33 · FIGURES 193–195 · MAP 152

Etymology. The species is named after Florida, the state where it is found.

Synopsis. The Florida Leaf Wing is the only leaf wing found in southern Florida. It has been considered by some to be a subspecies of the Tropical Leaf Wing (*Anaea aidea*), which ranges from extreme southern Texas south through Mexico and Central America.

Butterfly. Male forewing: $\overline{X}$ = 3.4 cm, range 3.3–3.7 cm; female forewing: $\overline{X}$ = 3.6 cm, range 3.3–3.9 cm. The Florida Leaf Wing is similar in size, shape, and appearance to the Goatweed Butterfly (*Anaea andria*), but has a more distinct dorsal pattern, and has a slightly serrate outer forewing margin. It has summer and winter forms that parallel those of the Goatweed Butterfly.

Range. This species is restricted to extreme southern peninsular Florida and the Keys.

Habitat. The Florida Leaf Wing is found only within pine-palmetto scrub or along its margins.

Life History. *Broods.* Two generations per year have been reported. Their timing is probably similar to that of the Goatweed Butterfly. The winter (dry-season) form is probably in reproductive diapause through most of the winter. Adults have been seen during every month in Dade County, Florida.

Early Stages. Early stages have been described by Matteson (1930). The caterpillar is green, with its body covered with many fine white tubercles. There are two lateral yellow stripes on the body, and the head has seven orange and two black tubercles or coronal horns. The chrysalis is pale green and similar in shape to that of the Monarch.

Adult Food. The Florida Leaf Wing may visit flowers on rare occasions, but it probably feeds at rotting fruit and dung most often.

Caterpillar Host Plant. Larvae use wooly croton (*Croton linearis*).

Goatweed Butterfly
Anaea andria Scudder, 1875
MAP 153

Etymology. The species name may be based upon the Greek *andria*, "a woman of Andros." Andros is one of the Cyclades islands in the Aegean Sea.

Synopsis. The Goatweed Butterfly is the only fairly widespread United States representative of a diverse tropical genus. Most species are cryptically patterned beneath and resemble dead leaves.

Butterfly. Male forewing: $\overline{X}$ = 3.1 cm, range 2.4–3.5 cm; female forewing: $\overline{X}$ = 3.4 cm, range 3.1–3.8 cm. This species is typified by the dorsal red-orange ground pattern, the falcate (hooked) forewing apex, and the short, pointed tail projecting rearward from each hind wing. The female is less red and has a distinct, irregular, pale-orange submarginal band. Like many tropically derived butterflies, the Goatweed Butterfly has two seasonal forms (Riley 1980). The male of the summer or wet-season form (April–September) is slightly less reddish and has a barely produced forewing apex and a short tail projecting from the hind wing. The winter or dry-season form male has a strongly produced forewing apex and a much longer tail than the summer form. In addition, the marginal areas above are broadly bordered in black, and there is a medium-black dorsal hind-wing band and a strong dorsal forewing black cell spot. The females of the two forms do not differ as strongly. The Florida Leaf Wing is very similar, but is more patterned and has a slightly serrate forewing outer margin.

Range. The Goatweed Butterfly ranges from West Virginia west to Nebraska and south through the Mississippi

River Drainage to the Gulf Coast. It ranges east along the Gulf to Georgia and the Florida panhandle and south and west to Vera Cruz, Mexico.

Habitat. The Goatweed's habitat is low-lying deciduous woodlands, particularly those along rivers and streams.

Life History. *Behavior.* This is a perching species. Males usually sit in clearings on stones or logs near the ground or on branch tips up to 3 m in height. They also concentrate along ridgetops. Mate-locating activity may take place during most daylight hours. At other times the butterflies perch on tree trunks or limbs with their wings tightly closed. The flight of Goatweed Butterflies is normally rapid and erratic.

Broods. The Goatweed Butterfly is bivoltine. The summer form emerges in early July and is almost immediately repro-ductive. The winter-form generation emerges in early to mid-August and hibernates and remains in reproductive diapause until the next spring.

Early Stages. The caterpillar lives in a folded leaf shelter during its later instars. The mature caterpillar is gray-green and its body is covered with small warty tubercles. Tubercles on the top of the head are slightly longer and are orange. The chrysalis is stout and is reminiscent of the Monarch's. It is pale green and granulated with white.

Adult Food. In general, leaf wings feed on rotting fruit, tree sap, bird droppings, or dung.

Caterpillar Host Plants. The larvae feed on goatweed (*Croton capitatum*), Texas croton (*C. texensis*), and prairie tea (*C. monanthogynus*).

Hackberry Butterfly
Asterocampa celtis (Boisduval and LeConte), 1834
PLATE 32 · FIGURES 191 & 192 · MAP 154

Etymology. The genus name is based upon a combination of the Latin *aster,* "star," and *campa,* "caterpillar." The species is named after *Celtis,* a genus of trees in the elm family, which serves as the butterfly's caterpillar host.

Synopsis. The hackberry butterflies, genus *Asterocampa,* are closely related to those of the American tropical genus *Doxocopa.* They are rapid-flying forest butterflies often found near their host plant, hackberry (*Celtis*). They will often alight on a person to feed on sweat, presumably for its salt content.

Butterfly. Male forewing: $\overline{X}$ = 2.4 cm, range 2.2–2.5 cm; female forewing: $\overline{X}$ = 3.0 cm, range 2.7–3.0 cm. The Hackberry Butterfly is sexually dimorphic, particularly in wing shape, the female having larger, broader wings. The forewing has a single black submarginal spot and there is a pattern of white marks in the apical portion. The Tawny Emperor (*Asterocampa clyton*) lacks the forewing spot and markings, and has a dorsal dark-orange ground color. The Hackberry Butterfly is variable geographically, and some have asserted that there are two separate species in the eastern United States. The other purported entity, the Empress Alicia (*Asterocampa alicia*), is larger and paler, but may be no more than a clinal subspecies. It is found in peninsular Florida.

Range. The Hackberry Butterfly ranges from central New England west to Minnesota and south through the entire

152. *Anaea floridalis*

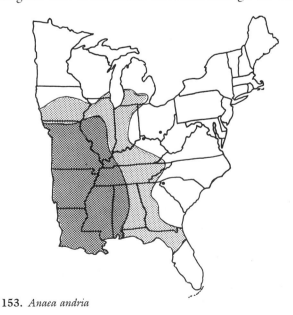

153. *Anaea andria*

eastern United States. It also ranges west along the Gulf through Texas to Arizona and northern Mexico.

Habitat. The usual habitat is along wooded streams, river-edge forest, and woodland glades, but the butterfly also occurs in upland forest if hackberry trees are present.

Life History. *Behavior.* Male Hackberry Butterflies usually perch on foliage of their hosts from early afternoon until dusk, but they may perch on other plants, logs, or rocks. Perches as much as 5 m above ground are utilized. The flight is swift and erratic. When inactive, Hackberry Butterflies perch upside down on tree trunks. Eggs are laid during the early afternoon.

Broods. The Hackberry Butterfly is bivoltine throughout its range. In the lower Delaware Valley of Pennsylvania, the generations occur in July and from mid-August to mid-September, while in Virginia the butterfly flies from the end of May to mid-July and from the end of July to early October. In Georgia the extreme dates for the two broods are May 16–July 10 and August 4–October 6. Probably not all caterpillars from the first generation develop into adults the same year, since at least some enter larval diapause.

Early Stages. The spherical white eggs are laid singly or in small groups under host leaves. Most larvae resulting from the first brood develop directly in about a month and pupate under host leaves. In contrast, the remainder of the first-brood larvae and all larvae from the second brood cease feeding when about half grown, attach themselves under host leaves, and turn brown. The caterpillars overwinter on the leaves under the host tree. The following spring they must crawl back up the tree and complete their development. The mature larvae are yellow-green and taper at both ends, with the posterior forked. They have a series of yellow spots along the midline and three yellow lines along each side. The head is surmounted by a pair of spinose, forked horns.

Adult Food. Hackberry Butterflies almost never visit flowers, although once, in western Iowa, we observed several adults taking nectar at common milkweed. Normally, the adults are attracted to dead animals, dung, tree sap, or rotting fruit. They may also be seen taking moisture at wet spots along roads or streams.

Caterpillar Host Plants. Various species of hackberry (*Celtis*) are the only larval hosts. In the East, sugarberry (*Celtis laevigata*) and hackberry (*C. occidentalis*) are eaten.

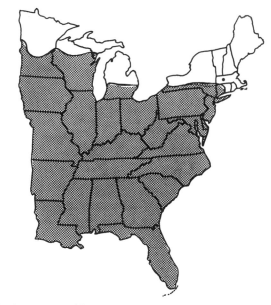

154. *Asterocampa celtis*

Tawny Emperor
Asterocampa clyton (Boisduval and Leconte), 1833
PLATE 33 · FIGURES 196–198 · MAP 155

(subspecies *flora* shown by darker shading in the extreme Southeast)

Etymology. The species name is derived from the Greek *klytos,* "famous" or "renowned."

Synopsis. The Tawny Emperor is more variable geographically than the Hackberry Butterfly. A number of separate forms are treated as species, but reproductive isolation has not been demonstrated and these may be no more than differentiated subspecies.

Butterfly. Male forewing: $\bar{X}$ = 2.4 cm, range 2.3–2.7 cm; female forewing: $\bar{X}$ = 3.1 cm, range 2.7–3.4 cm. The distinguishing features of the Tawny Emperor are mentioned in

the account of the Hackberry Butterfly. Both sexes are dimorphic. In one form, the dorsal hind wing surfaces are orange with prominent submarginal black spots, while the other has predominantly black hind wings. The significance of these morphs, if any, has not been explained. The southeastern form, *flora,* here considered a subspecies, is larger and paler and has reduced black markings. In addition, the black hindwing form is rare among *flora* populations.

Range. The Tawny Emperor ranges from southern New England west to Nebraska and south to Florida and the Gulf of Mexico. It extends west and south to southern Texas and northeastern Mexico. The species is rare in the northern portion of its range.

Habitat. The rarer Tawny Emperor seems to be more restricted to densely wooded riparian habitats than the Hackberry Butterfly, but is often found together with it.

Life History. *Behavior.* This butterfly is apparently similar to the Hackberry Butterfly in its activity patterns, but it tends to perch higher in the trees.

Broods. In the northern part of its range, the Tawny Emperor has only a single generation, as for example, in New York (July 15–28) and Illinois (June 18–August 30). Farther south (e.g., in Missouri), the species may have a partial second generation, while there are probably two full broods in the Deep South. The first brood normally appears in June, followed by the second in August. In Florida, the species flies from March to November.

Early Stages. The female lays clusters of 200–500 eggs in two or more layers on hackberry bark or leaves. The caterpillars hatch about a week later and feed communally until after the third stadium. If the larvae occur where the butterfly is univoltine, they become brownish and seek shelter for the winter. Overwintering larvae have been found in tight groups under dead leaves. Larvae of the first of two broods complete their development, and larvae of the succeeding generation enter diapause, with feeding completed the following spring. The larva is yellow-green, with a narrow, dark-green or indigo blue dorsal stripe. The head is white or black or a combination of the two colors and has paired, many-branched, spinose horns. The chrysalis is pale green with longitudinal yellow lines.

Adult Food. The Tawny Emperor's dietary habits are probably similar to those of the Hackberry Butterfly. In Iowa a single individual was found with milkweed pollinia attached to its legs. Common milkweed was common nearby.

Caterpillar Host Plants. Hackberries are the sole food of the Tawny Emperor. Sugarberry (*Celtis laevigata*) and hackberry (*C. occidentalis*) are fed upon in the eastern United States.

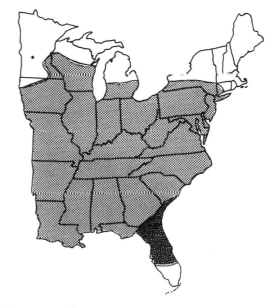

155. *Asterocampa clyton*

Subfamily Satyrinae: Satyrs and Wood Nymphs

The satyrs are generally medium-sized brown butterflies. Submarginal eyespots, or ocelli, are found on the ventral wing surface of most species. Both sexes have reduced forelegs that are used for substrate sensing instead of walking. The veins of the forewing are often swollen at their bases, and the males often have patches of specialized scales on the fore- or hind wings.

Except for the pearly eyes (*Enodia*) and arctics (*Oeneis*), males patrol likely habitats in search of receptive females. Mating usually occurs in the morning or late afternoon, and oviposition often takes place in the afternoon. The species seldom sit with open wings, except for occasional dorsal basking. The satyrs have a low, erratic, skipping flight. The adults have short proboscises and almost never visit flowers, but prefer sap flows, fermenting fruit, or dung instead.

Almost all satyrs feed on grasses or sedges as caterpillars. The female lays single eggs on host leaves or places them on the ground near the host. The eggs are drum-shaped, being somewhat taller than broad. The posterior end of the larva is forked (bifid) and often there is a hornlike structure on the vertex (summit) of the head capsule. The larvae often have tan, brown, or green bodies with several longitudinal stripes.

The caterpillars of most species hide at the base of their host by day and feed on leaves at night. Most of our species overwinter as larvae, with some *Oeneis* and *Erebia* requiring 2 years to complete development. The pupae hang down vertically from a silk button or are formed near the ground in debris.

Pearly Eye
Enodia portlandia (Fabricius), 1781
PLATE 34 · FIGURE 199 · MAP 156

Etymology. The species name is derived from Portland, a name for several cities.

Synopsis. This large brown satyrine is a denizen of swamp forests with cane brakes found along the southern Atlantic and Gulf Coasts, as well as of the southern Mississippi Valley. The butterfly was long confounded with the Northern Pearly Eye (*Enodia anthedon*), and its female confused with that of the Creole Pearly Eye (*E. creola*), until careful studies pointed out the differences (Gatrelle 1971). The Pearly Eye often occurs in the same habitat as the Creole Pearly Eye.

Butterfly. Male forewing: $\bar{X}$ = 2.9 cm, range 2.6–3.3 cm; female forewing: $\bar{X}$ = 2.9 cm, range 2.5–3.2 cm. The adults are warm brown dorsally, with a row of oval black spots irregularly bordered with yellow along the outer margins of both wings. Ventrally, the wings are purple-brown and the pattern is more complex. The black spots have blue pupils and are narrowly ringed with yellow-orange, while the spot field is surrounded by diffuse white. Where the Pearly Eye and Northern Pearly Eye occur together, they may be distinguished by the color of the antennal clubs, which are orange in the former and black in the latter. In addition, the ventral forewing spot row is curved on the Pearly Eye and straight on the Northern Pearly Eye. The male Pearly Eye may be easily distinguished from the male Creole Pearly Eye, since the latter has distinctive patches of raised scent scales on the dorsal forewings that are entirely lacking in the former. Females are more similar and must be studied carefully. The spot row on the ventral forewing of female Pearly Eyes has only four spots, while the Creole Pearly Eye usually has five. Florida Pearly Eye females may have five. More critically, the dark line just inward of the spot row on the ventral forewing is almost straight or slightly zig-zag in the Pearly Eye, while it has a distinct arcuate extension on the Creole Pearly Eye.

Range. This species ranges from the south Atlantic coastal plain of Maryland south to Florida, and from the Gulf Coast east to Louisiana and the Mississippi Valley north to Kentucky.

Habitat. The Pearly Eye inhabits moist, shady, wooded areas with cane, near small streams, which usualy lead into swamps.

Life History. *Behavior.* The Pearly Eye seldom strays far from patches of its host, maiden cane. Males perch on tree trunks and probably do not patrol. Gatrelle (1971) reports that males are found in dense woods, while females frequent more open areas. At the Great Dismal Swamp in Virginia, the Clarks (1951) found the species most active in the early morning or at dusk. Harris (1972) reported that courtship occurs at nightfall. Oviposition has been observed in late afternoon (1630 hr).

Broods. The species is triple-brooded in most of its range. In the North, flight dates extend from mid-May to early October (Virginia), while in Georgia, collection records for three putative broods are April 19 to May 25, July 2 to August 8, and September 10 to October 18. Farther south, in Mississippi, flight extremes are March 23 to October 26. Finally, in Florida, the butterflies may be found from March to December.

Early Stages. The early stages and their timing are imperfectly known. Harris (1972) describes the mature caterpillar as yellow-green with two sets of red-tipped horns, one set on the head and the other on the last abdominal segment.

Adult Food. The Pearly Eye feeds on oozing tree sap, as does the Northern Pearly Eye.

Caterpillar Host Plants. The butterfly has been closely associated with giant cane (*Arundinaria gigantea*) in Mississippi, North Carolina, and Virginia, while switch cane (*Arundinaria tecta*) seems to be the host in Georgia.

Northern Pearly Eye
Enodia anthedon Clark, 1936
PLATE 34 · FIGURES 200 & 201 · MAP 157

Etymology. The species name is derived from Anthedon, a city or village in Euboea (Ovid).

Synopsis. The Northern Pearly Eye, long confused with its close southern relative, the Pearly Eye (*Enodia portlandia*), inhabits damp deciduous woods, and is more of a woodland dweller than any other satyrine in the northern states. It is usually found in hilly or mountainous terrain, unlike its southern sibling (Heitzman and dos Passos 1974).

Butterfly. Male forewing: $\bar{X}$ = 2.7 cm, range 2.4–2.8 cm; female forewing: $\bar{X}$ = 2.8 cm, range 2.5–3.1 cm. The adults are extremely similar to those of the Pearly Eye (see that species' account). In addition to the cited differences, the Northern Pearly Eye's ventral wing surfaces have darker ground color.

Range. The Northern Pearly Eye may be found from southern Quebec west to southern Manitoba and south to

Kansas, southern Arkansas, northeastern Alabama, and northern Georgia.

Habitat. This butterfly is found in damp deciduous woods, especially near streams, in rolling or mountainous terrain. The Northern Pearly Eye is most often found in the Transition Life Zone, but is also found on occasion in the Upper Austral or Canadian Life Zones.

Life History. *Behavior.* The butterflies are most often found within woodland areas, where they perch on tree trunks, but they may occasionally be seen in small clearings, along woodland roads, or along small streams. During early morning and possibly late afternoon, the males take up perches on tree trunks or foliage at the edge of small clearings anywhere from near ground level up to 3 m above ground. From this perch they will dart out at passing butterflies and rapidly course back and forth before returning to another perch. A mating pair was observed at 1530 hr (Cardé et al. 1970).

Broods. The Northern Pearly Eye is univoltine in the northern portion of its range. For example in Maine the flight dates range from July 1 to August 9, and in Michigan's upper peninsula from June 20 to August 5. In the more southern parts of its range, it is bivoltine. In Virginia, flight dates range from May 20 to August 4, while in Illinois the species is found from May 24 to September 14, with adult emergences during early June and mid-August.

Early Stages. The green-white, squat, barrel-shaped eggs are laid singly on the host. After hatching in the fall, the young larvae overwinter. The mature caterpillar has a yellow-green head with red-tipped coronal horns, while its body is yellow-green, striped with dark green and yellow. The terminal segment is cleft and the resultant "tails" are pink-tipped. The chrysalis is glossy green or blue-green and the top of the head and the edges of the wing cases are cream.

Adult Food. The butterflies have been reported to feed at willow and poplar sap flows in Maine, while they have been observed at birch sap flows in Pennsylvania.

Caterpillar Host Plants. All known or suspected host plants are grasses. The following have been reported: plume-grass (*Erianthus*) in Maryland, white grass (*Leersia virginica*) in North Carolina, *Brachyelytrum erectum* in New York, and broadleaf uniola (*Uniola latifolia*) in Arkansas and Missouri. The Northern Pearly Eye is associated with bottlebrush (*Hystrix patula*) in Virginia.

156. *Enodia portlandia*

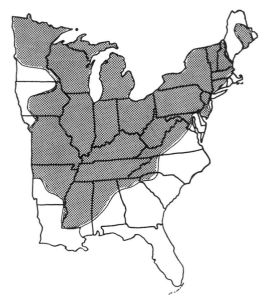

157. *Enodia anthedon*

Creole Pearly Eye
Enodia creola (Skinner), 1897
PLATE 34 · FIGURES 202 & 203 · MAP 158

Etymology. The species name is derived from Creole, meaning a person born in the West Indies but of European ancestry.

Synopsis. This southeastern endemic often flies with the Pearly Eye (*Enodia portlandia*), and has long been thought of as a rare species. One must enter dense swamps or dank woods with stands of cane, an unlikely butterfly habitat, to find the Creole Pearly Eye, but there the species is often found in abundance.

Butterfly.   Male forewing: $\overline{X}$ = 2.9 cm, range 2.7–3.2 cm; female forewing: $\overline{X}$ = 3.0 cm, range 2.7–3.1 cm. This is the largest *Enodia* in North America, and its male is the most distinctive, with its raised scent-scale patches on the forewings. Its females are distinguished from those of the Pearly Eye only with great difficulty. A discussion of the pertinent differences is presented in the account of the Pearly Eye.

Range.   This butterfly is generally endemic to the southeastern states from southern Virginia, Kentucky, southern Illinois, and southern Missouri southward (Irwin 1970). It is not found in Florida, and earlier records from that state probably refer to the Pearly Eye.

Habitat.   In Virginia, the Creole Pearly Eye is found in dense, moist woods with cane (*Arundinaria*), usually on the periphery of large swamps, and is found in similar situations throughout its range, except in the southern Appalachians, where it may be found in dense cove forests.

Life History.   *Behavior.* Adults are often found perching on tree trunks during the day, often up to 3 m above ground.

Although Harris (1972) reported that courtship takes place mainly in the daytime, the Clarks (1951) found the species to be predominantly crepuscular. In the vicinity of the Dismal Swamp, Virginia, we have found the species most active in late afternoon or on cloudy days. There, on an extremely hot day when the temperature exceeded 104°F, adults of both sexes were found resting on moist sand along an intermittent creek in dense woodland. In South Carolina, males were found in more open woods than females.

Broods. In the northern part of the range there are two broods; in Virginia, flight dates extend from early June to early July and from mid-August to mid-September, while in southern Illinois dates range from June 20 to July 4 and from August 1 to September 1. Flight extremes in Mississippi (April 26–September 11) are indicative of three generations in the South.

Early Stages. The larva is similar to but lighter in color than that of the Pearly Eye.

Caterpillar Host Plant.   The Creole Pearly Eye has been found to feed on or be associated with only switch cane (*Arundinaria tecta*). Such observations have been made in Virginia, South Carolina, and Georgia.

Northern Eyed Brown
Satyrodes eurydice (Johansson), 1763
MAP 159

Etymology.   The genus name resembles the Greek word for "similar to a satyr." The species name is Eurydice, the wife of Orpheus (Greek mythology).

Synopsis.   A satyrine of open marshes, the Northern Eyed Brown was long confused with the Appalachian Eyed Brown until the landmark study by Cardé et al. (1970) clearly demonstrated differences between the two.

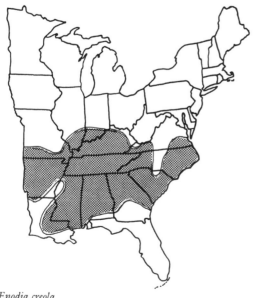

158. *Enodia creola*

159. *Satyrodes eurydice*

Butterfly. Male forewing: $\overline{X}$ = 2.4 cm, range 2.0–2.5 cm; female forewing: $\overline{X}$ = 2.4 cm, range 2.1–2.7 cm. This medium-sized brown satyrine has a series of small black spots on the outer margins of both the fore- and hind wings above. Ventrally, the spots are centered with yellow and ringed outwardly with orange. In typical *S. eurydice,* the eyespots on the ventral forewing are approximately equal in size, while the postmedian line below is deeply indented on both wings.
Range. The butterfly occurs from the Maritime Provinces and southern Hudson Bay west to Alberta and south to Delaware, Pennsylvania, northern Ohio, Illinois, Minnesota, and North Dakota.
Habitat. The Northern Eyed Brown is a butterfly of open sedge meadows and the margins of freshwater cattail marshes. It is rarely found in bogs or drier meadows.
Life History. *Behavior.* The Northern Eyed Brown flies weakly just above or within low sedge growth or short scrubby vegetation. Individuals perch frequently and engage in both lateral and dorsal basking behavior. Males patrol in search of females by periodically dipping low among vegetation (Angevine and Brussard 1979). Mating of typical *eurydice* has been observed from 1230 to 1500 hr.
Broods. The Northern Eyed Brown is univoltine throughout the majority of its range. In northern New York, the flight

extends from June 24 to September 8, in Pennsylvania from mid-June to early August, and in Illinois from June 14 to August 1. Cardé et al. (1970) report a small emergence of *eurydice* during the first half of August that represents either a partial second brood or an unrecognized sibling species.
Early Stages. Eggs may not always be deposited on the proper host, as is the case for several nymphalines. The caterpillars are yellow-green with red lateral stripes. A pair of pointed, red, hornlike structures arise from atop the head. Feeding begins in late summer, and the larvae turn straw yellow and enter diapause for the winter in the third or fourth stadium. Feeding resumes and development is completed during the following spring. The chrysalis is green and striped with buff, while the head case is buff and is more produced than that of the Pearly Eyes.
Adult Food. The normal food of Northern Eyed Browns is probably sap or bird droppings, although Shapiro (1973) found that they occasionally nectar from flowers of swamp milkweed or joe-pye weed.
Caterpillar Host Plants. In New York, where most biological study of the Northern Eyed Brown has been carried out, the host plants are all sedges (*Carex stricta, C. lupulina, C. bromoides,* and *C. trichocarpa*), although one population appears to be associated with a grass.

Smoky Eyed Brown
Satyrodes fumosa Leussler, 1916
PLATE 35 · FIGURES 205 & 206 · MAP 160

Etymology. The species name is derived from the Latin *fumosus,* "smoky."
Synopsis. The Smoky Eyed Brown is a denizen of freshwater marshes within the tall-grass prairie biome. The butterfly has lost much of its former haunt through conversion of the original prairies to corn and soybean fields.
Butterfly. Male forewing: $\overline{X}$ = 2.6 cm, range 2.5–2.6 cm; female forewing: $\overline{X}$ = 2.8 cm, range 2.7–2.9 cm. This butterfly has been treated as a subspecies of the Northern Eyed Brown by most recent workers. Cardé et al. (1970) cited minor genital and pattern differences. It is larger than either the Appalachian Eyed Brown or Northern Eyed Brown and has a pattern most like the latter. It differs from the Northern Eyed Brown in often having five ocelli on the ventral forewing, and in having the black areas in the ventral hind-wing ocelli more pronounced and the bands on the ventral wing surfaces darker and sharper, and in that the male is darker brown dorsally, with the forewing postmedian patch more extensive.
Range. Originally, the Smoky Eyed Brown occurred from northeastern Colorado east in a relatively narrow band along the Platte and Missouri River drainages to eastern Nebraska, Iowa, and northern Illinois.

160. *Satyrodes fumosa*

Habitat. The species frequents freshwater marsh edges and cord grass swales in the tall-grass prairie biome.

Life History. *Behavior.* The flight and patrolling behavior are similar to those of the Northern Eyed Brown. At Lake Okoboji, Iowa, a female was observed ovipositing at 1215 hr. A single egg was laid on each occasion, but the leaves of blackberry and other broadleaf plants were the chosen substrate, not sedges, the likely hosts. Adults were most active at midday.

Broods. The Smoky Eyed Brown is univoltine. Recorded flight dates range from June 19 to July 22 in Iowa and Illinois.

Appalachian Eyed Brown
Satyrodes appalachia (R. L. Chermock), 1947
PLATE 34 · FIGURE 204 · MAP 161

Etymology. The species name is derived from the Appalachian mountain chain or region.

Synopsis. It was not until the classical study of Cardé et al. (1970) that the "Eyed Brown" was demonstrated to be at least two closely related yet distinct species. The Appalachian Eyed Brown, which may occur within a few feet of its sibling in their overlap zone, differs from the Northern Eyed Brown (*Satyrodes eurydice*) in adult color pattern, genitalia, behavior, and habitat.

Butterfly. Male forewing: $\overline{X}$ = 2.5 cm, range 2.2–2.8 cm; female forewing: $\overline{X}$ = 2.6 cm, range 2.5–2.7 cm. The Appalachian Eyed Brown is a darker butterfly above and below than the Northern Eyed Brown. The pattern is the same in both species, except that the ventral forewing eyespots are unequal in size on the Appalachian Eyed Brown, with the top and bottom ones usually the largest, and the postmedian line on the ventral surface is only slightly wavy, not jagged as in the Northern Eyed Brown.

Range. The Appalachian Eyed Brown occurs from central New England and southern Quebec west to eastern Minnesota and south through the Appalachians and coastal plain to Florida, Alabama, and Mississippi.

Habitat. The Appalachian Eyed Brown is found in a variety of wet, wooded habitats: swamp forest, shrub swamp, forest-edge ecotone, wet woods and swamps with maiden cane (Virginia coastal plain), bottomland near streams, clearings along slow-moving streams, and water oak–bald cypress swamps.

Life History. *Behavior.* The Appalachian Eyed Brown inhabits small openings or forest edges, where the males patrol slowly back and forth, flying low over vegetation and periodically resting on sedges or other vegetation. On one occasion, at midday two interacting males were seen in a rapid circular chasing flight about 1.5 m in the air within the shade of a young forest. Unlike the Northern Eyed Brown, the Appalachian Eyed Brown will fly readily through shade, although it usually is found in small, sunlit openings.

A mating pair of freshly emerged individuals was observed during 1980 in Fairfax County, Virginia, at 1040 hr. The female was the carrier and rested on blades of grass while the male hung quietly. Oviposition occurs during low light intensity in late afternoon (1600 hr).

Broods. In the northern part of its range the Appalachian Eyed Brown is univoltine, as in New York from July 15 to September 8 and Michigan from June 20 to July 24. As far north as Virginia there are two full flights each year (June 6 to July 28 and August 4 to September 11). In Georgia, dates for the two broods range from June 6 to July 4 and July 18 to September 9, while individuals of a second brood were found in west-central Florida as late as October 8.

Early Stages. Caterpillars of the Appalachian Eyed Brown are much like those of its northern counterpart, but the red stripes on the head do not extend below the horns. In the laboratory, the entire life cycle required about 60 days for completion at 24°C.

Adult Food. Adults have been seen feeding at sap flows.

Caterpillar Host Plants. Sedge (*Carex lacustris*) in New York and giant sedge (*Rhynchospora inundata*) in Florida are the only two larval hosts reported.

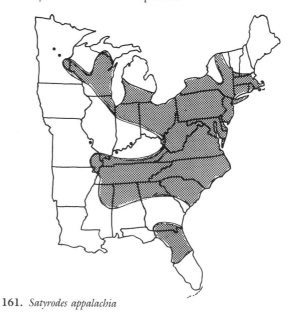

161. *Satyrodes appalachia*

Gemmed Satyr
Cyllopsis gemma (Hübner), 1818

PLATE 35 · FIGURE 207 · MAP 162

Etymology. The genus name is derived from the Greek *kyllē,* "palm of the hand," and *opsis,* "appearance." The species name is derived from the Latin *gemma,* "bud," and may refer to the iridescent patch on the outer margin of the ventral hind wings.

Synopsis. Most *Cyllopsis* species are restricted to mountain ranges in Central America and Mexico. In addition, several species are found in the mountain ranges of the southwestern United States. The Gemmed Satyr is therefore unusual, not only because it is the only eastern representative, but also because it occurs in lowland habitats.

Butterfly. Male forewing: $\bar{X}$ = 1.7 cm, range 1.6–1.8 cm; female forewing: $\bar{X}$ = 1.8 cm, range 1.7–2.0 cm. The Gemmed Satyr, instead of having eyespots, as do all of its relatives in our area, has a ventral patch of iridescent black scales along the outer hind wing margin.

Range. The Gemmed Satyr ranges westward from Virginia and Maryland (rare) to southern Missouri, Oklahoma, and Texas, thence southward to central Florida, the Gulf Coast, and Mexico.

Habitat. Gemmed Satyrs are usually found in flat or gently rolling terrain near wet, open woodland. Most often, these areas are in long-leaf pine forests (Virginia, Florida, and Louisiana), but they may be in mixed live oak–sweet gum forest (Louisiana) or near streams and ponds (Georgia).

Life History. *Broods.* There are probably three broods throughout the range of the Gemmed Satyr, although the proportion of the year through which it flies increases southward. In Virginia, the broods occur in late April–early May, late June–early July, and August–September. In Mississippi, flight extremes range from early March to late November, while in Florida individuals have been found as early as late February.

Early Stages. The larvae have pairs of long gray-brown tubercles on both the head and anal segment. Summer caterpillars are yellow-green with dark green longitudinal stripes, while those resulting from the fall brood are brown with darker brown longitudinal stripes. The chrysalis also has blue-green and brown forms.

Caterpillar Host Plant. The Gemmed Satyr has been closely associated with Bermuda grass (*Cynodon dactylon*) in the field by Kendall (1964), who found that captive females readily oviposit on the same plant.

162. *Cyllopsis gemma*

Carolina Satyr
Hermeuptychia hermes sosybius (Fabricius), 1793

PLATE 35 · FIGURE 208 · MAP 163

Etymology. The genus name is derived from the species name and the Greek *eu,* "well," plus *ptykhios,* "folded." The species name is derived from Hermes, god of merchants and messenger of the gods (Greek mythology). The subspecies name is derived from Sosibius, a historian from the third century B.C.

Synopsis. The Carolina Satyr is the most common and widespread satyrine of the Southeast and much of tropical America.

Butterfly. Male forewing: $\bar{X}$ = 1.6 cm, range 1.5–1.7 cm; female forewing: $\bar{X}$ = 1.7 cm, range 1.6–1.8 cm. In our area, the Carolina Satyr is the only satyrine with a combination of plain brown unmarked wings dorsally and a series of small, black, yellow-rimmed eyespots ventrally. The most similar butterfly in the East is the Little Wood Satyr (*Megisto cymela*). The latter is larger, has only two eyespots on the ventral forewings, and, most importantly, has the eyespots repeated on the dorsal wing surfaces.

Range. This butterfly is found in southern New Jersey and the entire southeastern area south and west to southern Texas and on into most of the tropical and temperate portions of the American tropics.

Habitat. The Carolina Satyr is found in a variety of woodland habitats, ranging from pinewoods and oak forest to wooded river bottoms.

Life History. *Behavior.* The Carolina Satyr is a patrolling species with a slower, less active flight than that of the Little Wood Satyr (*Megisto cymela*). It is most often seen flying weakly or perching on grass or leaf litter in forest understory. The butterflies seem to rest during most of the day but have increased activity in late afternoon. Courtship has been observed during most daylight hours but is seen primarily between 0800 and 1000 hr. The male lands behind the female on a horizontal grass blade, walks rapidly up until parallel to her left side and curves his abdomen around in a U-shaped copulatory position. A mated pair was found at 1330 hr by Miller and Clench (1968). The female was the active flying member of the pair. Oviposition occurs form 1215 to 1540 hr. Survival and longevity of the Carolina Satyr were studied in Florida by Kilduff (1972). He found maximum life expectancy during May to be 6.25 days, but it is expected to be longer during the dry season.

Broods. There are two broods in the northernmost portion of the Carolina Satyr's range, and three to six in much of the South. In Virginia, three broods have been reported, i.e., late April–early June, late June–early August, and late August––early October. Farther south, flight dates range more widely, and extremes have been reported as: Georgia, January 10 to October 30; Louisiana, February 10 to December 25. In Florida, the Carolina Satyr is found throughout the year. In northern Florida, adults undergo reproductive diapause between November and January (Emmel and Eliazar, unpublished).

Early Stages. The egg is green-white and semi-ovoid, while the caterpillar is light green with darker green longitudinal stripes. Its body is covered with setose yellow tubercles. The chrysalis is green, with the exception of the yellow-green abdomen, and there is a pair of small ridges along the abdominal dorsum and three black dots along each side.

Caterpillar Host Plants. In Louisiana, the Carolina Satyr oviposits on carpet grass (*Axonopus compressus*), while in North Carolina the females select centipede grass (*Eremochloa ophiuroides*). The butterfly has laid eggs on St. Augustine grass (*Stenotaphrum secundatum*) and Kentucky bluegrass (*Poa pratensis*) in captivity. Other grasses are probably used in nature.

Georgia Satyr
Neonympha areolatus (J. E. Smith), 1797
PLATE 35 · FIGURE 209 · MAP 164

Etymology. The genus name is derived from the Greek *neo,* "new, recent," and *nymphé,* one of the inferior divinities of nature (Greek and Latin myth). The species name is derived from the Latin *areolatus,* "having small rings."

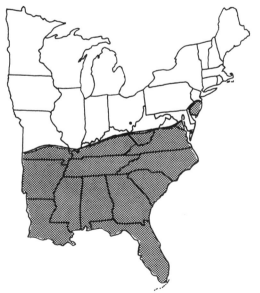

163. *Hermeuptychia hermes sosybius*

164. *Neonympha areolatus*

Synopsis. This southern coastal plain and Piedmont butterfly is found in open, sandy pinewoods and is related to Mitchell's Satyr, which is a species of northern tamarack–poison sumac bogs. Little is known of its habits or life history.

Butterfly. Male forewing: $\bar{X}$ = 1.8 cm, range 1.7–2.0 cm; female forewing: $\bar{X}$ = 2.0 cm, range 1.8–2.4 cm. The Georgia Satyr is characterized by the submarginal row of elongated eyespots on the ventral hind wings, together with the red-orange lines that cross the wings below.

Range. The Georgia Satyr is found on the coastal plain and Piedmont from coastal New Jersey south to the Florida Keys and Gulf states, thence westward to Texas.

Habitat. The butterfly is found in open grassy areas in sandy pinewoods or pine barrens.

Life History. *Behavior.* This satyr has a very slow, bobbing flight. It rests frequently.

Broods. The Georgia Satyr is univoltine in New Jersey (July), and bivoltine throughout most of the South, e.g., late April–May and August–early September in Virginia, and mid-May–June and September in Louisiana. In Florida, there may be more than two broods, since Kimball (1965) reports records from almost every month.

Early Stages. Descriptions of the caterpillar are based on paintings by Abbot. The larva is yellow-green with narrow yellow longitudinal stripes and a green head. The caudal projections are red-brown. The chrysalis is green, with the head end and wing cases creamy white.

Caterpillar Host Plants. Grasses have been reported to be larval hosts, although there are no definitive records available. The fact that Mitchell's Satyr has been reared on sedges strongly suggests that the Georgia Satyr also feeds on sedges, not grasses.

Mitchell's Satyr
Neonympha mitchellii French, 1889
MAP 165

Synopsis. Mitchell's Satyr inhabits tamarack–poison sumac bogs in the upper Midwest and New Jersey and is a close relative of the Georgia Satyr. Even after its discovery in an area it has been lost for long periods lacking a lepidopterist bold enough to slog into its favored marshes, within which poison sumac, biting flies, and pygmy rattlesnakes abound.

Butterfly. Male forewing: $\bar{X}$ = 1.7 cm, range 1.6–1.8 cm; female forewing: $\bar{X}$ = 1.9 cm, range 1.8–2.1 cm. Adult Mitchell's Satyrs are fragile, mothlike creatures. They may be identified by the rows of black, yellow-ringed eyespots in the submarginal areas of the ventral hind wings, and which are dotted with small patches of yellow scales. In addition, there are two orange lines bordering both fore- and hind-wing margins below. The only other close relative that occurs in its range, the Little Wood Satyr, has fewer eyespots and they appear on both surfaces of the wings.

Range. The historical range of Mitchell's Satyr is curious, with populations restricted to bogs and swamps in southern Michigan, northern Indiana, and Ohio, as well as a few localities in northern New Jersey. The butterfly has not been found in Ohio for many decades, and has a tenuous existence in New Jersey.

Habitat. In Michigan and Indiana, Mitchell's Satyr is found within tamarack–poison sumac swamps that have open areas in the center with tall grass and sedge (Badger 1958). In New Jersey, the species has been reported from a swamp that lacks both tamarack and poison sumac. There, it has been found flying along narrow grassy strips bordering small watercourses (Rutkowski 1966). Fairly complete vegetation descriptions have been given for several Michigan habitats by McAlpine et al. (1960).

Life History. *Behavior.* Mitchell's Satyr is a patrolling species and flies during sunlit periods throughout the day. Individuals fly with a slow, bobbing flight no more than a foot above the grasses in the marshes where they occur.

165. *Neonympha mitchellii*

Although this butterfly is rare in collections, it may be common when found, since population estimates of from 500–5,000 have been made on single days in its favored Michigan bog habitats. A female followed closely in the field repeatedly landed on sedge plants (*Carex stricta*) but did not oviposit. From July 9–16, a captive female laid 107 eggs on sedge stems between 1030 and 1300 hr (McAlpine et al. 1960). A maximum lifespan estimate of 8 days is based upon this captive female.

Broods. Mitchell's Satyr has a single midsummer generation, with adults flying from June 21 to July 28 (peak July 1–14) in Michigan and Indiana and from July 7–19 in New Jersey. Males emerge several days earlier than females, on average.

Early Stages. The life history and early stages of Mitchell's Satyr have been described in detail by McAlpine et al. (1960). Eggs are laid in mid-July and hatch in about 10 days. The caterpillars undergo three molts during summer and fall, with the first three stadia averaging 16 days, 12 days, and 24 days, respectively. Feeding ceases in late September and the fourth-instar larva spends the winter attached to a silken pad on a stem or leaf 1 or 2 inches above the ground. In nature, feeding probably resumes in late May, and two more larval molts occur. Finally, in June, pupation occurs, and adults emerge 10–15 days later.

The spheroid or rounded cubical egg (0.8–1.0 mm diameter) is whitish green, changing to pale brown prior to hatching. The third- to sixth-instar larvae are similar, being lime green and densely covered with whitish papillae. There are several longitudinal whitish lines running the length of the body. The pupa is stout and truncate at the anterior end, tapers posteriorly, and is light lime green with some areas that are darker and more bluish.

Caterpillar Host Plants. The plants utilized in nature by Mitchell's Satyr larvae are not known but are believed to be several different sedges, as females oviposit readily upon *Carex stricta* in captivity and the resultant larvae ate foxtail sedges (*Carex alspecoidea* and *C. cephalophora*), and bulrush (*Scirpus atrovirens*).

Little Wood Satyr
Megisto cymela (Cramer), 1777
PLATE 36 · FIGURES 211 & 212 · MAP 166

Synopsis. The Little Wood Satyr is the most widespread satyrine, and usually the most common, in the area covered by this book. It is able to occupy a wide variety of woodlands, including some parks in metropolitan areas. There is a strong indication that this butterfly may actually represent a complex of two or three sibling species.

Butterfly. Male forewing: $\overline{X}$ = 2.0 cm, range 1.8–2.1 cm; female forewing: $\overline{X}$ = 2.1 cm, range 1.9–2.2 cm. This pale-brown butterfly is distinguished by the pair of black, yellow-rimmed eyespots that occur on the submarginal area of each wing above and below. Only the Carolina Satyr and Mitchell's Satyr are similar to the Little Wood Satyr, and neither has eyespots on the dorsal wing sufaces.

Range. This wood satyr occurs throughout the eastern United States except for northernmost New England and the Florida Keys. Outside our area, it ranges west in river forests to extreme eastern Colorado.

Habitat. The butterfly frequents woodland and woodland edges in many habitat types, including deciduous woods, pinelands, and young second-growth brush. The Little Wood Satyr is most common on limy or basic soils and least common in barren habitats or other woods on acid soils.

Life History. *Behavior.* Male Little Wood Satyrs patrol along forest edges and in sunlit glades in search of receptive females during most daylight hours. Particularly strong male–male interactions occur early in the morning. Courtship has been observed in northern Virginia at 0710 and 1305 hr, while mating pairs have been seen from 0810 to 1510 hr. Females carry the inactive males during mating. A female was seen laying eggs at 1220.

During early morning and late afternoon, Little Wood Satyrs bask with their wings spread while perched on tree leaves or on leaf litter on the forest floor. Unlike other euptychiine satyrs, which seldom rise more than a few feet above the ground, Little Wood Satyrs frequently patrol the crowns of trees along forest edges and occasionally over the tops of tall trees.

Broods. The Little Wood Satyr is univoltine over most of its range, but two or three broods may be the mode in the Deep South. At univoltine localities, some sample flight extremes are: Maine, June 5 to July 7; Michigan (upper peninsula), June 20 to July 26; and New York (Ithaca), May 27 to July 14. Flight extremes of April 1 to early September for Georgia and mid-March to late September for Mississippi suggest two or three broods. The two emergences in Virginia and Pennsylvania (late May through June and then early July to mid-August) and some other locations that follow each other so closely strongly indicate that two cryptic siblings must be involved.

Early Stages. The ovoid yellow-green eggs are deposited singly on grass blades. The caterpillars hibernate partially grown. The mature caterpillar has a dirty white head mottled with dark brown, and a pale-brown body slightly tinged with green. In addition, the body is covered with tubercles, each

bearing a red-brown hair, and has a median black stripe together with a series of lateral brown patches. The anal forks are dirty white. The chrysalis is yellow-brown with two brown stripes and two rows of brown dots on the abdomen.
Adult Foods. The Little Wood Satyr probably feeds on tree sap, as do other forest satyrines. Adults have been seen feeding on aphid honeydew on tulip poplar leaves. Adults visit flowers rarely, and usually in late afternoon (1625–1805 hr). Visits to white or pale flowers, including common milkweed, white sweet clover, ox-eye daisy, dewberry, viburnum, privet, and staghorn sumac, have been observed.
Caterpillar Host Plants. In New York, larvae utilize orchard grass (*Dactylis glomerata*), while centipede grass (*Eremochloa ophiuroides*) is eaten in North Carolina. Females in captivity have laid eggs on St. Augustine grass (*Stenotaphrum secundatum*).

Viola's Wood Satyr
Megisto viola (Maynard), 1891
MAP 167

Synopsis. The Viola's Wood Satyr has been recognized as a separate species since it occurs in the same localities as the Little Wood Satyr without hybridizing (Oliver 1982).
Butterfly. Male forewing: $\bar{X}$ = 2.1 cm, range 2.0–2.2 cm; female forewing: $\bar{X}$ = 2.2 cm, range 2.1–2.3 cm. Viola's Wood Satyr may be distinguished from the Little Wood Satyr by its larger size, larger eyespots above and below, and its more strongly marked ventral surface, including strongly bowed postmedian lines.
Range. Viola's Wood Satyr is best known from northern and central Florida, but it also occurs westward at least to central Texas and north along the lower Mississippi drainage to central Arkansas and possibly eastern Kansas.
Life History. *Broods.* Viola's Wood Satyr has only a single brood, flying from early to late April in peninsular Florida. It occurs later in other parts of its range.
Early Stages. The larvae of Viola's Wood Satyr develop very slowly and unlike northern populations of the Little Wood Satyr, may not undergo diapause. Its larvae are similar to those of its close relative but are a lighter shade of brown (C. G. Oliver, personal communication).

166. *Megisto cymela*

167. *Megisto viola*

Inornate Ringlet
Coenonympha inornata Edwards, 1861

PLATE 35 · FIGURE 210 · MAP 168

Etymology. The genus name is derived from the Greek *coeno,* "common or new," and 'nymph,' while the species name is derived from the Latin *inornatus,* "unadorned, plain."

Synopsis. The ringlets in our area have been referred to as *Coenonympha inornata* and *C. nipisquit* by Brown (1961). It has been stated that the former flies in June and the latter only in August. Furthermore, it has been asserted that, where the two fly together, e.g., the St. Lawrence River islands, the fall-flying *nipisquit* is not the offspring of the early-summer *inornata.* Nevertheless, uncertainty exists, and the species relationships of this baffling group are not clear. For example, a Michigan population flies in July, when neither species should be present.

Butterfly. Male forewing: $\overline{X}$ = 1.7 cm, range 1.6–1.8 cm; female forewing: $\overline{X}$ = 1.8 cm, range 1.7–1.8 cm. The ringlets found in our area are orange-brown to ochre-brown dorsally, usually with a black spot near the forewing apex. Ventrally, the forewing is more reddish, with one or two submarginal eyespots and a partial whitish submarginal line. The ventral hind wing is green-gray with a sinuous incomplete whitish line. Much variation in color and pattern may be expected.

Range. In North America ringlets of the "tullia" complex occur west to Arizona, southern California, and north-westward to Alaska. They are found in the Rocky Mountains, the prairie provinces of Canada, and northward to the Maritimes (Brown 1955). In our area they occur in northern New England, northern New York, and the northern portions of the Great Lakes states. Most of the populations in our area are recent colonizations from the West and North. For example, ringlets were first found in Maine in 1968.

Habitat. Tullia ringlets are associated with open grassy areas in fields, marsh edges, and occasionally within young second-growth woods.

Life History. *Behavior.* The behavior of ringlets in our area is unreported, but is probably much like tullia ringlets elsewhere in North America. The males patrol grassy areas most of the day in search of receptive females and fly a jerky, bouncing flight just above the grass tops. Mating pairs of western subspecies have been seen in the morning (0906–1220 hr) and the female is the carrying sex.

Broods. A single generation occurs in Michigan's upper peninsula (June 18 to August 2) and Maine (June 26 to July 12), while two generations have been found in New York (June 15 to July 12, August 23 to September 15). Brown (1961) refers to the New York broods as single generations of *Coenonympha inornata* and *C. nipisquit,* respectively.

Early Stages. The egg is pale green when newly laid and is somewhat barrel-shaped, with 36 vertical ribs and many transverse ridges. There are five larval instars. Fourth-instar caterpillars from New York are dark green with pale-green lateral lines and a white ventral band. The head is green with white papillae. The caterpillars hibernate in either the first or fourth instar. Development is completed the following spring and summer.

Adult Nectar Sources. Ringlets are unusual among the satyrines, in that the adults are avid flower visitors. We have no flower visitation records for the East, but they are particularly fond of yellow composites in the West.

Caterpillar Host Plants. Larval hosts are unreported for any population in the eastern United States. Various grasses are hosts for western relatives.

Common Wood Nymph
Cercyonis pegala (Fabricius), 1775

PLATE 36 · FIGURES 213–215 · MAP 169

Etymology. The genus name is derived from Cercyon, the son of Poseidon (Greek mythology).

Synopsis. Next to the Little Wood Satyr (*Megisto cymela*), the Common Wood Nymph is the most frequently encountered satyrine in the eastern United States. It is also the most variable, ranging greatly in size, development of eyespots, and extent of yellow on the forewings.

Butterfly. Male forewing: $\overline{X}$ = 3.0 cm, range 2.8–3.2 cm (typical *pegala*); $\overline{X}$ = 2.4 cm, range 2.2–2.6 cm (*nephele*); female forewing: $\overline{X}$ = 3.4 cm, range 3.0–3.6 cm (typical *pegala*); $\overline{X}$ = 2.4 cm (*nephele*). The Common Wood Nymph is the most variable satyrine in North America. Some of its geographic forms had been previously considered full species until Thomas Emmel's (1969) revisionary study. With its two large eyespots on the outer portion of each forewing, this butterfly cannot be confused with any other in our territory. There is usually a series of unequal eyespots on the ventral hind wings, the lower of which is usually repeated above. There are two basic color forms in the East that replace each other geographically. The "pegala" type has a large rectangular yellow or orange-yellow patch surrounding the two large forewing eyespots; the "nephele" form lacks the yellow patch, and tends to be smaller (see above). The pegala type is found in coastal and more southern portions of the East, while the latter is more northern and more inland in its occurrence.

Range. The Common Wood Nymph occurs throughout the eastern United States except for portions of southern Florida and northern Maine.

Habitat. Preferred habitats always have at least some fairly large, sunlit, grassy areas and range from prairies and old fields to open, grassy pinewoods, damp meadows near streams, and open bogs.

Life History. *Behavior.* The males randomly patrol in search of females for much of the day, with a low dipping flight through grassland or scrubby vegetation. When shrubs are present, the Common Wood Nymph has a habit of either changing flight direction when passing behind a shrub or landing there momentarily. Females exhibit very different behavior, flying less actively and resting in the shade much of the time. Mating usually occurs in late morning or early afternoon (1000–1500 hr), but pairs have been seen as early as 0905 and as late as 1720 hr. The female acts as the carrier for the nuptial pair, although the male does not always hang limply.

Broods. The Common Wood Nymph is univoltine wherever found, with flights beginning in late June or early July in the North (New York, Michigan, and Maine) and as early as mid-June in the South (Illinois, Georgia, and Louisiana).

The males are relatively short-lived, living no more than 2 or 3 weeks, while the females, emerging a few days to a week later, have the potential to live several months. Some females are found in late September and even into early October. Their longevity seems to be an adaptation that defers oviposition until as late in the summer as possible and is similar to the adaptation of some fritillaries.

Early Stages. The female lays single, waxy yellow eggs on host leaves in late summer. The caterpillars hatch in 14 to 25 days and then enter diapause for the winter. Feeding takes place in the spring. Mature caterpillars have green heads and are green anteriorly, becoming yellow-green posteriorly. There are four paler longitudinal stripes and the anal fork is reddish. The chrysalis is green with paler mottlings and yellow edging or is yellow-green with white marks.

Adult Food. As is true for most satyrines, floral nectar is not usually a major food source. Males are seldom seen feeding at all. In Virginia, females have been seen feeding at moisture from rotting blackberries and exudates from a puff-ball. During late summer in the East, females may occasionally take nectar from New York ironweed, fleabane, mint, virgin's bower, and sunflowers. Purple coneflower is utilized on native tall-grass prairie in Iowa.

Caterpillar Host Plants. Various grasses are believed to be larval hosts. Purpletop (*Tridens flavus*) has been reported for the Delaware Valley.

168. *Coenonympha inornata*

169. *Cercyonis pegala*

Disa Alpine
Erebia disa mancinus Doubleday and Hewitson, 1851

Synopsis. The Disa Alpine is an inhabitant of northern spruce bogs and has been captured only once in northern Minnesota. The butterfly is brown-black dorsally with four black submarginal forewing spots surrounded by yellow-orange. A conspicuous white patch in the center of the ventral hind wing is distinctive. The Disa Alpine is crepuscular in its activities, and is found in acid bogs with dense stands of spruce. It is holarctic, being an inhabitant of subarctic zones in both Eurasia and North America. In North America it ranges from northern Minnesota north to Hudson Bay and northern Alaska and west to southern British Columbia. The species has a single brood during June every other year in at least part of its range.

Red-disked Alpine
Erebia discoidalis (Kirby), 1837
MAP 170

Etymology. The generic name is derived from the Greek *Erebos*, god of the darkness and son of Chaos.
Butterfly. Male forewing: $\bar{X}$ = 2.1 cm, range 2.0–2.3 cm. *Erebia discoidalis* is a distinctive blackish brown butterfly with a large orange-red patch taking up most of the forewing. Unlike most of our other satyrines, it lacks spots of any sort.
Range. This species occurs in the Canadian Life Zone in Central Asia and in North America from the northern Great Lake states northward and westward through Ontario and Churchill, Manitoba, to central Alaska. In the eastern United States, the Red-disked Alpine occurs in the upper peninsula of Michigan, northern Wisconsin, and northeastern Minnesota.
Habitat. *Erebia discoidalis* prefer large open bogs with abundant grass and a few spruce trees. They are found near bog edges or near trees, except during the early and late afternoon, when they move into the large, open, grassy areas (Masters 1970).
Life History. *Behavior.* Red-disked Alpines are in flight in early morning (before 1000 hr) or in late afternoon (after 1600 hr), but will also fly during midday on overcast or partly cloudy days. The butterflies have a weak, slow flight close to the ground. They normally alight low in the grass or sphagnum. Males search for females with a weak form of patrolling flight. Adults have never been observed at flowers.

Broods. The Red-disked Alpine is univoltine, flying from May 10 to June 18, with a peak during the last week of May.
Caterpillar Host Plant. The larval host was recorded in Manitoba as Canby bluegrass (*Poa canbyi*), a grass of acid soils.

170. *Erebia discoidalis*

Macoun's Arctic
Oeneis macounii (Edwards), 1885
MAP 171

Etymology. The genus is named after Oeneus, king of Calydon, an ancient city in western Greece. Oeneus is believed to be the first man to have grown grapes.
Synopsis. Eleven arctics inhabit North America, and six of them are found in the eastern United States. Of these six, only two species inhabit true arctic-alpine habitats. Only the Macoun's and Nevada Arctics are forest species, although the Jutta Arctic inhabits forest bogs.

Butterfly. Male forewing: $\overline{X}$ = 3.0 cm, range 2.8–3.2 cm; female forewing: $\overline{X}$ = 3.2 cm, range 3.1–3.4 cm. This butterfly, the largest arctic in the eastern United States, is reddish orange with black margins above and with two small black submarginal spots on the forewing (the uppermost being white-centered), and one near the anal angle of the hind wing. The ventral hind wing is blackish gray with a dark mesial band, while, in contrast, the hind wings of the Chryxus and Uhler's Arctics are marbled.

Range. The butterfly occurs in northern Minnesota and Michigan as well as in southern Ontario west to Alberta.

Habitat. In northern Minnesota, the habitat is jack pine forest growing on ridges that has interspersed growth of bracken (*Pteridium*) and blueberry (*Vaccinium*).

Life History. *Behavior.* Males perch in glades within jack pine forest and await receptive females. In flight, males resemble the Viceroy. Males usually perch on sunlit tree branches or shrubs that have a vantage point over their clearing. If a male is caught, another male will replace him within an hour or two, often taking up the previous male's perch. Females fly through the pine forest without apparent direction. On cool days (68°F), Macoun's Arctics sit with their wings fully spread, presumably to increase their thoracic temperature (dorsal basking), while on hot days (80°F) they land with their wings closed and lean over, in order to reduce sun exposure as much as possible (Masters and Sorensen 1968).

Broods. One adult flight occurs during late June every other year. In the Great Lakes states, the Macoun's Arctic flies in even-numbered years, but in the Riding Mountains of Manitoba, the adults are found only in odd-numbered years (Masters 1974).

Adult Food Sources. Masters (1972a) reports that the adults rarely visit flowers.

Caterpillar Host Plants. Masters (1972a) suggests that the larval host is probably a sedge (*Carex* sp.).

171. *Oeneis macounii*

Chryxus Arctic
Oeneis chryxus strigulosa McDunnough, 1934

MAP 172

Etymology. The subspecies name is based on the Latin *strigula*, "swath, windrow, bristle."

Synopsis. The Chryxus Arctic flies in a variety of habitats throughout its extensive range, exhibits a broad spectrum of phenotypes, and might actually consist of more than one biological species.

Butterfly. Male forewing: $\overline{X}$ = 2.3 cm, range 2.2–2.5 cm; female forewing: $\overline{X}$ = 2.4 cm, range 2.3–2.5 cm. The wings are ochre-orange dorsally, with three or four small black submarginal spots on the forewing and one or two on the outer margin of the hind wing. Ventrally, the hind wings are marbled gray-brown with white scaling along the veins.

Range. The Chryxus Arctic occurs from the Gaspé Penninsula, southern Quebec, and northern portions of Michigan and Wisconsin west through the Rockies, Sierra Nevada, and Cascade Mountains. It occurs through western Canada north to Alaska and the Yukon Territory. In the eastern United States it occurs only in northern Michigan and Wisconsin.

172. *Oeneis chryxus strigulosa*

Habitat. In northern Michigan, this species is found on sand prairies or prairielike habitats between jack pine forests.
Life History. *Behavior.* Males perch and occasionally patrol in search of receptive females. When on the ground or low vegetation, individuals of this species lean over on their sides with wings closed. This habit probably serves both as a thermoregulatory device and as camouflage by shadow elimination.
Broods. There is at most a single flight annually, and some populations may fly only every other year. In Michigan and Wisconsin, the Chryxus Arctic flies every year in late spring (May 24 to June 21), but numbers are greater in even-numbered years.
Early Stages. Little is known of the early stages except that winter is passed by the third- or early fourth-instar larva. The mature larva has a dark-brown head and a tan body that is laterally striped with brown. The pupa is tan and has a brown head and wing cases.
Caterpillar Host Plants. The Chryxus Arctic is believed to be a sedge feeder.

Uhler's Arctic
Oeneis uhleri (Reakirt), 1866
MAP 173

Etymology. The species is named in honor of Philip Reese Uhler (1835–1913), who was librarian at the Museum of Comparative Zoology at Harvard and President of the Maryland Entomological Society. He published papers on several insect orders, primarily Hemiptera, and collected briefly in Colorado with the Hayden Survey.
Synopsis. The Uhler's Arctic is a grassland butterfly that barely enters our area.
Butterfly. Male forewing: $\overline{X}$ = 2.3 cm, range 2.0–2.5 cm; female forewing: $\overline{X}$ = 2.5 cm, range 2.2–2.6 cm. This species is very similar to the Chryxus Arctic, particularly on the dorsal surface. It differs by its tendency to have two or more eyespots on the dorsal hind-wing margins, and the pattern on the ventral hind wing, which is marbled and lacks the median band and white veining of the Chryxus Arctic.
Range. Uhler's Arctic ranges from western Minnesota west across the northern plains to Colorado, thence north through the Rocky Mountains and Canada to the McKenzie River of the Northwest Territories.

Habitat. Dry, open, bunch-grass slopes or glades in pine forest are the Uhler's habitats.
Life History. *Behavior.* The reproductive behavior of the male Uhler's Arctic is characterized by perching and periodic patrolling or hovering. South-facing slopes in a bunch-grass community are favored, and the males space themselves 4 to 7 m apart just below ridge crests. They hover, 2 or 3 feet in the air, for 30 to 60 seconds, probably so that they can scan for females downslope.
Broods. This butterfly has a single generation each year during June and early July.
Early Stages. The larvae hibernate in the fourth instar and complete their feeding in the following year. The mature caterpillar is tan, striped longitudinally with dark brown. The head is brown with pale vertical stripes. The chrysalis is yellow-brown.
Caterpillar Host Plants. Larval hosts are reported to be grasses.

Jutta Arctic
Oeneis jutta (Hübner), 1805
PLATE 36 · FIGURE 216 · MAP 174

Etymology. Jutta is a German proper name for Jane or Joan.
Synopsis. This is the only North American arctic restricted to bogs.
Butterfly. Male forewing: $\overline{X}$ = 2.5 cm, range 2.3–2.7 cm (Maine); female forewing: $\overline{X}$ = 2.7 cm, range 2.6–2.8 cm (Maine). The Jutta Arctic is the only North American *Oeneis* that is brown above with black submarginal spots and distinct yellow markings.
Range. The Jutta Arctic is found in subarctic habitats in North America from Maine west to Colorado and northwestward to central Alaska. In the eastern United States it is found in northern Maine, northern Michigan, Wisconsin, and Minnesota. It also inhabits subarctic habitats across northern Eurasia.
Habitat. The Jutta Arctic is found in tamarack or black spruce–sphagnum bogs with glades containing leatherleaf, laurel, Labrador tea, and cotton grass (Masters and Sorensen 1973).
Life History. *Behavior.* Male Jutta Arctics perch with closed wings and periodically patrol glades within their bog habitats. Low vegetation or logs are favorite perch sites.

Female Juttas wander without apparent direction, and are more frequently encountered at bog edges (Masters and Sorensen 1969).

Broods. In Minnesota, the Jutta Arctic has a single generation only in odd-numbered years (June 16–July 18), and flies primarily in odd-numbered years in Wisconsin and Michigan (June 4–July 3). In Maine, it appears in about the same numbers every year (May 26 to June 29).

Early Stages. The Jutta Arctic overwinters as a third-instar caterpillar. In some areas, 2 years may be required to complete the life cycle. The yellow-white, barrel-shaped eggs are laid indiscriminately. The mature caterpillar has a green head

and a pale-green body with dark-brown lateral lines and a middorsal brown spot on each segment. The chrysalis has an amber head, green wing cases streaked with brown, and a pale yellow-green abdomen. Finally, there is a dark-green dorsal line.

Adult Food Sources. In Wisconsin, Jutta Arctics nectar at Labrador tea, while in Wyoming they take nectar from yellow flowers during late morning and early afternoon (1100–1400 hr).

Caterpillar Host Plants. "Hare's-tail" (*Eriophorum spissum*), a sedge, has been reported as the larval host in Michigan.

173. *Oeneis uhleri*

174. *Oeneis jutta*

White Mountain Butterfly
Oeneis melissa semidea (Say), 1828

MAP 175

Etymology. *Melissa* is "honeybee" in Greek, while the sub-species name is a combination of the Greek *semi*, "half," and the Latin *dea*, "goddess."

Synopsis. This butterfly is an Ice Age relict now limited to the windswept rocky alpine tundra of the White Mountains in New Hampshire.

Butterfly. Male forewing: $\bar{X}$ = 2.3 cm, range 2.1–2.4 cm; female forewing: $\bar{X}$ = 2.3 cm, range 2.2–2.4 cm. The adults are uniformly gray-brown dorsally, with translucent wings. Ventrally, there may be a faint hint of a small spot on the forewing apex, and the hind wing has a mottled blackish base and a grayish outer area.

Range. *Oeneis melissa* is holarctic, being found in arctic and

subarctic regions of North America and Eurasia. In North America it is found from Newfoundland to Alaska and south in the Rocky Mountains to Colorado and northern New Mexico. In the eastern United States, it is found only on the summits of the higher peaks in the White Mountains. Mount Washington (1930 m) is the best-known locality, since a paved road goes to its summit.

Habitat. The habitats are open, rocky summits and saddles in the Arctic-Alpine Life Zone.

Life History. *Behavior.* The White Mountain Butterfly rarely flies more than a half meter above the ground, but if disturbed may allow the wind to carry it several hundred meters downslope. The butterflies are not active at temperatures below 7° C or in winds greater than 40 mph. Males perch and patrol rocky areas on hills or ridge tops during most of the day (Anthony 1970).

Broods. The White Mountain Butterfly is univoltine, with adults flying from June 27 to July 22.

Early Stages. Females oviposit on the ground or in litter at the base of sedges. The caterpillars feed at night and pupate between rocks and mosses. The egg is pale yellow-white and is higher than it is broad. The mature caterpillar has a yellow-brown or dull green-brown head with seven black bands or spots. There are also separate lateral black and narrow dark-green stripes. The chrysalis is dull yellow-brown, with black head and black wing veins.

Caterpillar Host Plant. On Mt. Washington, the butterfly selects Bigelow's sedge (*Carex rigida*) as its caterpillar host.

Polixenes Arctic
Oeneis polixenes katahdin (Newcomb), 1901
MAP 176

Etymology. The species name is derived from Polyxena, the daughter of Priam and Hecuba, the bride of Achilles (Homerian epic). The subspecies name is derived from Mt. Katahdin, Maine. Furthermore, Katahdin means "big mountain" in the Algonquin language.

Synopsis. This butterfly is found in our area only on the high, alpine tundra of Mt. Katahdin, Maine.

Butterfly. Male forewing: $\overline{X}$ = 2.3 cm, range 2.0–2.5 cm; female forewing: $\overline{X}$ = 2.3 cm, range 2.1–2.5 cm. This butterfly's fragile wings are a warm brown and are unmarked dorsally in the male. The female may have two tiny black submarginal eyespots on the forewing, both above and below. Ventrally, the hind wings are mottled brown, gray, and black, with a dark brown mesial band, outlined in gray. There is also a faint row of submarginal white spots on the ventral hind wings.

Range. This New World species is found from Maine to Labrador east to the Colorado Rocky Mountains, thence north through Canada to arctic Alaska.

Habitat. On Mt. Katahdin this fragile-looking butterfly is found at altitudes between 1350–1500 m on short alpine tundra with tufts of sedges and grasses. It and the White Mountain Butterfly are the only *true* arctic-alpine butterflies in the eastern United States.

Life History. *Behavior.* In Colorado, Polixenes Arctic males have been observed to patrol all day but to perch occasionally in swales on grassy north-facing slopes.

Broods. The Polixenes Arctic is found between June 24 and July 28 on Mount Katahdin. Greater numbers are found in even-numbered years.

Early Stages. Early stages have been partially described by Edwards (1862–97). Larvae overwinter in the first or third instar, and 2 years may be required to complete the life cycle at some localities. The chalk-white subconical egg is laid on the host. The mature caterpillar has a brown head and a light-brown body with dark longitudinal stripes.

Caterpillar Host Plants. Larval hosts have not been reported, but are believed to be alpine grasses.

175. *Oeneis melissa semidea*

176. *Oeneis polyxenes katahdin*

Subfamily Danainae: Milkweed Butterflies

The milkweed butterflies are a largely tropical group of medium to large species. The males have either hindwing scent patches (all our species) or tufts of long scales ("hair pencil") near the end of the abdomen. The males of our species patrol for mates, and mating occurs predominantly in the afternoon. Females oviposit near midday. One species, the Monarch (*Danaus plexippus*), is well known for its long migrations, and other members of the subfamily are probably short-distance emigrants at tropical latitudes.

All stages are distasteful and emetic due to the cardiac glycosides contained in host leaves that are incorporated by the caterpillars. Our species have average-length proboscises, and the butterflies feed only on floral nectar. Milkweeds, dogbanes, and solanums are the larval hosts.

The caterpillars are naked except for fleshy tubercles on the posterior and anterior body segments. The cylindrical pupae are pendant from a silk button. No stage of these butterflies can withstand subfreezing winters.

Monarch
Danaus plexippus (Linnaeus), 1758
PLATE 37 · FIGURES 217–219 · MAP 177

Etymology. The genus name is derived from Danaus, the mythical king of Argos in ancient Greece.

Synopsis. The Monarch is perhaps the best-known American butterfly. Its incredible migrations of several thousand kilometers to central Mexico each winter, and its mimetic relationship with the Viceroy (*Limenitis archippus*) make it both the best example of regular insect migration and the best North American example of Batesian mimicry (Sheppard 1965).

Butterfly. Male forewing: $\bar{X}$ = 5.1 cm, range 4.3–5.9 cm; female forewing: $\bar{X}$ = 5.0 cm, range 4.0–5.9 cm (Mather 1955). The Monarch is a large orange butterfly with black scaling on the veins and a broad black margin on both wings above. The male is bright orange with an enlarged black scent patch on the second cubital vein. The female is a brownish orange with the black veining blurred.

Range. The Monarch occurs from southern Canada south through all of the United States, Central America, and most of South America (Urquhart 1960). Permanent populations are now established in Australia, Hawaii, and other Pacific islands. Resident populations in tropical areas are not migratory, but appear to undergo altitudinal movements during the dry season.

Habitat. The Monarch is found in a wide range of open habitats, often those disturbed by man. Milkweeds are common in open fields, pastures, marshes, and roadsides, and Monarchs may be found in these situations.

Migration. Throughout most of the United States and southern Canada, when fall arives, Monarchs vacate their breeding grounds and fly south to sites where winter temperatures are relatively mild. Usually, Monarchs follow river valleys or the sea coast, but they may also cross ridges or large bodies of water. During migration, their flight is high above the ground (5–500 m) and direct, although they usually pause to take nectar, particularly from fall-blooming composites, such as goldenrods, asters, ironweed, joe-pye weed, and bonesets. The Monarchs roost at night, sometimes by the thousands. The Monarchs' departure usually begins in mid-August, peaks during September, and continues through October in decreasing numbers. Monarchs from eastern North America winter in small numbers along the Atlantic Coast from South Carolina south to Florida and along the Gulf Coast of Florida. Some individuals winter on Cuba or other Caribbean islands, but the vast majority travel to the highlands of Mexico, where they form dramatic aggregations, sometimes numbering in the millions (Urquhart and Urquhart 1976a). These wintering Monarchs cling to pendant tree branches and each other in great festooned clusters. At the beginning of this period, the individuals have much stored fat in their bodies and are in reproductive diapause or arrest. As the winter months proceed, the Monarchs take moisture along streams and occasionally visit flowers on warm days. Their stored fat is gradually used up and their reproductive organs develop. The majority of the flock have mated by the time they leave the Mexican colonies and their stored reserves are nearly spent. They begin their northward flight during March and April. They arrive as early as March 5 in Mississippi and by late April in southern New York and Illinois, but not until May in Michigan. The returning individuals are few; perhaps only about one percent survive to return.

During the southward migration, some Monarchs either fly off course or more likely get caught in storms, as they are sporadically found in England and southern Europe, where no milkweeds grow.

Mimicry. The milkweeds upon which Monarch caterpillars

feed contain cardiac glycosides. These substances are incorporated into the body of the larva and adult, rendering them both distasteful and emetic to many vertebrates, especially birds. This, of course, constitutes a form of protection, and both larvae and adults have bright coloration to warn would-be predators. This strategy is known as "aposematic," or warning, coloration. Some milkweed populations lack cardiac glycosides and the Monarchs reared on them are perfectly edible. That edible Monarchs look exactly like distasteful ones has been termed "automimicry" by Brower et al. (1970), since the edible individuals presumably suffer less predation than their value as perfectly good bird food would otherwise dictate.

The Viceroy (*Limenitis archippus*), whose larvae feed on willow or poplar leaves, is perfectly edible, but its adult is very close to the Monarch in color and pattern, and very much unlike its close relative the Red-spotted Purple (*Limenitis arthemis astyanax*). Because of this resemblance, the Viceroy is presumed to be a Batesian mimic of the Monarch, and birds avoid it because it is so close in appearance to the Monarch. This concept has been demonstrated in laboratory feeding experiments with caged animals, although no field experiments have been attempted.

The Monarch and Viceroy have somewhat different habitats, the Monarch most often being found in dry situations and the Viceroy near streams and in marshy situations. In late summer, when Monarchs turn to swamp milkweed (*Asclepias incarnata*) in marshy situations, they are often seen with Viceroys. This is perhaps the situation in which ancestors of the Viceroy first gained a selective advantage by coming to look more and more like the Monarch.

Life History. *Behavior.* Male Monarchs patrol open fields, freshwater marshes, and gentle slopes in the vicinity of the milkweed hosts. The males interact not only with each other in high circling flights, but also with other large orange butterflies. Mating occurs all day (0725–1830 hr), but most pairs are seen in late afternoon (1400–1630 hr). Males are the carrying sex, and mated pairs often rest on foliage high up in a tree. The butterflies may remain coupled for more than an hour. Females fly closer to the ground than males and, upon encountering a suitable host, deposit a single egg on each of several leaves. Monarchs increase their body temperature by dorsal basking while perched 1 or more meters above the ground.

Broods. The Monarch has the potential to have continuous broods if weather and food-plant conditions permit, as they do in portions of the tropics. In the eastern United States, the length of the growing season limits the number of generations, which are seldom synchronized, as one goes north. In the South, there are as many as four to six broods (in Louisiana, Mississippi, and Virginia), and some Monarchs breed through the winter in southern Florida (Urquhart and Urquhart 1976b). In southern New York there are two to four broods, but in such places as northern New York, Maine,

and northern portions of the Great Lakes states there is time for only one to three generations. Overwintering Monarchs may live as long as 6 months, but Monarchs of the summer reproductive broods may not live much more than a month. *Early Stages.* The white, conical eggs are laid under host leaves. Development is always direct and requires about 6 weeks. The caterpillar has a white and black striped head. The body is transversely striped on each segment with yellow, black, and white, and has pairs of fleshy tubercles on segments 3 and 11. The chrysalis is glossy green, with metallic golden dots, and has a black and gold band between the second and third abdominal segments.

Adult Nectar Sources. Favored nectar sources of the Monarch are milkweeds (all *Asclepias* species) and composites, which they visit in the fall virtually to the exclusion of other plants. At that time, goldenrods, thistles, blazing stars, joe-pye weed, ironweed, and tickseed sunflower are favored. Earlier in the year, before many milkweeds flower, a variety of unrelated flowers, including red clover, dogbane, winter cress, buttonbush, and lilac, are utilized. In Florida, lantana, thistles, shepherd's needle, and other composites are visited. Although Monarchs are not the only visitors to milkweeds nor even the most numerous, they are constant when present, and probably contribute significantly to the plants' pollination. Thus they are one of several butterflies who pollinate the same plants they feed upon as larvae.

Caterpillar Host Plants. Throughout the eastern United States, Monarch caterpillars feed on various milkweeds (*Asclepias*). Common milkweed (*Asclepias syriaca*) and swamp milkweed (*A. incarnata*) are both widespread throughout the region and are frequently utilized. Butterflyweed (*Asclepias*

177. *Danaus plexippus*

tuberosa) is also widespread but is seldom selected by Monarchs, perhaps because of its leaf chemistry or its low stature. Others, such as poke milkweed (*Asclepias exalta*) and purple milkweed (*A. purpurascens*), are also selected. Larvae are occasionally found on dogbane (*Apocynum*), white vine (*Sarcostemma clausa*) or green milkweed (*Acerates*). In Florida, as well as in other tropical and subtropical areas, Monarchs often select bloodflowers (*Asclepias curassavica*), a common pantropical weed. In the American tropics, various genera of milkweed vines occur, and Monarchs have been reported to feed on *Matelea reticulata* in Texas and mudar (*Calotropis procera*) in the West Indies.

Queen
Danaus gilippus (Cramer), 1775
PLATE 37 · FIGURE 220 · MAP 178

Synopsis. Like the Monarch, the Queen is a member of a basically tropical subfamily, the Danainae. In the United States, two subspecies of the Viceroy, *Limenitis archippus floridensis* and *L. a. obsoletus*, are Batesian mimics of the peninsular Florida and Southwest Queen populations, respectively.

Butterfly. Male forewing: $\overline{X}$ = 4.2 cm, range 3.7–4.6 cm; female forewing: $\overline{X}$ = 4.2 cm, range 3.7–4.6 cm. Dorsally, the Queen is rich deep chestnut with small white apical and marginal spots. Both sexes are similar, but the male has a black androconial patch on the dorsal hind wing. The Queen is most similar to the Soldier (*Danaus eresimus tethys*); refer to the latter's species account for details of its distinguishing traits.

Range. The Queen occurs regularly in peninsular Florida and southern Georgia, as well as in the southern portions of the states bordering on Mexico, thence south through the West Indies and Central America to Argentina. Periodically, strays enter the plains and Midwest from the Southwest.

Habitat. The Queen favors open fields, pastures, roadsides, dunes, and other open, sunny areas where milkweeds grow.

Mimicry. As was mentioned above, the Monarch is a rare breeding species in our southern tier of states. For example, Kimball, in his treatment of Florida Lepidoptera, notes that the Queen is more common than the Monarch there. Queens are presumably just as distasteful and emetic as Monarchs, since their caterpillars also eat milkweeds, and Florida Viceroys developed a selective advantage by coming to mimic the more common Queen.

Life History. *Behavior.* The Queen is very similar to the Monarch in its general habits except that it does not undergo dramatic migrations. In areas of the tropics that have a distinct dry season, Queens fly from the lowlands to high elevations to wait out that unfavorable time of year. Courtship and mating are afternoon activities. Mated pairs have been seen from 1530 to 1715 hr. Males are the carrying sex. *Broods.* In tropical areas with no distinct dry season, the Queen may have continuous broods throughout the year. The species is found throughout the year in Florida, but those adults found in drier winter months may be in reproductive diapause (Brower 1962). There is no definite record of the number of broods anywhere. In portions of the eastern United States (Mississippi, Louisiana, Illinois) Queens appear as emigrants, normally in August, but occasionally during other months.

Early Stages. The early stages are mostly similar to those of the Monarch, but the caterpillar is brown-white with yellow and brown transverse stripes on each segment and has a yellow-green lateral stripe. There are three pairs of black fleshy tubercles along the back, on segments 2, 5, and 11. The pupa is similar to that of the Monarch, but has a blue band below the black abdominal band.

Adult Nectar Sources. In Florida, Queens take nectar from their milkweed hosts as well as from shepherd's needle and fogfruit.

Caterpillar Host Plants. Butterflyweed (*Asclepias tuberosa*) is eaten in Florida, and bloodflower (*Asclepias curassavica*) has been recorded as a larval host at several widely

178. *Danaus gilippus*

separated localities. Larvae were portrayed by Abbot on blunt-leaved milkweed (*Asclepias amplexicaulis*) and Riley (1975) reports honey vine (*Cynachum*) to be favored in the West Indies. Other reported hosts in the tropics are oleander (*Nerium*), angle pod (*Gonolobus*), white vine (*Sarcostemma*), and carrion flower (*Stapelia*).

Soldier
Danaus eresimus tethys Forbes, 1943
MAP 179

Etymology.   The subspecies is named for Tethys, a Greek sea goddess, wife of Oceanus.

Synopsis.   Until recently the Soldier was a rare vagrant from the West Indies, but it is now resident in southern Florida and the Keys.

Butterfly.   Male forewing: $\overline{X}$ = 4.0 cm, range 3.7–4.7 cm; female forewing: $\overline{X}$ = 3.7 cm, range 3.3–3.8 cm. Generally, the Soldier is much like the Queen in color and appearance, but may be distinguished by the dorsal black veining and the paler color outwardly, as well as by a pattern of pale blotchy spots on the outer area of the ventral hind wing.

Range.   The Soldier ranges from southern Florida and Texas south through the West Indies and Central America to Brazil.

Habitat.   This species frequents open pastures and fields as well as edges of seasonally dry tropical forest.

Life History.   *Behavior.* Behavior has not been reported but is believed to be similar to that of the Queen. Individuals emigrate to moister areas during the dry-season.

Broods. In Florida, adults may be found all year, but they are most common from October to December.

Early Stages. The egg is bright orange and looks much like an aphid that is common on milkweeds.

Adult Nectar Sources.   In Costa Rica, adults take nectar from bloodflower, a possible host, as well as from heliotrope and composites. Similar flowers, including shepherd's needle, are probably utilized in southern Florida.

Caterpillar Host Plant.   The only reported host in our area (Florida) is white vine (*Sarcostemma clausa*).

179. *Danaus eresimus tethys*

SUPERFAMILY HESPERIOIDEA: SKIPPERS

Family Hesperiidae

The skippers have three subfamilies in our area, the Pyrgines (Pyrginae), the Branded Skippers (Hesperiinae), and the Giant Yucca Skippers (Megathyminae). Skippers are usually small to medium (rarely, large), thick-bodied, dull-winged (orange, black, white) butterflies with six fully developed walking legs and elongated antennal clubs that are often hooked at the tip (except in the Megathyminae).

Skippers are incredibly diverse in the American tropics, but are very poor in species in arctic and subarctic habitats. Much of the information in this section was drawn from the authoritative treatment by MacNeill (1975).

Subfamily Pyrginae: Pyrgines

The Pyrgine skippers are often black or black and white. Sometimes they have iridescent scales or patches of other colors. The males lack forewing stigmata or sex brands, but may have specialized scales in costal folds or in specialized leg tufts.

Males of most Pyrgines are perchers, but some species patrol. Generally, Pyrgines mate in the afternoon, and females oviposit from midday to midafternoon. Pyrgines often rest with their wings wide open, and some species, especially in the tropics, roost upside down under leaves. Most species are multiple-brooded, and winter is usually passed by nearly mature larvae within folded leaf shelters. Pyrgines are avid flower visitors, although their proboscises are short relative to those of Branded Skippers. The larval hosts are a wide array of dicotyledonous plants. Females usually deposit single eggs on their host's leaves. Exceptions are the Gold-banded Skipper (*Autochton cellus*), which lays its eggs in groups, and the Silver-spotted Skipper (*Epargyreus clarus*), which may lay its eggs on plants adjacent to the true host. The eggs are ovoid or drum-shaped and often have strong, vertical ribs. The larvae have heart-shaped head capsules, which often have paired eyespots on the front. The caterpillars have thick bodies that taper gradually at either end so that there is a constricted "neck". The larva lives in a series of folded leaf shelters of increasing size and feeds at night. When fully grown it pupates in a loose cocoon within leaves that have been sewn together.

Mangrove Skipper
Phocides pigmalion okeechobee (Worthington), 1881
PLATE 38 · FIGURE 223 · MAP 180

Etymology. The genus name is possibly taken from Phocis, an ancient Greek state. Pygmalion was a sculptor who fell in love with a statue of a maiden. He was a grandson of Agenor, a king of Cyprus (Greek myth).

Synopsis. The Mangrove Skipper is the only butterfly in our territory that feeds on mangrove.

Butterfly. Male forewing: $\overline{X} = 2.6$ cm, range 2.5–2.8 cm; female forewing: $\overline{X} = 3.0$ cm, range 2.9–3.3 cm. This large skipper is brownish black with purplish highlights and a row of ill-defined submarginal iridescent blue spots on the ventral hind wings.

Range. The species occurs from southern Florida through the Bahamas and Greater Antilles to Argentina.

Habitat. The butterfly frequents coastal mangrove thickets in the intertidal area of the Tropical Life Zones.

Life History. *Behavior.* The flight is fast and powerful.

197

Broods. The number of generations is not documented. Adults have been found from November through August in southern Florida.

Early Stages. The mature larva is covered with a powdery white exudate and has a brownish head with two large orange or yellow spots in front. The pupa is creamy white with a greenish tinge on the thorax (Strohecker 1938).

Adult Nectar Sources. The Mangrove Skipper is partial to the flowers of its host, red mangrove. It also visits shepherd's needle, citrus, and bougainvillaea.

Caterpillar Host Plant. As mentioned above, red mangrove (*Rhizophora mangle*) is this butterfly's only host.

Mercurial Skipper
Proteides mercurius sanantonio (Lucas), 1857

MAP 181

Synopsis. Individuals of the Caribbean subspecies of this tropical skipper, which is superficially similar to the Silver-spotted Skipper (*Epargyreus clarus*), have been found twice in northern Florida and once in Louisiana. The Mercurial Skipper has a tropical distribution, occurring in the West Indies and from southern Texas south through Latin America to Argentina. Various woody tropical plants have been reported as hosts.

Zestos Skipper
Epargyreus zestos (Geyer), 1832

MAP 182

Synopsis. The Zestos Skipper is similar to the Silver-spotted Skipper but lacks the ventral silver spot. It is limited in range to southern Florida and the Lesser Antilles. It is found from January through September in Florida and little is known of its behavior or early stages.

180. *Phocides pigmalion okeechobee*

181. *Proteides mercurius sanantonio*

Silver-spotted Skipper
Epargyreus clarus (Cramer), 1775

PLATE 37 · FIGURES 221 & 222 · MAP 183

Etymology. Epargyreus is possibly derived from *argyros*, Greek for "silver." *Clarus* is Latin for "clear, bright, loud."

Butterfly. Male forewing: $\overline{X}$ = 2.7 cm, range 2.2–2.9 cm; female forewing: $\overline{X}$ = 2.9 cm, range 2.4–3.1 cm. This large brown-black skipper has pointed forewings with a large translucent gold patch in the center of the forewing, and a large metallic silver patch in the center of the ventral hind wing. Similar species are the Hoary Edge and Gold-banded Skipper.

Range. The Silver-spotted Skipper occupies a broad area extending from southern Quebec west to southern British Columbia, thence southward to Florida, the Gulf Coast, Texas, and northern Mexico.

Habitat. This skipper occurs in a variety of disturbed and open forest habitats. It does best in young, brushy second growth, where its chief caterpillar host, black locust, abounds. It is found from Tropical through Transition Life Zones.

Life History. *Behavior.* Males perch on projecting branches and tall weeds from 1–3 m above the ground during the morning and early afternoon (0730–1415 hr). They occasionally patrol back and forth 20–30 m in either direction from their perches before returning. They have frequent aerial encounters with other males or various large passing insects. Males may court females from their perches or interrupt their flower visits if a female is nearby. Mating occurs primarily at midday (1040–1315 hr), and females are the

carrying sex. Females oviposit near young locusts or other hosts. They have a slow, dipping flight in which they briefly touch various plants; after touching a correct host plant, the female then lands on another plant nearby to lay an egg. However, many reported hosts may not be true hosts, since it is the young caterpillar's chore to locate the correct host, which is presumably nearby.

At night the Silver-spotted Skippers perch upside down under large leaves. They may seek their roosts in mid-afternoon on hot or cloudy days.

Broods. Throughout most of the East, the Silver-spotted Skipper is bivoltine, usually in May to early July and then in late July to mid-September. In the northern part of its range, e.g., northern New York and northern Michigan, the skipper is not resident but may colonize and carry off a single generation in some years. At the other extreme, in Florida and the Gulf states, adults have been found from February to December and three or four broods may occur.

Early Stages. After hatching, the larvae must find their proper host, where they first make a single-leaf shelter but later make a shelter of several bunched leaves. Caterpillars resulting from eggs laid by first-brood females develop directly, but those resulting from the second brood overwinter as mature larvae in tightly spun leaf shelters. The egg is a green hemisphere with 16–19 minutely beaded vertical ridges. The mature caterpillar is yellow with horizontal cross lines. It is a red-brown with two large, round, red-orange spots low on the

182. *Epargyreus zestos*

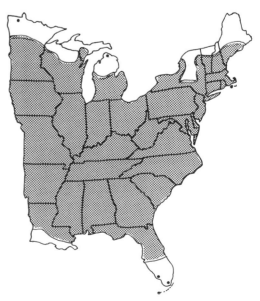

183. *Epargyreus clarus*

front. The pupa is brown with darker or lighter lines and blotches.

Adult Nectar Sources. The Silver-spotted Skipper has a predilection for blue, red, pink, or purple flowers, but also visits white or cream-colored flowers. It almost never visits yellow flowers, the only exceptions known to us being occasional visits to winter cress in early May. Favored nectar plants include purple vetch, everlasting pea, dogbane, common milkweed, Japanese honeysuckle, red clover, buttonbush, swamp milkweed, selfheal, blazing star, joe-pye weed, thistles, privet, and viper's bugloss.

Caterpillar Host Plants. Small, isolated individuals or suckers of various locust trees (*Robinia*) are the preferred larval hosts, among which black locust (*Robinia pseudacacia*) is used most widely. Many other legumes, including beggar's ticks (*Desmodium*), wisteria (*Wisteria*), hog peanut (*Amphicarpa bracteata*), and kudzu (*Pueraria thunbergii*), are also fed upon. Females oviposit on or near some plants upon which the young caterpillars cannot survive, i.e., coral bean (*Erythrina*).

Hammock Skipper
Polygonus leo (Gmelin), 1790
PLATE 38 · FIGURE 224 · MAP 184

Etymology. The genus name is taken from the Greek *polygonos*, "having several angles." *Leo* is Latin for lion.

Synopsis. This is a common butterfly of hardwood hammocks in southern Florida.

Butterfly. Male forewing: $\bar{X}$ = 2.5 cm, range 1.9–2.7 cm; female forewing: $\bar{X}$ = 2.4 cm, range 2.3–2.5 cm. The Hammock Skipper may be distinguished from Manuel's Skipper by its blue sheen and a prominent dark spot on the hind wing's subcostal area. The latter species probably does not occur in our area. The color pattern is similar to that of many tropical forest understory dwellers that fly through the sun-dappled shade, e.g., the purple wings (*Eunica*).

Range. The Hammock Skipper occurs from southern Florida, southern Texas, Arizona, and California south through the Caribbean, Mexico, and Central America to Peru and Argentina.

Habitat. In tropical Florida, the butterfly is found in shaded hardwood hammocks, being seen in openings and along trails.

Life History. *Behavior.* This butterfly perches upside down on the underside of large leaves. Like many large tropical skippers, it may be most active at dawn and at dusk. *Broods.* Broods are not documented, but adults may be found throughout the year in Florida. *Early Stages.* The egg is green and barrel-shaped. The caterpillar is translucent green, covered with dense, fine, short, white hairs, and has two lateral yellow stripes on each side. The head is green with two black spots on the upper front.

Caterpillar Host Plants. The Hammock Skipper feeds on the leaves of several woody legumes throughout its range. In Florida it utilizes Jamaican dogwood (*Piscidia piscipula*) and karum tree (*Pongamia pinnata*).

184. *Polygonus leo*

Long-tailed Skipper
Urbanus proteus (Linnaeus), 1758
PLATE 38 · FIGURES 225 & 226 · MAP 185

Etymology. The Latin word *urbanus* means "citizen of a town, city." Proteus was god of the sea, son of Poseidon (Greek myth).

Synopsis. This butterfly is one of the more attractive and widespread tropical skippers that occur north of our southern borders. It is occasionally a serious pest of beans.

Butterfly. Male forewing: $\overline{X}$ = 2.4 cm, range 2.3–2.7 cm; female forewing: $\overline{X}$ = 2.4 cm, range 2.1–2.7 cm. The Long-tailed Skipper is characterized by its long tails and the iridescent green sheen covering much of the wing's dorsal surfaces. Males and females are similar.

Range. This skipper ranges from the southern portions of the United States south through the Caribbean, Mexico, and Central America to Argentina. In the East, it is resident in Florida and on the coastal plain of Georgia and the Gulf states. It periodically moves north, especially along the immediate Atlantic Coast as far as Long Island, but its stages cannot survive prolonged freezing temperatures.

Habitat. The Long-tailed Skipper is a butterfly of open, disturbed habitats. It favors brushy fields and woodland margins. It is resident in the Tropical Life Zone.

Life History. *Behavior.* Males perch along forest edges or in sunlit glades from 1 to 2 m above ground. They roost upside down under leaves or under overhanging limbs, in the manner of many tropical skippers.

Broods. Adult Long-tailed Skippers are found throughout the year where the species is resident. There are probably two or three generations from spring to fall and a long winter generation of somewhat inactive adults in reproductive arrest.

Early Stages. The caterpillars live in rolled leaf shelters. The egg is a yellow hemisphere with about 12 vertical ridges. The mature caterpillar is yellow-green, with a thin black dorsal line and a yellow to reddish stripe with two thin green lines along each side. Its head is black, with a brown top and two large orange to yellow spots low on the front. The pupa is dark brown and is covered with a whitish powder.

Adult Nectar Sources. Long-tailed Skippers take nectar at lantana, bougainvillea, several composites, including shepherd's needles, and a variety of other flowers.

Caterpillar Host Plants. Caterpillars feed on a variety of climbing legumes. They prefer various beans (*Phaseolus*), but they also utilize wisteria (*Wisteria*), blue peas (*Clitoria*), beggar's ticks (*Desmodium*), and hog peanut (*Amphicarpa bracteata*). Records for crucifers and monocotyledonous plants, e.g., canna, are probably erroneous.

Dorantes Skipper
Urbanus dorantes (Stoll), 1790
PLATE 38 · FIGURE 227 · MAP 186

Synopsis. During 1969 this widespread skipper became established in southern Florida from Central American stock (Knudson 1974).

Butterfly. Male forewing: $\overline{X}$ = 2.2 cm, range 2.1–2.3 cm; female forewing: $\overline{X}$ = 2.3 cm, range 2.2–2.4 cm. The Dorantes Skipper closely resembles the Long-tailed Skipper (*Urbanus proteus*) but lacks the dorsal iridescent green on the head, thorax, and wings.

Range. This tropical skipper occurs from peninsular Florida, Texas, southern Arizona, and southern California

185. *Urbanus proteus*

186. *Urbanus dorantes*

southward through the Caribbean, Mexico, and Central America to Argentina. Strays have appeared north to Missouri and Tennessee.

Habitat. During spring, summer, and fall, the Dorantes Skipper may be found along roadsides, in overgrown fields, and at the edges of tropical hardwood groves. During the winter it is found along trails and in clearings within evergreen hardwood hammocks. It is restricted to the Tropical Life Zone.

Life History. *Behavior.* This skipper's behavior is very similar to that of the Long-tailed Skipper. During the winter months it feeds at flowers each day, but spends most of the time perched upside down under leaves.

Broods. Adults may be found in all months in Florida. There

are probably two or three generations during the warmer months, and one long late fall and winter generation that does not become reproductive until spring.

Early Stages. The egg is an iridescent green hemisphere with a flattened top. The mature larva has a black to brown head and a green to pale pink-orange body that becomes redbrown posteriorly. There is a dark red-brown dorsal line. The pupa is light brown.

Adult Nectar Sources. Lantana and shepherd's needle are the favored nectar flowers in Florida, but the adults also visit bougainvillea, ironweed, trilisa, and a variety of other plants.

Caterpillar Host Plants. Larval hosts are viny legumes, including both wild and cultivated beans (*Phaseolus*), blue peas (*Clitoria*), and others.

Gold-banded Skipper
Autochton cellus (Boisduval & Leconte), 1837
PLATE 38 · FIGURE 228 · MAP 187

Etymology. The genus comes from the Greek *autochton,* "sprung from the land itself," while the species name is based upon the Latin *cella,* "cell."

Synopsis. The Gold-banded Skipper has an extensive range, from southern Pennsylvania to northern Mexico, but it is always rare and local in its occurrence.

Butterfly. Male forewing: $\bar{X}$ = 1.8 cm, range 1.7–2.0 cm; female forewing: $\bar{X}$ = 2.3 cm, range 2.2–2.4 cm. The Goldbanded Skipper should be confused with no other species. It has a broad, golden yellow band crossing each forewing diagonally from costal margin to the termen. There are no white patches or bands on the ventral hind wings. The male has a white band on each antenna just below the club.

Range. The Gold-banded Skipper ranges from southeastern Pennsylvania and New Jersey (rarely) south to central Florida and the Gulf states, and thence southwestward across the eastern United States through central Texas and southern Arizona, and southward in the mountains to southern Mexico.

Habitat. The Gold-banded Skipper always seems to occur in damp, wooded ravines with permanent streams, bogs, or ponds. The species is limited to Lower and Upper Austral Life Zones.

Life History. *Behavior.* Males perch during the afternoon, on rocks or up to about 2 m above ground on vegetation.

Broods. In the most northern part of its range, in Pennsylvania and New Jersey, where the species may be only a periodic colonist, there is only a single generation in June. In the southeastern states the skipper is bivoltine, although the second flight is usually brief and incomplete. In Virginia, the flights occur from late May to early July, and then again in late July and August. In Georgia, the emergences are earlier, occurring from early April to mid-June and from early July to

early September. In Florida most adults are found from late February to mid-April.

Early Stages. The eggs are laid in strings of two to seven. (Virtually all other skippers lay single eggs.) The egg is yellow and flask-shaped, with 15–20 vertical ridges that are more prominent near the top. The mature caterpillar is bright yellow-green with scattered tiny yellow points and a wide yellow lateral line. The head is red-brown with two large round yellow patches on the front. The pupa is green and yellow at first, but turns brown after a few hours (Clark 1936).

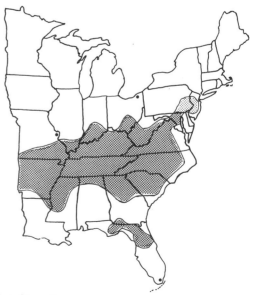

187. *Autochton cellus*

Adult Nectar Sources. Native plants whose flowers are visited include blackberry and trailing arbutus, while hollyhock and abelia are also visited in suburban situations.

Caterpillar Host Plant. The only recorded host is hog peanut (*Amphicarpa bracteata*).

Hoary Edge
Achalarus lyciades (Geyer), 1832
PLATE 39 · FIGURES 229 & 230 · MAP 188

Etymology. The genus name, *Achalarus*, is from the Greek *a*+*khalaros*, "not loosened." The species name, *lyciades*, may be taken from Lycia, an ancient country in S.W. Asia.

Synopsis. This skipper is the only eastern respresentative of a group whose species occur primarily in Mexico and Central America.

Butterfly. Male forewing: $\overline{X}$ = 2.1 cm, range 2.1–2.3 cm; female forewing: $\overline{X}$ = 2.2 cm, range 2.2–2.3 cm. The Hoary Edge is slightly smaller than the Silver-spotted Skipper (*Epargyreus clarus*) and has a broad white band along the outer portion of the hind wing below and less angular wings.

Range. The Hoary Edge occurs from southern New England west to southern Minnesota and south through almost all our area to northern Florida, the Gulf states, and eastern Texas.

Habitat. The Hoary Edge is at home in open woodland and adjacent brushy areas. It is especially prevalent in pine or oak woods with sandy soils. It occurs in the Lower Austral through Transition Life Zones.

Life History. *Behavior.* Males perch in forest glades or openings at the end of a twig or small branch about 1 m (occasionally 3 m) above ground. They will repeatedly chase other butterflies and large insects, returning to the same perch over and over again. Females oviposit during early afternoon (1214–1615 hr), laying a single egg under the leaflet of the host. Hosts in full or partial shade are usually selected.

Broods. The Hoary Edge is univoltine, usually flying from early or mid-June to late July or early August, in the northern part of its range (Michigan, northern New York). In the central portion, the first brood flies from mid-May through June and a partial second brood flies in late July and August. In the Deep South it is fully bivoltine, with flights in April–May and July to early September.

Early Stages. The egg is dull white with 13–15 vertical ridges. the mature caterpillar is dark green, with a blue-green dorsal line, a narrow orange lateral stripe, and a profuse scattering of yellow-orange dots. The head is black. The pupa is pale brown with scattered patches of black dots.

Adult Nectar Sources. Dogbane and common milkweed are visited most often, but flowers of Japanese honeysuckle, buttonbush, and New Jersey tea are also relied upon in some instances.

Caterpillar Host Plants. Large-leafed species of beggar's tick (*Desmodium*) are the usual hosts, but other legumes, such as false indigo (*Baptisia*) and bush clover (*Lespedeza*) may be eaten in some situations.

188. *Achalarus lyciades*

Southern Cloudy Wing
Thorybes bathyllus (J. E. Smith), 1797
PLATE 39 · FIGURES 231 & 232 · MAP 189

Etymology. *Thorybes* is an irregular formation from the Greek *thorybos*, "noise." Bathyllus was a poet and a contemporary of Vergilius.

Synopsis. The three cloudy wings that occur in our area are easily confused, and great care must be taken in the identification of many individuals. The Confused Cloudy Wing

(*Thorybes confusis*) is more restricted to the South than the present species, despite its common name.

Butterfly. Male forewing: $\bar{X}$ = 1.9 cm, range 1.7–2.2 cm; female forewing: $\bar{X}$ = 1.9 cm, range 1.8–2.0 cm. The Southern Cloudy Wing generally has a slightly elongated hindwing anal angle and its hyaline forewing spots are broader and more aligned than those of the other two species. Male Southern and Confused Cloudy Wings lack a costal fold.

Range. The Southern Cloudy Wing occurs from southern Maine, New Hampshire, and New York west across southern Ontario and Michigan to Minnesota, Nebraska, southeastern Colorado, and northern New Mexico. It ranges south to central Florida, the Gulf Coast, and Texas.

Habitat. The Southern Cloudy Wing is a skipper of open and scrubby habitats ranging from dry meadows and moist hayfields to power-line right-of-ways, burn scars, upland barrens, and dry prairie hills.

Life History. *Behavior.* Males perch 0.5–1.5 m above ground on ridge tops or small knolls or in open fields if no promontories are nearby. They are extremely faithful to particular sites: some males probably retain the same perch for most of their brief lives. One marked male in Virginia was observed repeatedly on the same perch for more than a week. Maximum lifespan is probably no more than 2 weeks. Females oviposit near midday (1120–1255 hr), laying a single egg on the underside of a host leaflet. Newly emerged males may imbibe moisture from wet areas along streams or trails. *Broods.* In the northernmost portions of its range, e.g., northern Michigan and central New York, where the species may be only a periodic colonist, the Southern Cloudy Wing is univoltine, with a flight in early summer (mid-June to mid-July). In most of its range this skipper is bivoltine and sometimes has a partial third flight. For example, in Virginia it flies during June, August, and sometimes late September to early October. In Florida, where adults have been found from mid-February to mid-October, there may be four generations.

Early Stages. The egg is a pale-green, broad hemisphere with about 15 slender vertical ridges and numerous cross-striations. The mature larva is dull mahogany-brown with paler dorsal and lateral lines and a deeply cleft black head covered with short golden brown hairs. The pupa is stout and is dull brown or green-brown.

Adult Nectar Sources. Southern Cloudy Wings seldom visit yellow, orange, or red flowers; white, pink, purple, or blue flowers are preferred. Adults nectar at dogbane, common milkweed, selfheal, red clover, crown vetch, purple vetch, Japanese honeysuckle, buttonbush, thistles, viper's bugloss, and others.

Caterpillar Host Plants. Various legumes are selected for oviposition and subsequent larval feeding, including beggar's tick (*Desmodium*), bush clover (*Lespedeza*), wild bean (*Glycine reticulata*), fuzzybean (*Strophostyles*), hoary pea (*Tephrosia*), butterfly pea (*Centrosema*), and milk vetch (*Astragalus*).

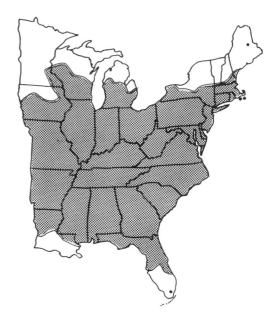

189. *Thorybes bathyllus*

Northern Cloudy Wing
Thorybes pylades (Scudder), 1870

PLATE 39 · FIGURES 233 & 234 · MAP 190

Etymology. The species is named after Pylades, son of Strophios. Strophios was king of Phocis, an ancient kingdom of central Greece.

Synopsis. The Northern Cloudy Wing is the most common, most widespread cloudy wing in North America. It is more likely to be found in woodlands than its two relatives.

Butterfly. Male forewing: $\bar{X}$ = 1.8 cm, range 1.6–2.0; female forewing: $\bar{X}$ = 2.0 cm, range 1.8–2.1 cm. Male Northern Cloudy Wings are distinctive in possessing a fold with specialized scales along the costal forewing margin. The hyaline marks on the forewing tend to be small, nonaligned, and triangular.

Range. The Northern Cloudy Wing ranges from Nova Scotia west across southern Canada to British Columbia, and south to Florida, the Gulf states, Texas, and central Mexico.

Habitat. The Northern Cloudy Wing is most common in

open or scrubby woodland, as well as along forest edges. It is occasionally found in open fields. It occurs from the Lower Austral Life Zone through lower portions of the Canadian Life Zone.

Life History. *Behavior.* Male Northern Cloudy Wings perch on or close to the ground in forest clearings, along forest roads, near woodland margins, and occasionally on hilltops if woods are nearby. Courtships are seen in late morning (1045–1100 hr), mated pairs are found in early afternoon (1220–1435 hr), and ovipositing females have been seen from late morning into the afternoon (1000–1730). Freshly emerged males readily visit moist spots along streams and forest trails.

Broods. In most of its range the Northern Cloudy Wing is univoltine with earlier flights to the south. For example, flight periods are June 4 to July 15 in Maine, May 14 to June 29 in Illinois, and March 28 to June 7 in Georgia. In the South there may be a rare partial second brood, since there are several records from late July to mid-September. Some of these records may represent misidentified Confused Cloudy Wings (*Thorybes confusis*).

Early Stages. The egg is pale green-white, with 12–15 ill-defined vertical ridges and numerous fine cross-lines. The mature larva is dark green, varying to maroon-green, with a brown or maroon dorsal line and two pale orange-pink lines along each side. The body is also covered with minute orange setiferous tubercles. The head varies from black to dark maroon and is covered with fine pale hairs. The pupa is dark brown, tan, and olive and is mottled with black.

Adult Nectar Sources. Flower preferences are similar to those of the Southern Cloudy Wing (*Thorybes bathyllus*), and the species was also observed nectaring at flowers of Deptford pink and hoary vervain on several occasions. No visits to yellow, orange, or red flowers were observed.

Caterpillar Host Plants. Beggar's tick (*Desmodium*) and bush clover (*Lespedeza*) are the primary larval hosts, but other legumes, including clover (*Trifolium*), alfalfa (*Medicago sativa*), lotus (*Hosackia*), milk vetch (*Astragalus*), lead plant (*Amorpha*), and snoutbean (*Rhynchosia*), have also been reported.

Confused Cloudy Wing
Thorybes confusis Bell, 1922
MAP 191

Etymology. The species name refers to the fact that this skipper is easily confused with the other eastern cloudy wings.

Synopsis. The Confused cloudy wing is little known, despite its wide range in the South.

Butterfly. Male forewing: $\overline{X}$ = 1.8 cm, range 1.7–1.8 cm; female forewing: $\overline{X}$ = 1.9 cm, range 1.8–2.2 cm. The Southern Cloudy Wing (*Thorybes bathyllus*), with which this species is often confused, lacks a costal fold in the male, and has a

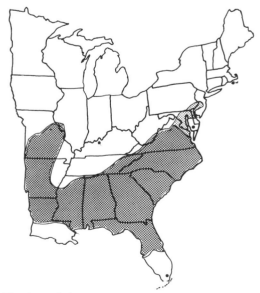

190. *Thorybes pylades*

191. *Thorybes confusis*

rounded forewing, elongate hyaline spots on the forewing, and a strong pattern on the ventral hind wing.

Range.   The Confused Cloudy Wing ranges from southeastern Pennsylvania south along the Atlantic Coastal plain to central Florida and then west along the Gulf Coast to Texas. The skipper also occurs in the Mississippi River drainage from Kansas and Missouri southward.

Life History.   *Broods.*   The Confused Cloudy Wing is bivoltine in most of its range, the second flight being only partial in many areas. There is a rare partial third generation in the Deep South. In Virginia, flights occur from mid-May to early June and from late July through August. In Mississippi the adults fly in April and May and then again in July and August.

Outis Skipper
Cogia outis (Skinner), 1894
MAP 192

Synopsis.   The Outis skipper, a close relative of the Mimosa Skipper (*Cogia calchas*), is known to be resident only in Texas. Strays have been taken on several occasions in northern

Arkansas and southwestern Missouri. The Outis Skipper has several generations, with adults flying from April to October. The caterpillars feed on leaves of several acacias.

Southern Sooty Wing
Staphylus hayhurstii (Edwards), 1870
PLATE 40 · FIGURE 235 · MAP 193

Etymology.   Staphylus was a Greek historian (century not mentioned). The species was named for Dr. L. K. Hayhurst of Sedalia, Missouri, who sent life-history information on several butterflies to W. H. Edwards.

Synopsis.   This unusual native, scallop-winged skipper has become more widespread because it now feeds principally on lamb's quarters, a widespread alien weed.

Butterfly.   Male forewing: $\overline{X}$ = 1.4 cm, range 1.3–1.5 cm;

female forewing: $\overline{X}$ = 1.3 cm, range 1.2–1.4 cm. This butterfly is superficially similar to the Common Sooty Wing (*Pholisora catullus*), but differs in its scalloped hind-wing margin, reduced white spotting, and two-toned wings with gold flecks. The female is brown with more contrast on the upper wing surfaces. There are several tropical American relatives.

Range.   The Southern Sooty Wing occurs from southern Pennsylvania and New Jersey (rare) westward across the

192. *Cogia outis*

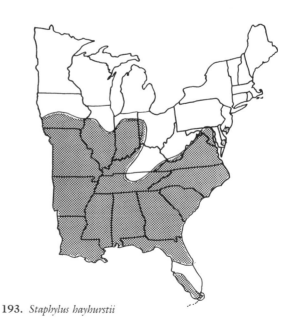

193. *Staphylus hayhurstii*

southern Midwest to eastern Nebraska and thence southward along the coastal plain, river valleys, and plains to Florida, the Gulf states, and central Texas.

Habitat. This skipper is found in vacant weedy lots, suburban gardens, and moist open wooded areas. It occurs primarily in the Lower Austral Life Zone.

Life History. *Behavior.* Males perch on low vegetation along woodland edges, often in broken sun or light spots. They move freely into open areas to visit flowers, and usually perch with their wings wide open. Oviposition occurs in late afternoon (1700–1800 hr), with the female laying a single egg on the underside of a host leaf.

Broods. In most of its range, the Southern Sooty Wing has two generations, usually with adults in May–June and then again in late July–August. In the Deep South there appear to be three flights. For example, in Mississippi flight dates are April 2–May 16, June 15–July 4, and August 1–September 14. In Florida this skipper flies from February on.

Early Stages. The egg is a flattened orange-brown hemisphere and has irregular creamy ridges that give it a starlike appearance when viewed from above. The caterpillars live in silk-lined folded leaf shelters and feed at night. The second-brood caterpillars overwinter in tightly sewn dried-leaf shelters and pupate the following spring. The mature caterpillar is deep green with an orange-pink cast posteriorly and a deep green dorsal line. The body is covered with fine white hair and the head is deep purplish brown. The pupa is pale orange-brown and light olive-brown, with the abdomen and head covered with short, stout orange hairs (Heitzman 1963).

Adult Nectar Sources. Southern Sooty Wings nectar at white clover, white sweet clover, dogbane, cucumber, wild marjoram, spearmint, knotweed, and marigold.

Caterpillar Host Plants. Lamb's quarters (*Chenopodium album*) is the chief host, but chaff flower (*Alternanthera*) is also fed upon in nature.

Sickle Winged Skipper
Achlyodes thraso (Hübner), 1807

Synopsis. A tropical American skipper, the Sickle Winged Skipper was captured once in Arkansas (October, 1973). It has a distinctively shaped forewing with a sinuous outer margin just below the pointed apex. The species is resident from southern Texas south to Argentina, and also occurs in the Caribbean area. Adults may be found all year in the tropics and the caterpillars feed on members of the citrus family.

Brown-banded Skipper
Timochares ruptifasciatus (Plötz), 1884

Synopsis. The Brown-banded Skipper is a tropical species that is resident from southern Texas through Mexico. A stray was once found in Illinois (August 29, 1941).

Florida Dusky Wing
Ephyriades brunnea floridensis Bell & Comstock, 1948
PLATE 40 · FIGURES 236 & 237 · MAP 194

Etymology. The Ephyriades were a race of Ephyra, the ancient name for Corinth. The species name, *brunnea* means "dark brown" in Latin.

Synopsis. The Florida Dusky Wing belongs to a small tropical American group found primarily in the Antilles, as well as in Central and South America.

Butterfly. Male forewing: $\bar{X} = 1.9$ cm, range 1.8–2.0 cm; female forewing: $\bar{X} = 2.0$ cm, range 1.9–2.2 cm. This skipper is extremely similar to the *Erynnis* dusky wings, but differs in having a more abruptly angled forewing apex, a lack of scattered gray wing scaling, and darker marginal areas on both wings dorsally. Dorsally, the male is a monotone black with very small white hyaline spots in the forewing apical region, while the female is distinctly two-toned, with larger hyaline spots.

Range. The Florida Dusky Wing is found only in southern Florida and the Keys. Other subspecies are found in the Antilles from the Bahamas south and east to Dominica.

Habitat. This species occurs in tropical pine–Sabal palm scrub.

Life History. *Behavior.* Males perch on bare twigs about a half meter above the ground in full sunlight.

Broods. In Florida, this skipper is found in every month except November.

Early Stages. The mature larva is green, with three white longitudinal stripes along the sides and an irregular translucent stripe along each side between the two uppermost white ones (Tamburo and Butcher 1955).

Caterpillar Host Plants. In Florida, the native host is Key Byrsonima (*Byrsonima lucida*), while caterpillars have also been found on the introduced Barbados cherry (*Malpighia glabra*).

Dreamy Dusky Wing
Erynnis icelus (Scudder & Burgess), 1870

PLATE 40 · FIGURE 238 · MAP 195

Etymology. Erynnis was one of the evening spirits who brought retribution for homicide (Greek myth). The species may be named after *Icelos*, a son of Somnus, god of dreams or sleep (Roman myth).

Synopsis. This skipper and the Sleepy Dusky Wing form a small American group closely related to the Old World dusky wings (Burns 1964).

Butterfly. Male forewing: $\overline{X}$ = 1.5 cm, range 1.3–1.6 cm; female forewing: $\overline{X}$ = 1.6 cm, range 1.5–1.7 cm. Both the Dreamy Dusky Wing and Sleepy Dusky Wing (*Erynnis brizo*) lack white spots on their forewings and have a discal series of black marks on the forewings in the form of opposed crescents, lending the band the semblance of a chain. The Dreamy Dusky Wing is smaller and has pointed antennal clubs and very long labial palpi.

Range. This small dusky wing occurs from Nova Scotia south to Arkansas, Alabama, and Georgia and west across southern Canada to British Columbia. In the western mountains it occurs south to New Mexico, Arizona, and central California.

Habitat. The Dreamy Dusky Wing inhabits a variety of open woodland or forest-edge habitats. In particular, it has been noted on acid birch and serpentine barrens, in rich open woods, and along the borders of damp woods in hilly or mountainous terrain.

Life History. *Behavior.* Males perch in flats or depressions adjacent to woods or in openings during most daylight hours. Courtship has been seen in the morning (1000–1030 hr). Freshly emerged males visit sandy or muddy areas along dirt roads and streams in early spring.

Broods. Except for the faint possibility of a very small second brood in the southern Appalachians during late July and early August, the Dreamy Dusky Wing is univoltine. In northern areas, the flight occurs in mid-May with stragglers to early July, while farther south the butterfly is on the wing from early April, with rare stragglers until late June.

Early Stages. Early stages are very similar in the entire dusky wing group. Eggs are laid singly on host leaves, and the caterpillars live within a shelter of one or more host leaves. Mature larvae overwinter in a leaf shelter and pupate the

194. *Ephyriades brunnea floridensis*

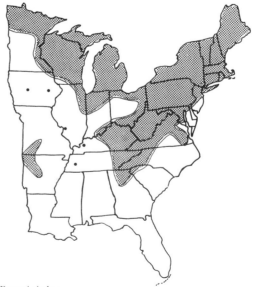

195. *Erynnis icelus*

following spring. The eggs are small, pale green, columnar, and have prominent vertical ribs. The mature caterpillar is pale green with white setiferous tubercles. The head is black with reddish and yellowish spots, is strongly angled, and is depressed on top.

Adult Nectar Sources. Adults favor flowers of blueberry, winter cress, wild strawberry, purple vetch, blackberry, lupine, Labrador tea, dogbane, and New Jersey tea for their nectar visits.

Caterpillar Host Plants. Larval hosts include willows, poplars, aspens, and occasionally birch. One record for black locust (*Robinia pseudacacia*) is unlikely.

Sleepy Dusky Wing
Erynnis brizo (Boisduval & Leconte), 1834
PLATE 40 · FIGURE 239 · MAP 196

Etymology. It is possible that *brizo* is taken from the Latin *brisa*, the word for the refuse of grapes left after they are pressed, or the grape skins.

Synopsis. The Sleepy Dusky Wing occurs from coast to coast, with different subspecies in the principal oak regions. It is found most predictably in association with shrubby oaks in well-drained habitats.

Butterfly. Male forewing: $\overline{X} = 1.8$ cm (smaller in New England and Florida), range 1.5–1.9 cm; female forewing: $\overline{X} = 1.8$ cm, range 1.6–2.0 cm. The Sleepy Dusky Wing differs from the Dreamy Dusky Wing in its larger size, shorter labial palpi, and blunt antennal clubs. The small subspecies *somnus* occurs in peninsular Florida.

Range. This dusky wing ranges from southern New Hampshire and Massachusetts west through the central Great Lakes area, the central Rocky Mountain states, and northern California. It ranges south to peninsular Florida, the Gulf states, Texas, and northern Mexico.

Habitat. In the East, the Sleepy Dusky Wing is a butterfly of oak or pine-oak scrub or barrens on sandy and shaly soils.

Life History. *Broods.* This skipper has one fairly brief flight period each spring. In northern Michigan, flight dates run from May 10 to June 15; in Virginia, from late March to late May; and in Florida, from early February to mid-April. *Early Stages.* Refer to Dreamy Dusky Wing.

Adult Nectar Sources. The Sleepy Dusky Wing favors the flowers of heaths, such as wild azalea and blueberry. It also nectars at blackberry and dandelion.

Caterpillar Host Plants. In the East, bear oak (*Quercus ilicifolia*) is the chief caterpillar host, although there is one report for American chestnut (*Castanea dentata*), and some other small oaks must be utilized in Florida and other areas where Bear Oak is absent.

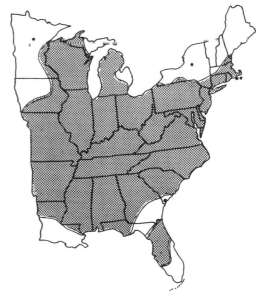

196. *Erynnis brizo*

Juvenal's Dusky Wing
Erynnis juvenalis (Fabricius), 1793
PLATE 41 · FIGURES 241–243 · MAP 197

Etymology. The species name *juvenalis* is derived from the name of Juvenal, who was a Roman satirical poet in the first century A.D.

Synopsis. This is the most common dusky wing in eastern North America, and in many localities may be the most common butterfly for several weeks during early spring.

Butterfly. Male forewing: $\overline{X} = 1.9$ cm, range 1.6–2.2 cm; female forewing: $\overline{X} = 2.0$ cm, range 1.7–2.2 cm. The two light spots on the apical third of the ventral hind wing serve to distingush Juvenal's Dusky Wing from the similar Horace's Dusky Wing (*Erynnis horatius*).

Range. Juvenal's Dusky Wing occurs from Nova Scotia

and central New England west to southern Manitoba, the Dakotas, and northeastern Wyoming. It then ranges south to Florida, the Gulf states, Texas, New Mexico, Arizona, and Mexico. It is not found in the Rocky Mountains.

Habitat. This skipper requires only oak woods or scrub and their adjacent edges and fields. In Nova Scotia, the butterfly is closely associated with beech stands. It occurs from the Lower Austral through the Transition Life Zone.

Life History. *Behavior.* Male Juvenal's Dusky Wings perch along forest edges or in openings during the afternoon (1300–1700 hr), usually on tips of bare exposed twigs 1–4 m above ground. From these perches they frequently patrol back and forth, attempting to overtake passing skippers in their tireless quest for receptive females. Mated pairs have been seen from midday through the afternoon (1140–1600 hr). The female is the carrying sex. Freshly emerged males commonly imbibe moisture along dirt roads and streams.

The usual sleeping posture of these butterflies is perching with their wings clasped about a twig and their antennae laid back in a most unbutterflylike posture.

Broods. Juvenal's Dusky Wing is univoltine, with a flight each spring; occasional individuals in late summer may represent a very partial second emergence or misidentified Horace's Dusky Wings. In northern areas, Juvenal's Dusky Wing emerges from mid-April to mid-May and flies until mid-June. In the Deep South it emerges in February to mid-March and flies until late May or early June.

Early Stages. Refer to Dreamy Dusky Wing.

Adult Nectar Sources. Winter cress, blueberry, dandelion, Carolina vetch, wild plum, redbud, wisteria, and lilac are favored for nectar visitation.

Caterpillar Host Plants. A variety of both white and red oaks is utilized for egg-laying and subsequent caterpillar feeding.

Horace's Dusky Wing
Erynnis horatius (Scudder & Burgess), 1870
PLATE 40 · FIGURE 240 · MAP 198

Etymology. There were two ancient Romans named Horatius. One was a legendary hero, the other a poet.

Synopsis. This oak-feeding skipper occurs in both spring and summer. Horace's Dusky Wing is extremely similar to and is found in the same localities as the Juvenal's Dusky Wing (*Erynnis juvenalis*), although the latter flies only in spring.

Butterfly. Male forewing: $\overline{X}$ = 1.8 cm, range 1.5–2.2 cm; female forewing: $\overline{X}$ = 1.9 cm, range 1.7–2.2 cm. Horace's

Dusky Wing is similar to Juvenal's Dusky Wing but is distinguished by its virtual lack of white overscaling and its more uniformly brown appearance above, as well as by the absence of subapical spots on the ventral hind-wing surface. There is a gradual increase in adult size as one proceeds southward, and second-generation individuals are significantly larger than those of the spring brood.

Range. This skipper occurs from Massachusetts west to Iowa and Minnesota and south through most of our area to

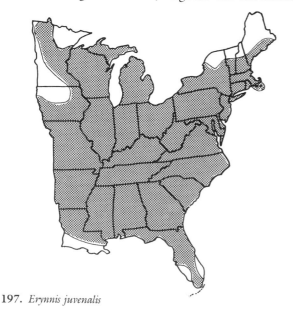

197. *Erynnis juvenalis*

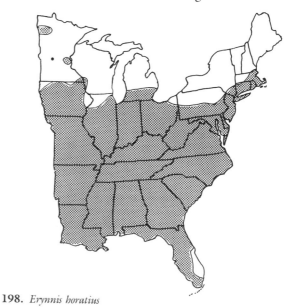

198. *Erynnis horatius*

Florida, the Gulf states, and eastern Texas. A second group of populations occurs in Colorado, northern New Mexico, and northeastern Arizona.

Habitat. Horace's Dusky Wing usually occurs in association with oaks, but it often wanders in search of nectar, particularly in late summer. It is found in open woods, woodland edges, power-line right-of-ways, wooded swamps, and open fields.

Life History. *Behavior.* Males perch on hilltops or slopes, about .3 m above ground at the tip of twigs or, rarely, on the ground itself. Mating has been observed near midday (1245–1345 hr). Females oviposit on young host leaves, often those of saplings, in mid- to late afternoon (1500–1700 hr). Freshly emerged males visit moist sand or mud along streams or roads.

Broods. There are two generations in the north (late April–May and July–August), while in the mid-South there are two broods and a partial third (March–early May, mid-June–July, and August–September). In the Gulf states there are three full broods (mid-February–April, May–mid-July, and mid-July–mid-October). In Florida, Horace's Dusky Wing occurs from January to early October, but the species is most common in two broods from January through May, and less common in the July and September generations.

Early Stages. Refer to Dreamy Dusky Wing. Caterpillars resulting from the last brood overwinter but those of other broods develop directly.

Adult Nectar Sources. Horace's Dusky Wing may visit flowers from ground level to 1.5 m high. White and yellow flowers are favored, including dogbane, peppermint, buttonbush, boneset, sneezeweed, winter cress, and goldenrod.

Caterpillar Host Plants. The red oak group is favored for larval hosts; species fed upon include willow oak (*Quercus phellos*), northern red oak (*Quercus velutina*), scrub oak (*Quercus ilicifolia*), and water oak (*Quercus nigra*). White oak group species selected include post oak (*Quercus stellata*) and live oak (*Quercus virginiana*).

Mottled Dusky Wing
Erynnis martialis (Scudder), 1869
PLATE 41 · FIGURE 244 · MAP 199

Etymology. The species name is derived from M. Valerius Martialis, a Roman poet.

Synopsis. This skipper's colonies are closely associated with New Jersey tea. The range coincides almost perfectly with the area covered by this book, with the exception of isolated Rocky Mountain and Black Hills populations.

Butterfly. Male forewing: $\overline{X}$ = 1.6 cm, range 1.3–1.8 cm; female forewing: $\overline{X}$ = 1.6 cm, range 1.5–1.6 cm. Both sexes are brown dorsally with black patches, lending the butterflies a strongly mottled or banded appearance. Fresh individuals have a purple-blue sheen. The spring-generation butterflies have more white scaling on the wings and are smaller than those of the second generation. This species is related to the Pacuvius Dusky Wing (*Erynnis pacuvius*) of western North America.

Range. The Mottled Dusky Wing occurs from Massachusetts and New York west across southern Ontario and the Great Lakes states to Minnesota and western Iowa, then south to Georgia, the Gulf states, and central Texas. Isolated populations occur in the Black Hills and in central Colorado.

Habitat. The Mottled Dusky Wing is usually confined to hilly country, often near woods or in open brushy fields. In the East, this skipper may be found in shale or serpentine barrens with acidic soils. In western Iowa, the skipper was found near the summits of prairie hills in the loess soil formation running east of the Missouri River.

Life History. *Behavior.* Males perch on hilltops or along ridges during most daylight hours; they sit on the ground or at the tip of small twigs.

Broods. The Mottled Dusky Wing is bivoltine everywhere it occurs, except in Colorado, where it is univoltine. Rare late-season males in the mid-South may represent a very

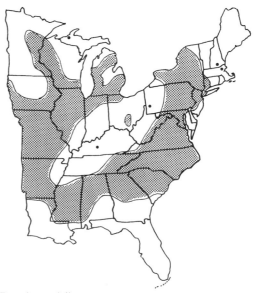

199. *Erynnis martialis*

sporadic third generation. Appearance of the first brood occurs progressively earlier southward, but the second brood appears during June or early July throughout the East. In Michigan the species first appears as early as April 24, in Georgia on March 27, and in Mississippi on March 4.

Early Stages. Refer to Dreamy Dusky Wing. Larvae resulting from the first brood develop directly.

Adult Nectar Sources. This species has been observed visiting bush houstonia, gromwell, and hoary vervain.

Caterpillar Host Plants. Although a variety of larval hosts were reported for this skipper in the past, it is now clear that the Mottled Dusky Wing selects only wild lilacs (*Ceanothus*, family Rhamnaceae) as hosts. New Jersey tea (*Ceanothus americanus*) is most widely used in the East, but adults were closely associated with redroot (*Ceanothus herbaceus* var. *pubescens*) in western Iowa.

Zarucco Dusky Wing
Erynnis zarucco (Lucas), 1857
MAP 200

Etymology. It is possible that the species name is a combination of the Italian *zar*, "Russian czar," and *uccio*, "bad, ugly, rotten."

Synopsis. This skipper is the most southern of the eastern dusky wings, and it is the only one that also occurs in the Caribbean.

Butterfly. Male forewing: $\bar{X}$ = 1.9 cm, range 1.7–2.0 cm; female forewing: $\bar{X}$ = 1.8 cm, range 1.7–1.9 cm. The Zarucco Dusky Wing may be most easily confused with the Wild Indigo Dusky Wing (*Erynnis baptisiae*), but it differs in being somewhat larger, having somewhat triangular hind wings, and possessing more pointed forewings. A few individuals on the Florida Keys have whitish hind-wing fringes. Individuals of the first generation are usually smaller, with more scattered white scaling above.

Range. This southern Skipper occurs from Maryland and Virginia east to western Arkansas and south to Florida and the Gulf states. Occasional wandering individuals may be found north to Pennsylvania and Connecticut, but the species is probably not resident north of the Carolinas. Populations also occur on Cuba and Hispaniola.

Habitat. This butterfly is found in open fields, scrubby areas, and wood edges within the southeastern coastal plain and Piedmont, being most abundant in the former. It is resident in the Tropical and Lower Austral Life Zones.

Life History. *Broods.* The Zarucco Dusky Wing has three poorly defined generations in the mid-South, and possibly four in Florida. In Mississippi, the skipper flies from early March through September, with occasional individuals found in November. In Florida, adults may be seen from mid-January through mid-October.

Early Stages. Refer to Dreamy Dusky Wing. Larvae from the last brood overwinter, but those from other broods develop directly.

Adult Nectar Sources. Adults often visit shepherd's needle in Florida.

Caterpillar Host Plants. Caterpillars have been found on short individuals of black locust (*Robinia pseudacacia*) on several occasions, and have been reliably recorded in the eastern states on hairy bush clover (*Lespedeza hirta*), *Sesbania longifolia*, and Colorado River hemp (*S. exaltata*), all legumes. The species probably feeds on several other legumes as well.

Funereal Dusky Wing
Erynnis funeralis (Scudder & Burgess), 1870
MAP 201

Synopsis. The Funereal Dusky Wing is a close relative of the Zarucco Dusky Wing, but it has blacker wings and a white hind-wing fringe. It occurs primarily in Texas and the Southwest, and south through Central America to Argentina. It may occasionally appear in western Arkansas, Missouri, Louisiana, or Florida. Its life history is similar to that of the Zarucco Dusky Wing; it has three generations each year, and its caterpillars feed on various legumes.

Columbine Dusky Wing
Erynnis lucilius (Scudder & Burgess), 1870

MAP 202

Etymology. Lucilius was a first-century (A.D.) Roman author of epigrams, short satirical poems.

Synopsis. The Columbine Dusky Wing is smaller than the other two eastern "persius" group skippers, the Wild Indigo and Persius Dusky Wings. It is usually found in close association with wild columbine in rocky habitats.

Butterfly. Male forewing: $\overline{X}$ = 1.5 cm, range 1.3–1.8 cm; female forewing: $\overline{X}$ = 1.5 cm, range 1.4–1.6 cm. The Columbine Dusky Wing is similar to the Wild Indigo Dusky Wing (*Erynnis baptisiae*), but is smaller, has a less distinct brown forewing patch, virtually lacks the hairlike scales on the dorsal forewing, and has two submarginal rows of distinct light spots on the ventral hind wing. This species occasionally hybridizes with the Wild Indigo Dusky Wing.

Range. This skipper occurs from southern Quebec and Ontario west to Michigan and Minnesota and south to New Jersey and Pennsylvania. The species also ranges southwest along the Appalachians to Virginia and Kentucky.

Habitat. The Columbine Dusky Wing is usually found in rich, rocky, deciduous or mixed woodland or along its edges. Rocky, wooded ravines and gullies are among its favored situations.

Life History. *Broods.* The Columbine Dusky Wing is bivoltine wherever found. The first generation flies from mid- to late April through early June, and the second flies from late July to early September.

Early Stages. Refer to Dreamy Dusky Wing. Larvae resulting from the second brood overwinter, while those from the first brood develop directly.

Caterpillar Host Plants. The sole native host is wild columbine (*Aquilegia canadensis*), although garden columbine (*Aquilegia vulgaris*) may be eaten on occasion.

Wild Indigo Dusky Wing
Erynnis baptisiae (Forbes), 1936

PLATE 41 · FIGURE 245 · MAP 203

Etymology. The species is named for *Baptisia*, the genus name of its larval host.

Synopsis. This is the most common, most widespread eastern member of the "persius" skippers (see Persius Dusky Wing). It should be sought near its host plants, wild indigo, lupine, or crown vetch.

Butterfly. Male forewing: $\overline{X}$ = 1.7 cm, range 1.6–1.8 cm; female forewing: $\overline{X}$ = 1.7 cm, range 1.6–1.8 cm. Throughout

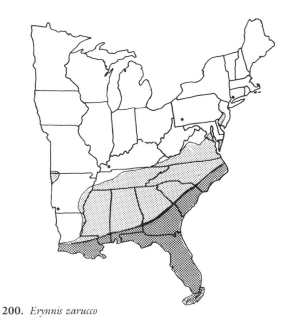

200. *Erynnis zarucco*

201. *Erynnis funeralis*

most of the Southeast, this is the only representative of the "persius" group. It is larger than the Columbine Dusky Wing, the basal half of its forewing is dark, and the apical half is lighter with dark bands. There is a pronounced brown patch near the costa in the postmedian area.

Range.　The Wild Indigo Dusky Wing occurs from Massachusetts and New York west to Iowa and south to Florida, the Gulf states, and central Texas.

Habitat.　This species is found in upland fields, in dry open woods, or on barrens in sandy, acid, or serpentine soils. Recently colonized habitats are highway margins and railroad banks with stands of crown vetch.

Life History.　*Behavior.* Males perch atop low shrubs in open areas during late morning and afternoon. During late afternoon, males have been seen engaged in a continuous, low, patrolling flight over an oval course perhaps 30 m in length. Egg-laying has been seen in the afternoon (1530 hr). Freshly emerged males will visit moist areas along dirt roads or streams.

Broods. The Wild Indigo Dusky Wing has two annual flights that become more widely separated southward. In New York, the adults appear in June and again in July, while in Virginia the two flights appear from late April to mid-June and again from early July to early September.

Early Stages. Refer to Dreamy Dusky Wing. Larvae from the first brood develop directly; those from the second brood overwinter.

Adult Nectar Sources.　The Wild Indigo is known to nectar at blackberry, sunflower, white sweet clover, crimson clover, and dogbane, and probably other plants.

Caterpillar Host Plants.　This skipper selects only a few members of the pea family as hosts. The usual host is wild indigo (*Baptisia tinctoria*), but others are also selected, including plains wild indigo (*Baptisia leucophaea* var. *laevicaulis*), wild blue indigo (*B. australis*), and white false indigo (*B. leucantha*), as well as lupine (*Lupinus perennis*), false lupine (*Thermopsis villosa*), and possibly rattlebox (*Crotalaria sagittalis*). In recent years this skipper has included crown vetch (*Coronilla varia*), an introduced alien that is widely planted along highway verges, among its oviposition plants, and the species is now more common and rapidly expanding the number of locations where it may be found (Shapiro 1979).

202. *Erynnis lucilius*

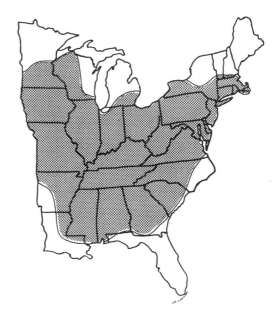

203. *Erynnis baptisiae*

Persius Dusky Wing
Erynnis persius (Scudder), 1863

MAP 204

Etymology.　Persius was a Roman poet of the third century A.D.

Synopsis.　In the East, the difficult "persius" group of skippers is composed of three species, the Persius, the Wild Indigo, and the Columbine Dusky Wings.

Butterfly.　Male forewing: $\overline{X}$ = 1.6 cm, range 1.4–1.8 cm;

female forewing: $\overline{X}$ = 1.7 cm, range 1.6–1.8 cm. This skipper is larger than the Columbine Dusky Wing (*Erynnis lucilius*) and its paler areas are covered with raised white hairlike scales. The pale postmedian patch along the forewing costa is not as brown as on the Wild Indigo Dusky Wing (*Erynnis baptisiae*).

Range. This skipper ranges from northern New England, Quebec, and the Great Lakes states west across Canada to British Columbia and north to Alaska. In the United States it ranges south in the Appalachians to Virginia and in the West to New Mexico, Arizona, and California (Burns, 1964).

Habitat. The Persius Dusky Wing is found in open areas within boreal forest. Marshes or seeps with small willows, sandy areas with small aspens, or open fields and boggy places near woods are situations where this rare eastern butterfly might be found.

Life History. *Behavior.* Males perch on ridges or hilltops during most daylight hours; they usually perch on the ground or on low projecting twigs. Females oviposit near midday. *Broods.* In the East there is a single spring flight. In Virginia flight dates extend from late April to early June, but the butterfly may be expected from late May through June in more northern states.

Early Stages. Refer to the Dreamy Dusky Wing.

Caterpillar Host Plants. In the eastern states, willows, poplar, and aspens are the chief hosts, but lupine (*Lupinus perennis*) is reported as an oviposition plant in at least two states.

Grizzled Skipper
Pyrgus centaureae wyandot (Edwards), 1863
PLATE 41 · FIGURE 246 · MAP 205

Etymology. The genus name is based upon the Greek *pyrgos*, "tower," while the species is derived from the Centaurs, a race of beings that were half man and half horse, the descendants of Ixion in Greek myth, or from *centaurea* (Knapweed), an Old World herb. Wyandot is a synonym for the Hurons, an Iroquoian Indian tribe.

Synopsis. The Grizzled Skipper belongs to a group of boreal species with relatives in western North America and Eurasia. This species, which ranges throughout the Northern Hemisphere, is usually found in barrens in the East.

Butterfly. Male forewing: $\overline{X}$ = 1.4 cm, range 1.4–1.5 cm; female forewing: $\overline{X}$ = 1.5 cm, range 1.3–1.5 cm. The Grizzled Skipper is largely black above, with small white checks, and the ventral forewing is predominantly black.

Range. The Grizzled Skipper is a boreal circumpolar species, being found in boreal regions of both North America and Eurasia. In North America, it occurs from Labrador west and north to British Columbia and Alaska. It ranges south in the Rocky Mountains to northern New Mexico. Our subspecies *wyandot* is found in northern portions of the Great Lakes states and ranges down the Appalachians to North Carolina and Kentucky.

Habitat. The Grizzled Skipper is generally found in open situations near woodlands. Specific habitats include hilltops

204. *Erynnis persius*

205. *Pyrgus centaureae wyandot*

in heath-shrub acid barrens (New York), grass hillsides and open pastures near woods (Pennsylvania), scrub oak openings (Michigan), and woodland clearings within shale barrens (Virginia). The species is restricted to the Transition and Canadian Life Zones.

Life History. *Behavior.* Males patrol all day in suitable open habitats but occasionally assume a perching posture at low temperatures.

Broods. The Grizzled Skipper is univoltine. In northern Michigan, flight dates extend from May 13 to 31; in New York from April 29 to May 19; and in southern Virginia from late March to mid-April.

Adult Nectar Sources. Adults nectar at flowers of low-growing plants, including Canadian cinquefoil, wild strawberry, and blueberry.

Caterpillar Host Plants. In Michigan, wild strawberry (*Fragaria virginiana*) has been found to be the host (M. C. Nielsen, personal communication).

Checkered Skipper
Pyrgus communis (Grote), 1872
PLATE 42 · FIGURES 247 & 248 · MAP 206

Etymology. The species name is taken from the Latin *communis*, "common."

Synopsis. Each year the Checkered Skipper expands its range far north of the area where it is resident and can survive the winter. It is the only common black-and-white patterned skipper in most of our area.

Butterfly. Male forewing: $\overline{X} = 1.5$ cm, range 1.4–1.6 cm; female forewing: $\overline{X} = 1.6$ cm, range 1.4–1.7 cm. The Checkered Skipper differs from the Tropical Checkered Skipper (*Pyrgus oileus*) in having a very small marginal row of white spots on the dorsal hind wing and the submarginal white spots large. In addition, the two bands on the ventral hind wing are dark gray or olive and contrast with the flat white ground.

Range. This skipper occurs from southeastern New York and southern New England northwest to southern Ontario and west across central Canada to the Pacific Coast. It then extends south through most of the United States, the lowlands of Mexico, and Central America to Argentina. It cannot survive very cold winters, and is therefore rare in New England and usually not resident north of the 40th parallel. Midwestern populations may be more cold-tolerant than those found east of the Applachians.

Habitat. This species is usually found in very open, sunny situations with some bare ground and low vegetation. Habitats include open pine forest, prairies, wet and dry meadows, farmyards, landfills, and highway verges. This species is resident in the Tropical through the Lower Austral Life Zones, but may colonize Upper Austral and Transition Life Zone areas each summer.

Life History. *Behavior.* Males patrol all day, but most actively in the afternoon, by flying rapidly back and forth over a beat about 30 m in length. They occasionally perch. Mating occurs from midday to midafternoon (1140–1400 hr), with females acting as carriers. Roosting begins in later afternoon (1700 hr), with the butterfly assuming an exposed perch atop some tall weed.

Broods. There appear to be three broods in most of the areas north of the Deep South. In Florida and the Gulf states, where the species may be found in most months, there may be four or five generations, although this has not been carefully documented. In the Deep South, the Checkered Skipper flies from February through October. Farther north there are records in various states beginning anywhere from late March to June, but the species does not become common, if at all, until August or September.

Early Stages. The egg is a pale-green hemisphere. The caterpillars make a shelter in the center of a host leaf by folding the leaf and spinning lots of silk. The mature larva is yellow-white to brownish with a gray-green or brown-green dorsal line and two prominant brown lines and two fine white lines along each side. The body is covered with small, white, setiferous tubercles. The head is black and has a dense cover-

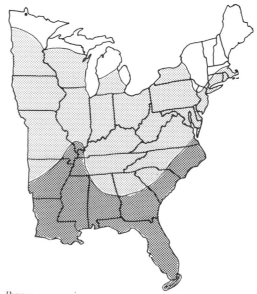

206. *Pyrgus communis*

ing of yellow-brown hairs. The pupa is partly light brown and partly light green, with black dots and streaks.

Adult Nectar Sources. Checkered Skippers nectar at white-flowered composites, such as fleabane, asters, and shepherd's needles, and at red clover, beggar's ticks, and knapweed.

Caterpillar Host Plants. A variety of plants in the mallow family are selected as larval hosts, including mallow (*Malva*), hollyhock (*Althaea*), velvet-leaf (*Abutilon*), poppy mallow (*Callirhoe*), Sidas (*Sida*), and globe mallows (*Sphaeralcea*). Females may lay eggs on lamb's quarters (*Chenopodium album*), but the caterpillars cannot survive on it.

Tropical Checkered Skipper
Pyrgus oileus (Linnaeus), 1767
PLATE 42 · FIGURES 249 & 250 · MAP 207

Etymology. Oileus was King of Locris and father of Aias and Medon. He is mentioned in histories of the Trojan war.

Synopsis. The widespread Tropical Checkered Skipper is resident within our area only in peninsular Florida and the immediate Gulf coast. It may be narrower in its selection of host plants than the Common Checkered Skipper.

Butterfly. Male forewing: $\overline{X}$ = 1.6 cm, range 1.4–1.7 cm; female forewing: $\overline{X}$ = 1.5 cm, range 1.4–1.6 cm. The male is distinctive, with its overlying mat of long blue-gray hairs on the dorsal wing surfaces, while both sexes have submarginal and marginal rows of equally large spots on the dorsal hind wing. There is often a strong infusion of brown on the ventral hind wing, and in any event there is not a strong contrast between the bands and ground color.

Range. The Tropical Checkered Skipper occurs from peninsular Florida, the Gulf Coast, and southern Texas south through the Antilles, Mexico, and Central America to Argentina (Burns and Kendall 1969).

Habitat. This skipper is found predominantly in wet or dry pastures, brushy fields, or other open, sunny areas in lowland Tropical Life Zones.

Life History. *Behavior.* The traits of this skipper are probably similar to those of the Common Checkered Skipper. Ovipositon takes place in late afternoon (1600–1700 hr), with the female placing single eggs on top of leaves of young hosts.

Broods. The number of generations has not been documented. The species is found all year in Florida. In Texas there appear to be four or five generations.

Early Stages. The egg is a pale yellow-white hemisphere. The caterpillar has a black head and a green body with a faint

dorsal line and numerous small, white, setiferous tubercles.

Adult Nectar Sources. The plants selected for nectaring are primarily Sidas and small-flowered composites, such as shepherd's needles.

Caterpillar Host Plants. Larval hosts include various mallows, especially Sidas, particularly axocatzin (*Sida rhombifolia*). Other reported hosts are hollyhock (*Althaea rosea*), velvet-leaf (*Abutilon*), mallow (*Malva*), and malva loca (*Malvastrum*).

207. *Pyrgus oileus*

Streaky Skipper
Celotes nessus (Edwards), 1877

Synopsis. A small skipper of dry foothill washes, the Streaky Skipper was collected once in Louisiana. It has long, narrow, dark, V-shaped markings projecting inward from the

margins. Its proper range is the arid Southwest and northern Mexico.

Common Sooty Wing
Pholisora catullus (Fabricius), 1793
PLATE 42 · FIGURE 251 · MAP 208

Etymology. The genus name *Pholisora* is possibly derived from the Greek *phōlis*, "lurking in a hole," and *ŏra*, "season, especially spring." Catullus was an adversary of Pompeius.

Synopsis. A butterfly of weedy lots and roadsides, the Common Sooty Wing is the most commonly encountered small black skipper in our area.

Butterfly. Male forewing: $\overline{X} = 1.3$ cm, range 1.2–1.4 cm; female forewing: $\overline{X} = 1.4$ cm, range 1.3–1.5 cm. This skipper has rounded black wings and two curved rows of tiny white spots on the outer portions of the forewings. There is sometimes a row of tiny white spots along the outer margin of the upper hind wing. It is most similar to the unrelated Southern Sooty Wing (*Staphylus hayhurstii*).

Range. The Common Sooty Wing may be found from southern Quebec west across southern Canada to southern British Columbia. It ranges south through most of the United States to northern Mexico, but is absent or rare in northern boreal regions and peninsular Florida.

Habitat. The original habitat of this skipper was probably sandy river banks and sand plains, but it is now abundant on land fills, roadsides, vacant lots, gardens, and farmyards.

Life History. *Behavior.* The males patrol by flying close to the ground in open sunny places during most daylight hours. Mating occurs from the morning to early afternoon (930–1430 hr), with the female carrying the male on short nuptial flights. Females lay eggs near midday, depositing a single egg on the upper surface of a host leaf, then fluttering off to the next plant. Freshly emerged males may imbibe moisture along dirt roads or trails.

Broods. This small skipper is bivoltine throughout its range. The two broods become more widely spaced southward. The first flight usually occurs from mid-May to mid-June, and the second from mid-July to mid-August. In Mississippi, the Common Sooty Wing flies from late March to early October.

Early Stages. The egg is white or creamy when laid on the host and has prominent vertical ribs and domelike tubercles at the top. Young caterpillars slice a leaf to the midrib and then fold part of it over to make a small shelter; older larvae fold over an entire leaf. Mature caterpillars of the second brood overwinter in their leaf shelters, which are lined with silk. They pupate and emerge in the following spring. The mature caterpillar is pale green with pale dots; its head is black and there is a black shield on the prothorax. The pupa is purple-brown with a glaucous bloom.

Adult Nectar Sources. The butterflies commonly nectar at white clover, dogbane, common milkweed, marjoram, peppermint, oxalis, cucumber, and melon flowers.

Caterpillar Host Plants. Hosts include members of the goosefoot, amaranth, and mint families. Lamb's quarters (*Chenopodium album*) is the usual food plant, but several amaranths (*Amaranthus*), cockscomb (*Celosia*), and a few mints, such as wild marjoram (*Origanum*), horehound (*Marrubium*), and others, may be fed upon.

Subfamily Hesperiinae: Branded Skippers

The skippers of the Hesperiinae subfamily are predominantly orange or black and usually have relatively pointed forewings. The males often have a stigma or brand (patches of specialized scales) on the dorsal forewing.

Male hesperiines are strong perchers, and only a few (*Ancyloxypha, Thymelicus, Oarisma*) employ patrolling to locate mates. Mating and egg-laying usually take place from mid- to late afternoon. Adults often rest with their hind wings separated from the forewings. In our area, a greater proportion of hesperiine skippers are univoltine. Hesperiinae have very long proboscises, often approaching or exceeding their forewings in length. They are avid flower visitors, and may take nectar from long-tubed flowers. All hesperiines select monocotyledonous plants as their caterpillar hosts. In our area, these are predominantly grasses and sedges, but other groups, such as palms, gingers, and aroids, are also utilized in the tropics. Females usually deposit single eggs on host leaves. The eggs are usually hemispherical and reticulate without prominent ribbing. The caterpillars have narrow, ovoid head capsules and relatively narrow, tapered bodies, which often have several longitudinal stripes. The caterpillars often feed at night and usually live in a shelter of tied, bent host leaves, or amid the base of the host clump. The larvae often overwinter, and pupation occurs within the shelter.

Arctic Skipper
Carterocephalus palaemon mandan (Edwards), 1863

MAP 209

Etymology. The genus name *Carterocephalus* is possibly derived from the Greek *Kephale*, "head" (e.g., hydrocephalus = head enlarged by fluid). The species is named for Palaemon, who was god of the sea in Greek mythology. He was the son of King Melikertès.

Synopsis. The Arctic Skipper and the Laurentian Skipper (*Hesperia comma laurentina*) are the only native North American hesperiine skippers that also occur in Eurasia. Both species range into the Arctic and probably crossed the Bering Straits during the Ice Ages.

Butterfly. Male forewing: $\overline{X}$ = 1.3 cm, range 1.2–1.4 cm; female forewing: $\overline{X}$ = 1.3 cm, range 1.2–1.4 cm. This distinctive skipper is black with orange spots dorsally, and has red-orange hind wings with several ovoid white patches ventrally.

Range. This skipper occurs in boreal regions of both North America and Eurasia. In North America it ranges from Nova Scotia west to the Pacific Coast and north to central Alaska. It occurs south to northern Pennsylvania, northern Michigan, northern Minnesota, Wyoming, and northern California.

Habitat. The Arctic Skipper is found within openings or glades in heavily forested areas. It occurs principally in the Canadian and Hudsonian Life Zones.

Life History. *Behavior.* Males perch atop grass stalks in small forest openings and occasionally patrol back and forth for about 20 m before returning to their perch.

Broods. The Arctic Skipper is univoltine, with adults on the wing from May 15 to July 7 in the eastern states.

Early Stages. The egg is a pale green-white hemisphere. The mature caterpillar has a whitish head and a blue-green to creamy body, with a dark-green dorsal stripe and a whitish or pale-yellow lateral stripe along each side. There is a line of black spots below each lateral stripe.

Adult Nectar Sources. In California, these skippers often nectar at iris flowers.

Caterpillar Host Plants. Various broadleaf grasses serve as food plants. In California, purple reedgrass (*Calamagrostis purpurascens*) is a known host, and in Europe brome grasses (*Bromus*) are preferred.

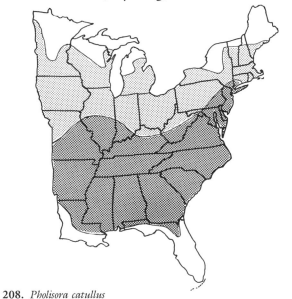

208. *Pholisora catullus*

209. *Carterocephalus palaemon mandan*

Swarthy Skipper
Nastra lherminier (Latreille), 1824

PLATE 42 · FIGURE 252 · MAP 210

Synopsis. This small, pale-brown skipper is often overlooked among tall grasses. It does not survive most northern winters and varies tremendously in abundance from year to year.

Butterfly. Male forewing: $\overline{X}$ = 1.3 cm, range 1.2–1.3 cm; female forewing: $\overline{X}$ = 1.2 cm, range 1.2–1.3 cm. This small, pale-brown skipper has virtually no markings dorsally and lightly veined brown ventral hind wings.

Range. The Swarthy Skipper may be found from Massachusetts and New York west to Missouri and then south through most of our area to Florida, the Gulf Coast, and eastern Texas. It is probably only a temporary colonist in the northern portion of its range.

Habitat. In most years the Swarthy Skipper is limited to dry or moist fields and meadows, sand barrens, and hillsides with good stands of its host. In years of abundance it appears in almost any open, grassy area, including suburban yards, vacant lots, landfills, and roadsides. It is primarily found in the Lower and Upper Austral Life Zones.

Life History. *Behavior.* Males perch close to the ground on grass stems. Courtship occurs primarily in late afternoon. *Broods.* In areas of permanent residence, the Swarthy Skipper is bivoltine. The first brood occurs progressively earlier southward, but the second occurs in August and September everywhere. In New York, the first adults appear in late June; in Virginia, in mid-May; in Georgia, in late April; and in Florida, by late March.
Early Stages. The egg is pearly white with a flattened summit. The first-instar larva is white with light-brown head.

Adult Nectar Sources. Low-growing white, pink, or blue flowers are preferred, including selfheal, peppermint, red clover, purple vetch, tick trefoil, and New Jersey tea.

Caterpillar Host Plants. Little bluestem (*Andropogon scoparius*) is the only recorded host.

Neamathla Skipper
Nastra neamathla (Skinner & Williams), 1923
MAP 211

Synopsis. The Neamathla Skipper is a wide-ranging but poorly known tropical and subtropical skipper.

Butterfly. There are usually definite but minute spots on the forewing. Identification of the Neamathla Skipper should be based only on examination of the genital features.

Range. The Neamathla Skipper occurs sporadically from Florida west to extreme southeastern California and then south through Mexico and Central America to at least Costa Rica.

Life History. *Broods.* There are probably three yearly flights in Florida (mid-February through March, May–June, and mid-August–mid-October).

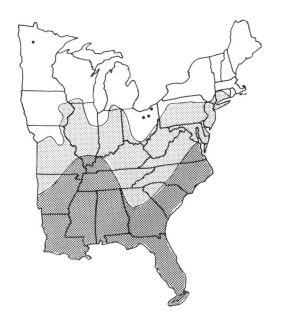

210. *Nastra lherminier*

211. *Nastra neamathla*

Three Spot Skipper
Cymaenes tripunctus (Herrich-Schaeffer), 1865
MAP 212

Etymology. The species name is derived from the Latin *tri*, "three," and *punctum*, "spot."

Synopsis. In our area this tropical American skipper is confined to southern Florida, where it may be confused with either the Eufala Skipper or the Neamathla Skipper.

Butterfly. Male forewing: $\overline{X}$ = 1.4 cm, range 1.3–1.5 cm; female forewing: $\overline{X}$ = 1.3 cm, range 1.3–1.4 cm. The Three Spot Skipper is a dull dark-brown skipper with a series of three tiny white hyaline spots on the costa near the forewing apex and two or three others at the end of the cell. This skipper's antennae are more than half of the forewing length, while those of the Eufala Skipper (*Lerodea eufala*) are less than half. The hyaline forewing spots should distinguish it from the Neamathla Skipper (*Nastra neamathla*).

Range. The Three Spot Skipper occurs in southern Florida, the Greater Antilles, and the mainland from southern Mexico to Brazil.

Life History. *Broods.* This skipper flies from March to October in Florida. The number of generations is unknown. *Early Stages.* The egg is white to light green and hemispherical. The caterpillar is blue-green with a gray-green dorsal stripe and light green lines separating green and gray-green longitudinal bands. Its head is variable, either white with brown marginal stripes or brown with two white stripes on the lateral portion of the front.

Caterpiller Host Plants. The larvae feed on Guinea grass (*Panicum maximum*) in nature, and will feed on sugarcane in captivity.

Clouded Skipper
Lerema accius (J. E. Smith), 1797
PLATE 43 · FIGURE 253 · MAP 213

Etymology. The genus name *Lerema* is taken from the Greek *lerema*, "silly discourse, chatter." Accius was a Roman poet (170–94 B.C.).

Synopsis. The Clouded Skipper is our only representative of a tropical American genus. It is resident on the southeastern coastal plain, but it wanders north irregularly in some years.

Butterfly. Male forewing: $\overline{X}$ = 1.7 cm, range 1.5–1.8 cm; female forewing: $\overline{X}$ = 1.9 cm, range 1.6–2.0 cm. Dorsally, the Clouded Skipper is brown-black with several white hyaline forewing spots and a black stigma for the male. Ventrally, the hind wing has a violet-blue iridescence, gray outer shading, and one or two dark bars parallel to the outer margin.

212. *Cymaenes tripunctus*

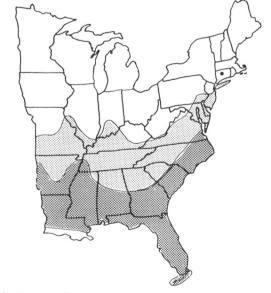

213. *Lerema accius*

Range. The Clouded Skipper is a permanent resident from Georgia south to Florida and east to Texas, then south again through Mexico and Central America to northern South America. It wanders north from its area of residence to at least Illinois and New York, and it may occasionally reproduce as far north as Virginia. In northern areas it usually appears in August or September.

Habitat. The Clouded Skipper is usually found near swamps and rivers at the edges of woods or in clearings. It may wander to nearby open fields or roadsides in search of nectar. It is resident in Tropical and Lower Austral Life Zones.

Life History. *Behavior.* Males perch within a few inches of the ground in semishaded areas along forest edges or in small clearings. A mated pair was seen at 0930 hr.

Broods. The Clouded Skipper is found throughout the year in Florida, where it may have three generations. It is barely possible that the species has a generation of long-lived dry-season adults. Farther north there are two annual flights. For example, in Georgia there is a flight from mid-August to mid-September.

Early Stages. The mature caterpillar is almost white and is mottled with black. Its head is white and is black margined, with three vertical black streaks on the front. The pupa is smooth, slender, and green-white.

Adult Nectar Sources. Several white, pink, or purple flowers are visited, including lantana, buttonbush, vervain, selfheal, and shepherd's needle.

Caterpillar Host Plants. Well-documented larval hosts are wooly beard grass (*Erianthus alopecturoides*), St. Augustine grass (*Stenotaphrum secundatum*), and *Echinochloa povietianum*. Abbot protrayed larvae on Indian corn (*Zea mays*), but it may not be a natural host.

Least Skipper
Ancyloxypha numitor (Fabricius), 1793
PLATE 43 · FIGURE 254 · MAP 214

Etymology. The genus name *Ancyloxypha* is derived from the Greek *ankylos*, "curved, hooked," and *xiphos*, "sword." Numitor, King of Alba, was grandfather of Romulus and Remus.

Synopsis. The Least Skipper may be incredibly abundant, especially in early summer, but its small size and low flight may cause it to be overlooked by the novice observer.

Butterfly. Male forewing: $\overline{X}$ = 1.1 cm, range 1.0–1.2 cm; female forewing: $\overline{X}$ = 1.2 cm, range 1.1–1.3 cm. In addition to its small size, the combination of a predominantly black forewing and a black-bordered orange dorsal hind wing, together with a yellow-orange ventral hind wing, should serve to distinguish the Least Skipper from all others in our area.

Range. This small skipper occurs from Nova Scotia and southern Quebec west to southern Saskatchewan and thence south through almost our entire area to Florida, the Gulf states, and Texas.

Habitat. The Least Skipper is found most predictably in wet or moist open areas with tall grasses. The borders of sluggish streams, ditches, or open freshwater marshes are often occupied, but individuals of summer broods often spread to other areas, such as old fields or even hillsides, if tall grasses are present.

Life History. *Behavior.* Males patrol by flying through grassy areas, usually less than 1 m above ground. The flight is usually slow, but may be quite rapid when in pursuit of a female or another male. Several attempted courtships were seen in late afternoon (1630 hr). One unusual rejection posture by females is to quickly flex their wings below the horizontal plane of their body. Mated pairs have been seen throughout the day (0920–1415 hr).

Broods. In most of the East, the Least Skipper has three full generations or two and a partial third. In Texas, there seem to be four broods. In Virginia, the broods extend from May 17 to July 1, July 15 to August 15, and August 29 to October 15. In the Deep South (Louisiana, Florida), flight dates range from mid-February to early December.

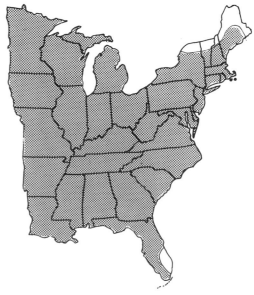

214. *Ancyloxypha numitor*

Early Stages. The egg is a shiny bright yellow when first laid, but soon acquires an orange-red band around the middle. The mature caterpillar is grass green and has a dark-brown head ringed with a white line on the front. The pupa is cream, with brown lines and patches.

Adult Nectar Sources. For its nectar visits, the Least Skipper favors low plants with small flowers, including wood sorrel, arrowhead, swamp verbena, mistflower, pickerel-weed, wild oregano, dogbane, buttonbush, chickory, and white clover.

Caterpillar Host Plants. Larval hosts are various grasses. Rice cutgrass (*Leersia oryzoides*), bluegrass (*Poa*), marsh millet (*Zizaniopsis miliacea*), and cultivated rice are documented hosts in nature.

Powesheik Skipper
Oarisma powesheik (Parker), 1870
PLATE 43 · FIGURE 255 · MAP 215

Etymology. In Greek, *oarisma* means "fond discourse" or "loving conversation." Powesheik was a Fox Indian chief during the Black Hawk Indian War in 1832. A county in Iowa was named for him.

Synopsis. Like many prairie denizens, the Powesheik Skipper has suffered tremendous losses of its native habitat. Soon it will be found only in prairie preserves and parks.

Butterfly. Male forewing: $\bar{X}$ = 1.3 cm, range 1.2–1.4 cm; female forewing: $\bar{X}$ = 1.3 cm, range 1.2–1.4 cm. The Powesheik is one of the more beautiful skippers. Its black ventral wings have an orange-lined costal area and veins, while the dorsal hind wings have the veins lined with white scaling.

Range. The former range of this skipper extended in an arc from southern Michigan across the southern terminus of Lake Michigan, then north and west through Iowa, western Minnesota, and the eastern Dakotas. Today, it remains locally abundant in Michigan (one site), Iowa (one site), Minnesota, and the Dakotas.

Habitat. The Powesheik Skipper is limited to relatively undisturbed, native tall-grass prairie remnants.

Life History. *Behavior.* Males patrol close to the ground with a fairly rapid flight.

Broods. The Powesheik Skipper is univoltine, with adult flight dates extending from the last week of June to mid-July. Occasional individuals are seen until the first week of August.

Early Stages. The pale, yellow-green, hemispherical egg is laid on grass. The caterpillars feed from mid-July to the end of September while passing to the fifth instar. Winter is passed in this stage, and feeding is resumed in April of the succeeding year. The mature larva is pale green with a dark-green dorsal band that tapers posteriorly. The band is outlined by cream lines and there are six fine cream lines along each side (McAlpine 1973).

Adult Nectar Sources. In Iowa, adults nectar at purple coneflower, ox-eye, daisy, and stiff-leaved coreopsis. McAlpine (1973) reports that they visit white clover and "yellow daisy" in Michigan, while the skipper takes nectar from black-eyed susans in Wisconsin.

Caterpillar Host Plant. The only recorded natural larval host is spikerush, *Eleocharis elliptica*, upon which eggs are laid in Michigan.

215. *Oarisma powesheik*

Garita Skipper
Oarisma garita (Reakirt), 1866

Synopsis. The Garita Skipper has been collected only once in our area, in western Minnesota, despite the fact that it is the most abundant skipper in neighboring North Dakota. An inhabitant of short-grass prairie knolls, the Garita Skipper ranges on the high plains from Alberta and Manitoba south to Texas. It may be easily distinguished from the larger Powesheik Skipper (*Oarisma powesheik*) by the largely orange anal area on its ventral hind wing. The skipper has a single flight each year from early June to early July.

Orange Skipperling
Copaeodes aurantiaca (Hewitson), 1868

Synopsis. The Orange Skipperling is similar to the Southern Skipperling (*Copaeodes minima*) but is larger and lacks the white ventral hind-wing stripe. Its usual range is from the southwestern states south to Panama. A male was once found in Little Rock, Arkansas. Like Southern Skipperling larvae, its caterpillars feed on Bermuda grass.

Southern Skipperling
Copaeodes minima (Edwards), 1870

PLATE 43 · FIGURE 256 · MAP 216

Etymology. *Copaeodes* is derived from the name for Copae, an ancient town in Boeotia, now called Topolia. The species name, *minima*, is taken from the Latin *minimus,* "smallest, least."

Synopsis. The Southern Skipperling is the smallest skipper in North America. It is one of three *Copaeodes,* which are primarily or completely tropical in occurrence.

Butterfly. Male forewing: $\overline{X}$ = 0.9 cm, range 0.8–0.9 cm; female forewing: $\overline{X}$ = 1.0 cm, range 0.9–1.0 cm. The small size, orange color, and white stripe on the ventral hind wing should permit easy identification of this skipper.

Range. The species occurs from North Carolina west to Arkansas and south through Florida, the Gulf states, Texas, and Mexico to Panama.

Habitat. The Southern Skipperling is found in open sunny fields in the Subtropical and Lower Austral Life Zones.

Life History. *Broods.* This skipper is bivoltine in Georgia, where there are two generations, one in May–June and the other in August–early October. In Louisiana the Southern Skipperling is in flight from March 2 to October 18, and in Florida it may be seen throughout the year.

Adult Nectar Sources. The only recorded nectar flower is fine-leaved sneezeweed (*Helenium tenuifolium*).

Caterpillar Host Plant. Bermuda grass (*Cynodon dactylon*) is the only known caterpillar host.

216. *Copaeodes minima*

European Skipper
Thymelicus lineola (Ochsenheimer), 1808

PLATE 43 · FIGURE 257 · MAP 217

Etymology. The genus name is derived from the Greek *thymele,* "altar, orchestra." Thymelici are theatrical musicians. The Latin word *lineola* means "small line."

Synopsis. The small, all-orange European Skipper was first found in North America in London, Ontario, in 1910. Since then, it has spread over much of our area. It may be the most common skipper in some localities.

Butterfly. Male forewing: $\overline{X}$ = 1.2 cm, range 1.1–1.3 cm; female forewing: $\overline{X}$ = 1.3 cm, range 1.2–1.3 cm. Dorsally, this skipper is burnt orange with a brassy sheen, its wings have a narrow black border, and the terminal portion of the veins are outlined in black. The male has a narrow black forewing stigma. A pale form occurs in some areas. The Southern Skipperling (*Copaeodes minima*) is similar but is smaller and more yellow-orange. At present the ranges of these two species do not overlap.

Range. The European Skipper now occurs from New Brunswick west to northern Minnesota, and south to Virginia, North Carolina, Tennessee, and Missouri. The species is continuing to expand its range. In the West, a second introduced population occurs in central British Columbia. Burns (1966a), who documented the early spread of the European Skipper, felt that transport of its eggs in hay shipments was the most likely explanation for much of this skipper's range expansion.

Habitat. Open grassy fields, including pastures and abandoned homesteads, and grassy road verges are typical situations in which the European Skipper may be found. It is most common in the Transition and Canadian Life Zones, but it is expanding into the Upper Austral Zone as it becomes acclimated.

Life History. *Behavior.* Males patrol during most of the day, with a low, steady, meandering flight through grassy stands. They pause periodically only to visit flowers or to court females. Mated pairs are seen near midday (1100–1415 hr). The skippers roost for the night by clinging to grassy stalks, and warm up in the morning by spreading their wings at about a 70° angle.

Broods. Although Shapiro (1966) reports two generations in Pennsylvania, there seems to be only one through most of the species' introduced range. Usually the European Skipper emerges in the first week of June and flies to mid-July. In northern Virginia it has been found as early as May 20, but no later than June 21.

Early Stages. The white, low, hemispherical eggs are laid on grass stems, and although the first-stage caterpillar soon develops, it does not hatch until the following spring. The mature larva is green with a dark dorsal stripe and has a light-brown head with white or yellow longitudinal stripes on the front. The pupa is green and yellow-green, with a longitudinal dark dorsal stripe, and also has a down-curved horn in front.

Adult Nectar Sources. European Skippers generally nectar at flowers of low stature, including ox-eye daisy, orange hawkweed, fleabane, red clover, white clover, selfheal, thistles, Deptford pink, swamp milkweed, and common milkweed.

Caterpillar Host Plants. Timothy (*Phleum pratense*) is the usual host, but other grasses are also eaten.

Fiery Skipper
Hylephila phyleus (Drury), 1773
PLATE 44 · FIGURES 259–261 · MAP 218

Etymology. The genus name is derived from the Greek *hylē*, "forest," and *philos*, "loving." Phyleus is a Latin proper name.

Synopsis. The Fiery Skipper is a lawn pest in parts of the West. It is the only North American representative of its genus, most of whose species are found in South America.

Butterfly. Male forewing: $\overline{X}$ = 1.5 cm, range 1.4–1.6 cm; female forewing: $\overline{X}$ = 1.6 cm, range 1.5–1.7 cm. This species may be readily distinguished by its short antennae, which are only as long as the head width. The males are bright yellow-orange above, with a narrow, black, dentate border and a wide black stigma, as well as small black spots scattered on the

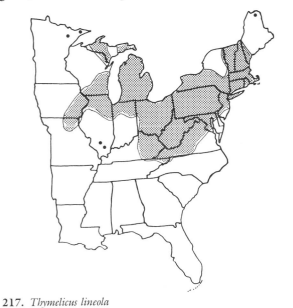

217. *Thymelicus lineola*

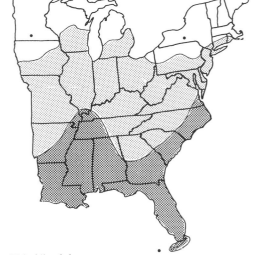

218. *Hylephila phyleus*

ventral hind wings. The female is similar to that of the Satchem but lacks hyaline spots on the forewings.

Range. The Fiery Skipper may be found from Connecticut west to California and south through the Caribbean, Mexico, and Central America to Chile. It cannot overwinter in any stage anywhere in the eastern United States where there are harsh winters. It is probably resident south from the coastal Carolinas and lower Mississippi Valley. It regularly colonizes more northern areas each summer.

Habitat. The Fiery Skipper is most often found in open, sunny areas, including dry fields, lawns, roadsides, levees, and second-growth scrub. It is resident only in the Tropical and Lower Austral Life Zones.

Life History. *Behavior.* Courtship and mating behavior were investigated by Shapiro (1975). Males perch all day, close to the ground on projecting objects, such as twigs or grass blades. The fluttering of the wings of passing insects elicits approach flight by males. In successful courtships, the male lands behind the female, thrusts his head between the female's wings, and then moves laterally to her while rapidly fluttering his wings. Then he bends his abdomen toward her and coupling takes place. Males mate primarily with virgin females, and coupling usually lasts about 40 minutes. If a pair is disturbed, the female is the carrying sex. Mated pairs have been seen in late afternoon.

Broods. In areas where the species is resident, there seem to be three to five generations annually. In Florida the Fiery Skipper may be seen throughout the year, while in Mississippi adults are present in all months except January and February. In more northern areas, the butterflies are usually not seen until May, and are not common until late August.

Early Stages. The caterpillars live in a shelter amidst grass sod that lies horizontal to the surface. The mature caterpillar has a black head with red-brown stripes, and its body varies in color from dark gray-brown to yellow-brown with three darker longitudinal stripes. The pupa is variably colored yellow-brown, reddish or greenish. The tongue case is free posterior to the wing cases.

Adult Nectar Sources. Flowers of many plants, including thistles, ironweed, knapweed, sneezeweed, asters, swamp milkweed, and sweet pepperbush, are visited for nectar.

Caterpillar Host Plants. Weedy grasses, especially crabgrass (*Digitaria*) and Bermuda grass (*Cynodon dactylon*), are selected by females for ovipositing, but others, such as St. Augustine grass (*Stenotaphrum secundatum*), sugar cane, and bent grass (*Agrostis*), are also fed upon.

Uncas Skipper
Hesperia uncas Edwards, 1863
MAP 219

Synopsis. In our area, the Uncas Skipper has been found only sparingly in Iowa and western Minnesota. It differs from other *Hesperia* chiefly on the ventral hind wings, on which the macular white spots are prolonged along the veins and contrast with the black patches between the basal spots and median macular band. The butterfly is a short-grass prairie inhabitant found on the high plains from Saskatchewan south to below Mexico City and west through the Great Basin to eastern California. Its caterpillars feed on blue grama grass (*Bouteloua gracilis*) and needlegrass (*Stipa*). Unlike our other eastern *Hesperia*, the Uncas Skipper has at least two generations and is in flight from May to September.

Laurentian Skipper
Hesperia comma laurentina (Lyman), 1892
MAP 220

Etymology. Hesperia, a nymph, was daughter of the river god, Kebren (Greek myth). The species name stands for a punctuation mark, while the subspecies name refers to the region surrounding the St. Lawrence River.

Synopsis. *Hesperia comma* occurs in both North America and Eurasia. The seventeen other *Hesperia* are entirely American in their occurrence. The Laurentian is slowly spreading southward.

Butterfly. Male forewing: $\overline{X}$ = 1.5 cm, range 1.3–1.5 cm; female forewing: $\overline{X}$ = 1.6 cm, range 1.4–1.7 cm. This northern skipper may be distinguished by the ventral hind wings, which have a pattern of clear white spots on a golden to golden green ground color.

Range. *Hesperia comma* is found from Labrador and New Brunswick west to the Pacific Coast, north to Alaska, and south to Maine, New Hampshire, northern Michigan, Wisconsin, Minnesota, western Texas, New Mexico, Arizona, and Baja California. The species also ranges across Eurasia. The Laurentian Skipper, a subspecies, occurs from Maine and New Brunswick west to southern Manitoba and northern Minnesota. It first appeared in Maine during the early 1930s.

Habitat. The Laurentian Skipper is always found in open, sunny areas, and presumably favors open fields and meadows.
Life History. *Behavior.* Males perch on bare ground, small twigs, or hilltops during most daylight hours and await the passage of receptive females. Mating occurs near midday, with females acting as carriers.
Broods. This skipper is univoltine, with flight dates from July 19 to as late as September 9.
Early Stages. The early stages of *Hesperia* are similar enough to permit one general description to apply to all (MacNeill 1964). The eggs are over 1 mm in diameter and are low, white hemispheres. The mature larvae are olive green with a black or brown head. The caterpillar lives within a silken shelter at the base of grass clumps, but leaves it to feed. The larvae overwinter at various stages and pupate in a loose cocoon amid debris during the following year. The pupae are various shades of brown, often with a waxy bloom.
Adult Nectar Sources. Blazing star and goldenrods are visited by adults in search of nectar.
Caterpillar Host Plant. The known hosts for other subspecies are various perennial bunch grasses, including needlegrass (*Stipa*), blue grass (*Poa*), fescue (*Festuca*), and brome grass (*Bromus*) (McGuire 1982a).

Ottoe Skipper
Hesperia ottoe Edwards, 1866
PLATE 44 · FIGURES 262 & 263 · MAP 221

Etymology. The Oto, or Otoe, were one of the three Sioux tribes forming the Chiwere group.
Synopsis. The Ottoe Skipper is a large, "bold" prairie inhabitant whose range spans almost the entire tall-grass prairie biome. Like other native prairie obligates, it has had great reductions in its habitat due to agricultural conversion.
Butterfly. Male forewing: $\overline{X} = 1.7$ cm, range 1.5–1.7 cm; female forewing: $\overline{X} = 1.8$ cm, range 1.6–1.9 cm. Male Ottoe Skippers have "gray felt" within the forewing stigma and clear orange-yellow ventral hind wings. Ventrally, the females are ochraceous, with vaguely defined markings. The forewings are more pointed than the similarly marked Dakota Skipper (*Hesperia dacotae*).

Range. This skipper occurs in southern Michigan, Illinois, and Wisconsin and then again from western Minnesota west to eastern Montana and south to Texas and Colorado.
Habitat. The Ottoe Skipper occurs in tall-grass prairie, often on hills or slopes, and requires a mixture of suitable nectar plants.
Life History. *Behavior.* Males perch during warm daylight hours (1000–1800 hr) on bare ground or low vegetation. Their interactions with other males may take them away on low chases of 100 m or more. Courtships were seen in late afternoon (1800 hr) in western Iowa. The male forces down the female while in flight, then lands behind her and gradually moves up beside her while quivering his wings. Females also

219. *Hesperia uncas*

220. *Hesperia comma laurentina*

oviposit in the afternoon (1300–1700 hr). They land on an area of bare soil, then walk toward a host grass and lay a single egg at the base of a stem. They may also lay eggs on the flower heads of purple coneflower.

Broods. The Ottoe Skipper is univoltine, with the flight beginning in our area about the third week in June and peaking in late June or early July. Some individuals may still be seen until the first week in August.

Early Stages. Refer to Laurentian Skipper.

Adult Nectar Sources. Adults visit yellow prickly pear, green and common milkweeds, vetch, blazing star, sunflower, alfalfa, compassplant, leadplant, purple coneflower, and bush houstonia.

Caterpillar Host Plants. In Michigan, fall witchgrass (*Leptoloma cognatum*) is fed upon (Nielsen 1958); in Iowa, several females oviposited on little bluestem (*Andropogon scoparius*).

Leonard's Skipper
Hesperia leonardus Harris, 1862
PLATE 43 · FIGURE 258 · MAP 222

Etymology. The species was named in honor of the Reverend L. W. Leonard of Dublin, New Hampshire, who corresponded with T. W. Harris, the species' describer.

Synopsis. The Leonard's Skipper is the only univoltine fall-flying butterfly in much of the eastern United States. Recently it was realized that the western Pawnee Skipper is actually a subspecies of Leonard's Skipper.

Butterfly. Male forewing: $\overline{X}$ = 1.7 cm, range 1.6–1.8 cm; female forewing: $\overline{X}$ = 1.8 cm, range 1.6–2.0 cm. Dorsally, Leonard's Skipper is red-orange with broad, dark borders. The male forewing stigma has yellow "felt" within. Ventrally, the hind wings are brick red with a median band of white, cream, or yellow spots. In Minnesota, individuals intermediate to the more western Pawnee Skipper are found.

Range. Leonard's Skipper occurs from Nova Scotia and Maine west across southern Ontario and the Greak Lakes states to Minnesota. It then occurs south to North Carolina,

Louisiana, and Missouri. The Pawnee Skipper, now recognized as a subspecies of Leonard's Skipper (Scott and Stanford, personal communication), ranges west to Montana and south to Kansas and Colorado.

Habitat. Leonard's Skipper is found in open grassy areas, including low-lying wet fields or meadows, grassy slopes, and pine–oak barrens. It occurs primarily in the Upper Austral and Transition Life Zones.

Life History. *Behavior.* Males may perch during most daylight hours on knolls or flats near blazing stars or may repeatedly patrol a beat that includes both nectar sources and places where females might be located. Courtships have been seen in the afternoon (1300–1515 hr).

Broods. There is only a single late-summer flight each year. In contrast to most other butterflies, the Leonard's Skipper emerges earlier at higher elevations and more northern latitudes. For example, in New York's Finger Lake region,

221. *Hesperia ottoe*

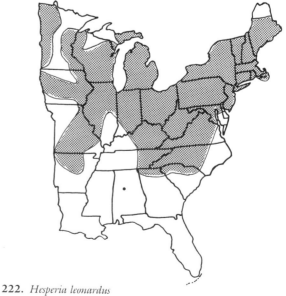

222. *Hesperia leonardus*

Leonard's Skipper flies from August 5 to September 10, while in lowland northern Virginia it is on the wing from August 28 to October 10. In the mountains of Virginia and West Virginia, its flight begins mid-August.

Early Stages. Refer to Laurentian Skipper.

Adult Nectar Sources. Purple or pink flowers are favored by this skipper. These include joe-pye weed, blazing star, New York ironweed, purple boneset, asters, teasel, selfheal, thistles, and trailing bush clover. Rare brief visits to yellow flowers such as tick-seed sunflower are sometimes witnessed.

Caterpillar Host Plants. Grasses, such as switch grass (*Panicum virgatum*), poverty oatgrass (*Danthonia spicata*), and bent grass (*Agrostis*), have been reported as larval hosts (McGuire 1982a). Leonard's skipper seems to be associated with the small stands of bluestem (*Andropogon*) where the Dusted Skipper (*Atrytonopsis hianna*) may be seen in spring.

Pahaska Skipper
Hesperia pahaska Leussler, 1938

Synopsis. The western Pahaska Skipper has been collected once in Clay County, Minnesota. It is similar to the Laurentian Skipper (*Hesperia comma laurentina*). The Pahaska Skipper is larger and flies earlier in the year. It ranges from Saskatchewan south to northern Mexico and west to Arizona and eastern California. The skipper is univoltine, with a single annual flight from early June to early July from Colorado northward, while in the South and West it is bivoltine. Like other *Hesperia*, its larvae feed on grasses. One host is blue grama grass (*Bouteloua gracilis*).

Cobweb Skipper
Hesperia metea Scudder, 1864

PLATE 44 · FIGURE 264 · MAP 223

Etymology. *Metea* is a variant spelling of Mitia, a Potawatomi Indian chief. There is a town in Indiana with the same spelling.

Synopsis. This small, very dark skipper is the earliest *Hesperia* to fly each year throughout most of our area. Its dark color probably helps it warm up to and maintain flight temperatures. It is often found with the Dusted Skipper (*Atrytonopsis hianna*).

Butterfly. Male forewing: $\bar{X}$ = 1.4 cm, range 1.3–1.5 cm; female forewing: $\bar{X}$ = 1.5 cm, range 1.5–1.6 cm. The Cobweb is very dark brown or blackish. The dorsal light areas are very restricted, and the light spots on the ventral hind wings are extended along the veins, giving them a cobweb effect. Populations in Arkansas and Texas (subspecies *licinus*) are almost immaculate above and below, and some individuals elsewhere in the Deep South are nearly so.

Range. The Cobweb Skipper occurs from southern Maine west to Wisconsin and south to the Gulf States and eastern Texas.

Habitat. This small skipper is found on dry, acid, sandy, or rocky sites, including pine–oak sand barrens, shale barrens, serpentine barrens, and pine prairies. Old pastures, burn areas, and dry, rocky hillsides are other habitats in which it may be found (Shapiro 1965).

Life History. *Behavior.* Males perch on or near the ground and periodically leave to visit low flowers.

Broods. There is a single brief flight of the Cobweb Skipper each spring. Emergence takes place earlier to the south. In the lower Delaware Valley, adults are found from May 6 to June 1, while in Mississippi, they fly from March 29 to April 21.

Early Stages. Refer to Laurentian Skipper.

Adult Nectar Sources. Low-growing plants, such as bird's-foot violet, Labrador tea, wild strawberry, winter cress, blackberry, and red clover, receive most nectar visits,

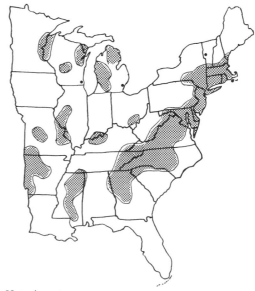

223. *Hesperia metea*

but wild hyacinth, Carolina larkspur, vervains, and lilac are visited on some occasions.

Caterpillar Host Plants. The Cobweb female selects primarily beard grasses (genus *Andropogon*), including little bluestem (*A. scoparius*) and its variety *glomeratus,* as well as big bluestem (*A. gerardi*).

Dotted Skipper
Hesperia attalus (Edwards), 1871

MAP 224

Etymology. *Attalus* "Grammaticus" was a grammarian in the second century B.C.

Synopsis. The Dotted Skipper is a very widespread but rarely seen skipper whose biological features are only recently becoming known.

Butterfly. Male forewing: $\overline{X}$ = 1.6 cm, range 1.5–1.6 cm; female forewing: $\overline{X}$ = 1.7 cm, range 1.5–1.8 cm. This skipper is so variable it is difficult to characterize. Ventrally, the hind wings may have a series of small dots or they may be immaculate. The forewing stigma has black "felt" within, and the forewings are very pointed.

Range. The Dotted Skipper is sparsely distributed where found, and occurs from Massachusetts and New Jersey south along the Atlantic Coastal Plain and Piedmont to Florida, thence west through the Gulf states to central Texas. The species is found rarely on the Plains.

Habitat. The Dotted Skipper frequents pine barrens or pine savannas in the east and southeast. It is found in sandy or serpentine barrens, but has not been reported from the Appalachian shale barrens. West of our area, the skipper is found on short-grass prairie.

Life History. *Broods.* This species appears to be bivoltine in most of its range, usually flying in May–June and August–September. In Florida, it flies from February to mid-May and from August through October.

Early Stages. Refer to Laurentian Skipper.

Adult Nectar Sources. The Dotted Skipper favors thistles, prickly pear cactus, alfalfa, and purple coneflower for its nectar visits.

Caterpillar Host Plants. Larval hosts are not documented, but Shapiro (1973) found the skipper associated with extensive stands of switchgrass (*Panicum virgatum*) in New Jersey. Studies by McGuire (1982b) note several other host plants for more western populations, including fall witchgrass (*Leptoloma cognatum*).

224. *Hesperia attalus*

Meske's Skipper
Hesperia meskei (Edwards), 1877

PLATE 45 · FIGURE 265 · MAP 225

Etymology. The species name may possbily be derived from *meska,* the Algonquin Indian term for spruce tree.

Synopsis. This is the only *Hesperia* that is limited to the Deep South. Its biological traits are almost completely unknown.

Butterfly. Male forewing: $\overline{X}$ = 1.5 cm, range 1.4–1.6 cm; female forewing: $\overline{X}$ = 1.6 cm, range 1.5–1.7 cm. Meske's Skipper may be distinguished from other *Hesperia* in its range

by its rich orange or reddish olive ventral hind wings, which usually have a band of faint pale orange spots. Dorsally, the wings are largely dark, but have restricted areas of bright orange.

Range. Meske's Skipper occurs from North Carolina south through Florida, including some of the Keys, and then west to Arkansas and Texas.

Habitat. The Meske's habitat has been described as sparse open woods.

Life History. *Broods.* Meske's Skipper has two or three yearly generations. In Georgia, it flies from late May to mid-June and from late September to early October. In Florida, adults are found from mid-April to mid-June, in September and October, and again in December.

Adult Nectar Sources. Meske's Skippers visit pickerel-weed and blazing star for nectar, but may visit other plants as well.

Dakota Skipper
Hesperia dacotae (Skinner), 1911
PLATE 45 · FIGURE 266 · MAP 226

Etymology. The Dacotas constituted the largest division of the Sioux.

Synopsis. The Dakota Skipper population is now much reduced, since its only habitat, native tall-grass prairie, has been largely converted to the production of corn and soybeans.

Butterfly. Male forewing: $\overline{X}$ = 1.4 cm, range 1.4–1.5 cm; female forewing: X = 1.5 cm, range 1.5–1.6 cm. The Dakota Skipper is a close relative of the Indian Skipper (*Hesperia sassacus*), with which it shares relatively stubby wings, a black interior felt patch in the male, and a poorly defined macular band of yellow spots on the ventral hind wing. The Dakota Skipper is small for a *Hesperia*. Males resemble Ottoe Skippers (*Hesperia ottoe*), and the females share with that species a hyaline spot below the end of the forewing cell. Females are distinctly gray beneath.

Range. The Dakota Skipper once occurred from extreme southern Manitoba south and east in a curving band through the eastern Dakotas, western Minnesota, Iowa, and northern Illinois. It is now limited to one site in Iowa, together with several in Minnesota and the Dakotas. A number of the remaining colonies occur on native prairie preserves.

Habitat. This rare species is limited to native tall-grass prairies on rolling hills in the Transition Life Zone.

Life History. *Behavior.* Males perch on low vegetation, including heads of composites. Perch sites are often on leeward slopes. When disturbed, they may fly 100 m or more until they land again. They are difficult to follow visually, due to their low, rapid flight and the constant prairie breeze. Adults mate on their first day of emergence, and both sexes may mate more than once (McCabe 1981)

Broods. The Dakota Skipper is univoltine, with a flight in early summer. Flight dates range from June 16 to August 2, but the species flies for only about 5 weeks at any locality in a given year. Unlike most butterflies, both sexes emerge at about the same time.

Early Stages. Females oviposit on broad-leaved plants, not their grass hosts. The eggs hatch in 7–20 days, and the young larvae climb to the ground and web two leaves together. As the larvae grow, they build a silken tube lined with grass

225. *Hesperia meskei*

226. *Hesperia dacotae*

blades and do most of their feeding at night. Winter is passed by fourth-instar larvae and feeding is completed the following spring.

Adult Nectar Sources. Dakota Skippers nectar at a variety of flowers, especially composites, with great regularity. Espe-cially favored are purple coneflower, prairie coneflower, flea-bane, blanket flower, black-eyed susans, ox-eye, and harebell.

Caterpillar Host Plant. The native host is unknown, but the larvae will feed on a variety of grasses in the laboratory.

Indian Skipper
Hesperia sassacus Harris, 1862
PLATE 46 · FIGURE 271 · MAP 227

Etymology. Sassacus was last chief of the Pequot Indians. He was killed in 1637 by the Mohawks in New York.

Synopsis. The Indian Skipper is the most widespread common *Hesperia* in the Northeast. In some places it is the most abundant skipper.

Butterfly. Male forewing: $\bar{X}$ = 1.4 cm, range 1.4–1.5 cm; female forewing: $\bar{X}$ = 1.5 cm, range 1.4–1.6 cm. The Indian Skipper is fairly small, is bright orange above and below, and has a somewhat dentate black border on the outer margins of both wings dorsally. The interior "felt" of the male stigma is black, and there are vague light spots on the ventral hind wings.

Range. The Indian Skipper occurs from Maine west across southern Ontario to Minnesota, and thence south in the mountains to Virginia and North Carolina (rarely).

Habitat. This skipper occurs in brushy old fields, wood-land clearings, pastures, and rocky headlands.

Life History. *Broods.* This late-spring skipper is uni-voltine. In Maine, adults may be found from the end of May to mid-July. Further south, in Virginia, they are found from early May to late June.

Early Stages. Refer to Laurentian Skipper.

Adult Nectar Sources. Indian Skippers visit flowers of henbit, orange hawkweed, blackberry, lithospermum, phlox, and viper's bugloss.

Caterpillar Host Plants. Panic grass (*Panicum*), little bluestem (*Andropogon scoparius*), and red fescue (*Festuca rubra*) are confirmed larval hosts, while other grasses prob-ably also are used.

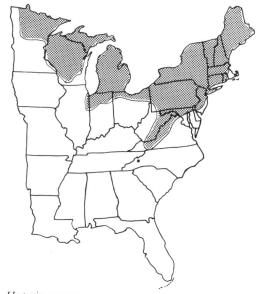

227. *Hesperia sassacus*

Peck's Skipper
Polites coras (Cramer), 1775
PLATE 45 · FIGURES 267 & 268 · MAP 228

Etymology. The genus was probably derived from 'polite' (French: *politesse*). The species is probably not classical.

Synopsis. The Peck's Skipper is a distinctive *Polites* with no obvious close relative. It is one of our most common skippers.

Butterfly. Male forewing: $\bar{X}$ = 1.2 cm, range 1.1–1.3 cm; female forewing: $\bar{X}$ = 1.3 cm, range 1.2–1.4 cm. Peck's Skipper is easily distinguished by its small size and the pattern of the ventral hind wing, which is composed of a central patch of yellow spots surrounded by a border of orange-brown. Individuals from the Atlantic coastal plain populations are a bit larger and have a more strongly defined hind-wing patch than those of inland populations.

Range. Peck's Skipper occurs from the Canadian Maritime Provinces west across southern Canada to British Columbia and south to Georgia, Texas, Colorado, Idaho, and northern California. It is rare in the South.

Habitat. This skipper is found in a wide variety of open, grassy habitats, including damp meadows, power-line cuts, lawns, landfills, roadsides, freshwater marshes, and vacant lots.

Life History. *Behavior.* Males perch during most daylight hours on low vegetation in open, sunny areas. Courtships may be seen during most of the day (0940–1740 hr). Mated pairs have been seen mainly in late afternoon (1520–1625 hr), but are occasionally seen as early as 0915 hr. As is usual in hesperiine skippers, the female is the carrier if the pair takes flight.

Broods. Peck's Skipper is bivoltine in most of the eastern United States, although the second may be only partial near the Canadian border, and a partial third occurs in the South. In Virginia, flights occur from late May to late June and again from late July through early September. There, a partial third brood is seen in late September and early October.

Early Stages. The egg is a pale-green globose hemisphere. The mature caterpillar is dark maroon with light-brown mottling and has a black head with two white vertical streaks on the upper front and two white patches on the lower front. **Adult Nectar Sources.** The native skipper has an inordinate predilection for red clover, especially during its first flight. This plant is an introduced exotic. Other plants whose flowers are visited freely include purple vetch, common milkweed, swamp milkweed, dogbane, New Jersey tea, self-heal, New York ironweed, thistles, and blue vervain. **Caterpillar Host Plants.** Shapiro (1974) reports one host to be rice cutgrass (*Leersia oryzoides*). Other grasses are probably fed upon as well.

Baracoa Skipper
Polites baracoa (Lucas), 1857
MAP 229

Etymology. The species was named after the Baracoa Indians, an original Cuban tribe.

Synopsis. The Baracoa Skipper is one of two Antillean *Polites*. It is like a small, dark Tawny-edged Skipper.

Butterfly. Male forewing: $\overline{X}$ = 1.1 cm, range 1.0–1.2 cm; female forewing: $\overline{X}$ = 1.2 cm, range 1.1–1.3 cm. The Baracoa Skipper is very similar to the Tawny-edged Skipper (*Polites themistocles*), but it differs in being smaller and darker, and in having a definite series of light spots on the ventral surface of the hind wing. The males have a narrow short stigma on the dorsal forewing.

Range. This species is resident in peninsular Florida, Cuba,

and Hispaniola. Temporary breeding populations are occasionally established in southern Georgia.

Habitat. The Baracoa Skipper is found in moist grassy areas near streams and marshes. It inhabits the Tropical Life Zone.

Life History. *Broods.* Broods are not well documented. Adults may be found from March to November in Florida. *Early Stages.* The egg is a tall yellow hemisphere. The mature caterpillar is brown with a darker brown dorsal stripe and a broad dark-brown line along each side. The head is dull gold with two white stripes on the upper front and two white spots on the lower front.

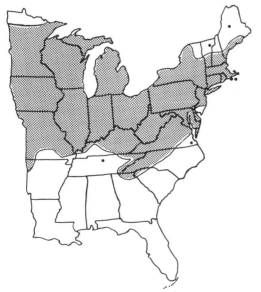

228. *Polites coras*

229. *Polites baracoa*

Tawny-edged Skipper
Polites themistocles (Latreille), 1824

PLATE 45 · FIGURES 269 & 270 · MAP 230

Etymology. Themistocles, son of Neocles, was an Athenian statesman.

Synopsis. The Tawny-edged Skipper is the most widespread *Polites* in North America. Its habitats are moister than those of the Cross Line Skipper.

Butterfly. Male forewing: $\overline{X}$ = 1.4 cm, range 1.0–1.6 cm; female forewing: $\overline{X}$ = 1.4 cm, range 1.3–1.6 cm. This common skipper is most similar to the Baracoa Skipper (*Polites baracoa*), but the two seldom overlap in range. The Tawny-edged Skipper overlaps broadly with the Cross Line Skipper (*Polites origenes*) and is often confused with it. The Tawny-edged Skipper may be identified by its generally smaller size, more rounded wings, and generally grayer ventral hind wings. Dorsally, the male has a broader, more sinuous forewing stigma and is brighter orange. The female has more extensive orange on the forewing costa.

Range. The Tawny-edged Skipper is found from Nova Scotia west across Canada to southern British Columbia. It then ranges south to Florida, the Gulf states, central Texas, central Arizona, and extreme northern California. It is rarer in the southern part of its range.

Habitat. This species favors situations moister than those favored by the Cross Line Skipper. For example, this skipper may occur in moist tall-grass prairie swales, damp meadows, pastures, lawns, vacant lots, and roadsides.

Life History. *Behavior.* Males perch low in open grassy areas during most warm daylight hours and await the passage of receptive females. Mated pairs have been seen in mid-afternoon (1325 to 1400 hr).

Broods. This species has one yearly flight in the northern portion of its range (e.g., Maine—June 13 to August 9), while two generations are the rule everywhere southward, except Florida, where the skipper is found all year. In the Delaware Valley, the Tawny-edged Skipper flies from early June to mid-July and again from early August to mid-September. In Georgia, the flights are from April 25 to June 7 and again from mid-July to early September.

Early Stages. Early stages are similar to those of the Cross Line Skipper except that the egg reticulations are finer and the caterpillar is more varied in color, with white vertical stripes on its "face."

Adult Nectar Sources. This skipper visits an array of flowers similar to those favored by the Cross Line Skipper, including dogbane, red clover, shrub houstonia, alfalfa, thistles, purple coneflower, and chicory.

Caterpillar Host Plants. Panic grasses (*Panicum*) are the larval hosts, and smaller species are preferred, although Shapiro (1966) found deertongue grass (*P. clandestinum*) was selected on occasion. Oviposition was observed on finger grass (*Digitaria filiformis*) in Illinois.

230. *Polites themistocles*

Cross Line Skipper
Polites origenes (Fabricius), 1793

PLATE 46 · FIGURES 272–274 · MAP 231

Etymology. The species name, *origenes,* was possibly derived from Origenes Adamantius, an Alexandrian writer (185–254 A.D.), or from a third-century A.D. Latin theologian of the same name.

Synopsis. The Cross Line Skipper and its close relative, the Tawny-edged Skipper (*Polites themistocles*), are found throughout most of the eastern United States. The Cross Line Skipper tends to occur in drier situations, although the two are often found together.

Butterfly. Male forewing: $\overline{X}$ = 1.4 cm, range 1.3–1.5 cm; female forewing: $\overline{X}$ = 1.5 cm, range 1.3–1.7 cm. The male Cross Line Skipper has a narrow forewing stigma, and the orange on its dorsal forewing is dull and brassy compared to that of the Tawny-edged Skipper. Ventrally, the hind wings are yellow-orange to orange-brown, usually with faint spots. Both wings tend to be relatively elongated or pointed. The female tends to be darker, with less orange above, and often has a squarish spot at the end of the forewing cell.

Range. This skipper occurs from southern Maine west across southern Ontario, the Great Lakes states, and plains to eastern Wyoming and the Colorado Rocky Mountain foothills. It ranges south to northern Florida, the Gulf states, central Texas, and northern New Mexico.

Habitat. The Cross Line Skipper is found in open grassy areas, often in less disturbed, drier situations than those favored by the Tawny-edged Skipper. Typical situations include old fields, meadows, brushlands, grassy openings in pine flatwoods, prairie hills, serpentine or sand barrens, and power-line cuts.

Life History. *Behavior.* Males perch within a meter of the ground in open areas during most warm daylight hours.

Courtship and mating take place from midday to mid-afternoon (1055–1530 hr).

Broods. The Cross Line Skipper is univoltine in the most northern portion of its range, e.g., northern Michigan (June 21–August 9), but everywhere else it is bivoltine, with occasional fresh individuals at the very end of the season. The gap between the broods becomes broader southward. For example, in northern Virginia, flights occur from the end of May through June and from early August to early September, while in Georgia they occur from May 2 to June 19 and August 29 to September 25.

Early Stages. The egg is a pale-green, flattened, reticulate hemisphere. The mature caterpillar is dark brown with dirty white mottling and a dull black head.

Adult Nectar Sources. Favored nectar sources include dogbane, purple vetch, red clover, selfheal, New Jersey tea, shrub houstonia, and New York ironweed. Yellow flowers, such as tick-seed sunflower or marigold, are visited only when the preferred white, pink, or purple flowers are scarce.

Caterpillar Host Plants. The caterpillars feed on purpletop (*Tridens flavus*), and probably other grasses as well.

Long Dash
Polites mystic (Edwards), 1863
PLATE 47 · FIGURES 277–279 · MAP 232

Etymology. The species name *Mystic* is taken from the Latin *mysticus,* "mystical, mysterious."

Synopsis. The Long Dash is a butterfly of wet marshy areas. It is named for the male's black stigma, which connects

231. *Polites origenes*

232. *Polites mystic*

with the subapical forewing patch, thus forming a "long dash."

Butterfly. Male forewing: $\overline{X}$ = 1.5 cm, range 1.3–1.5 cm; female forewing: $\overline{X}$ = 1.6 cm, range 1.5–1.7 cm. The Long Dash is larger than our other *Polites* and is more similar to the Indian Skipper (*Hesperia sassacus*) than to its relatives, but compared to the latter it has a broader stigma, is less orange above, and has a less angled spot band on the ventral hind wing. Prairie populations are lighter and have been given the subspecies name *dacotah*.

Range. The Long Dash occurs from the Maritimes west across southern Canada to British Columbia and south to New Jersey, Virginia (mountains), West Virginia (mountains), northern Illinois, Iowa, Nebraska, Colorado, Idaho, and Washington.

Habitat. This skipper is found in moist, open situations, including prairie swales, wet meadows, moist areas along streams, marshes, and wood edges.

Life History. *Behavior.* Males perch in low spots during most sunny daylight hours, but courtships and male-male interactions take place from mid- to late afternoon (1430–1730 hr). One mated pair was reported at 1655 hr. *Broods.* There is only a single annual generation in most of this skipper's range, flying from the end of May to mid-July or early August. Along the Atlantic coastal plain in New York and New Jersey, there are two broods, primarily in June and August.

Early Stages. The egg is a pale-green, finely reticulate hemisphere. The mature caterpillar is chocolate brown with dull white mottling and a dull black head.

Adult Nectar Sources. Flowers visited include selfheal, common milkweed, mountain laurel, and tick trefoil.

Caterpillar Host Plant. Shapiro (1966) reports the larval host in Pennsylvania to be blue grass (*Poa*). Other grasses may be fed upon as well.

Whirlabout
Polites vibex (Geyer), 1832
PLATE 46 · FIGURES 275 & 276 · MAP 233

Etymology. *Vibex* is the Latin word for "the mark of a blow," e.g., a linear subcutaneous extravasation of blood.

Synopsis. The Whirlabout is a common skipper in the Deep South. The male looks like that of the Fiery Skipper, but the female is mostly brown-black and looks like a completely different species.

Butterfly. Male forewing: $\overline{X}$ = 1.4 cm, range 1.3–1.8 cm; female forewing: $\overline{X}$ = 1.6 cm, range 1.4–1.7 cm. The male may be separated from the Fiery Skipper (*Hylephila phyleus*) by its broader black forewing stigma, and smoother black margins dorsally, as well as by larger black dots on the ventral hind wing. Females are brown-black dorsally and gray-brown on the ventral hind wing, with a pale central patch outlined with dark scaling.

Range. The Whirlabout is resident on the Atlantic coastal plain and Piedmont from South Carolina south through Florida and thence west along the Gulf Coast to Texas. This skipper also ranges farther south through Mexico and Central America to Argentina. In the eastern United States, emigrants occasionally range north along the immediate coast to Connecticut and up the Mississippi Valley to Iowa.

Habitat. The Whirlabout occupies a variety of open or scrubby habitats, including beach dunes, pine flats, open fields, and forest openings. The species is resident in the Tropical and Lower Austral Life Zones.

Life History. *Behavior.* Males perch and interact with other males during the afternoon. Their flight is rapid and darting. Females are more likely to be found along wood edges or in forest openings.

Broods. In Florida, the Whirlabout may be found all year, while in Georgia there are two generations (April 20–June 2 and July 6–September 27).

Early Stages. The egg is a smooth, white hemisphere. The mature caterpillar is pale with a faint stripe along each side. Its head is black with two yellow white stripes on the upper front and two yellow-white patches on the sides of the face. The

233. *Polites vibex*

pupa is pale green with a whitish abdomen and a long tongue case that extends to the tip of the abdomen.

Adult Nectar Sources.   In Florida, lantana and shepherd's needle are both visited freely.

Caterpillar Host Plants.   Recorded larval food plants are *Paspalum setaceum,* Bermuda grass (*Cynodon dactylon*), and St. Augustine grass (*Stenotaphrum secundatum*).

Broken Dash
Wallengrenia otho (J. E. Smith), 1797
PLATE 47 · FIGURE 280 · MAP 234

Etymology.   The species name, *otho* is possibly derived from Otto the First, Emperor of the Holy Roman Empire.

Synopsis.   The range of the Broken Dash overlaps with that of the Northern Broken Dash along the south Atlantic coastal plain and in Texas. The two species have different flight phenology, behaviors, and habitat preferences.

Butterfly.   Male forewing: $\overline{X}$ = 1.3 cm, range 1.1–1.5 cm; female forewing: $\overline{X}$ = 1.4 cm, range 1.1–1.6 cm. The Broken Dash is distinctly smaller than the Northern Broken Dash (*Wallengrenia egeremet*). Above, the male has more extensive pale areas, including the costa, and they are more orange than white or pale yellow. Below, the hind wing is tawny orange or red-orange with a medial series of small pale spots. John M. Burns (Smithsonian Institution) is presently completing a study of both species (personal communication).

Range.   The Broken Dash occurs from the coastal plain of Maryland (rarely) and Virginia south through peninsular Florida, then west through the Gulf states to Texas. Vagrant individuals have been found as far north as Missouri and Kentucky. On the Caribbean islands there are populations of skippers similar to this species.

Habitat.   Along the Atlantic coastal plain the Broken Dash is found in more or less wet situations near woodland associated with swamps or large rivers. In other parts of its range, it may be found in drier habitats.

Life History.   *Behavior.* Males perch within a half meter of the ground in sunny areas amid low vegetation at forest edges. Most mate-seeking activity has been observed in early morning.

Broods. The Broken Dash is bivoltine, but sometimes has a partial third. Flight dates extend from mid-March to late September. In areas where the two *Wallengrenia* occur together, the Northern Broken Dash never has more than a rudimentary second brood (J. Burns, personal communication).

Early Stages. The caterpillars are case bearers: they cut out circular pieces of material, which they carry over themselves.

Adult Nectar Sources.   Favored flowers of the Broken Dash include pickerelweed, sweet pepperbush, and selfheal.

Caterpillar Host Plants.   The native larval hosts of the Broken Dash are not definitely known. The caterpillars will feed on St. Augustine grass (*Stenotaphrum secundatum*) in captivity.

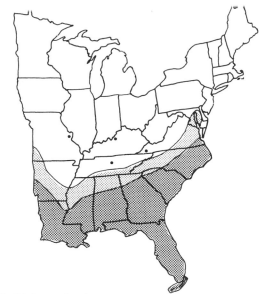

234. *Wallengrenia otho*

Northern Broken Dash
Wallengrenia egeremet (Scudder), 1864
PLATE 47 · FIGURES 281 & 282 · MAP 235

Etymology.   The species name was probably derived from Egeria, a nymph in Roman mythology.

Synopsis.   Until about 1950, this skipper and the Broken Dash were confused, but then it was realized that two similar yet separate species were involved. The common name is derived from the complex male stigma, which appears to be divided into two sections.

Butterfly.   Male forewing: $\overline{X}$ = 1.4 cm, range 1.1–1.6 cm;

female forewing: $\overline{X}$ = 1.5 cm, range 1.2–1.7 cm. The Northern Broken Dash is very similar to the Broken Dash (*Wallengrenia otho*) but is larger and has restricted amounts of yellow or white scaling dorsally, and its ventral hind wing is a lustrous purplish or reddish brown with a row of faint spots. Females are more likely confused with those of the Little Glassy Wing (*Pompeius verna*) or Dun Skipper (*Euphyes ruricola*). Florida specimens are larger than those from other populations (Burns, personal communication).

Range.   The Northern Broken Dash occurs from coastal Maine, New York, and southern Quebec west to northern Minnesota, Nebraska, and Kansas. It then ranges south through most of our area to peninsular Florida, the Gulf states, and Texas.

Habitat.   This species is found in open areas near woods or in brushy second-growth fields. It seems to avoid the driest situations, and is usually not found far from woods or brush.

Life History.   *Behavior.* Males perch on exposed twigs or leaves from 1 to 2 m above ground, (higher than its sibling, *W. otho*) and are most active in their mate-seeking behavior during the morning (0730–1020 hr) in direct contrast to most related skippers. A mated pair was seen at 1215 hr.

Broods. North of Virginia, the Northern Broken Dash is univoltine, usually flying from mid-June or early July to mid-August. Further south there are two broad flight periods. For example, in Mississippi, adults may be found from early April to late June and then again in August and September.

Early Stages. The larva of this species may be a case-bearer, as has been reported for the Broken Dash. The egg is a pale-green or yellow-white hemisphere. The mature caterpillar is pale green with darker green mottling and indistinct green and yellow stripes along the sides. The head is dark brown and has a dark central stripe and pale vertical stripes on the face.

Adult Nectar Sources.   This species prefers white, pink, or purple flowers. Favored nectar plants include dogbane, New Jersey tea, red clover, and sweet pepperbush. A variety of other flowers are visited.

Caterpillar Host Plants.   Large species of panic grass, such as deertongue grass (*Panicum clandestinum*), are selected as larval hosts.

Little Glassy Wing
Pompeius verna (Edwards), 1862
PLATE 48 · FIGURES 283 & 284 · MAP 236

Etymology.   Pompeius was a Roman triumvir with Caesar and Crassus. The Latin *verna* means "slave, born in the master's house."

Synopsis.   The Little Glassy Wing belongs to a group of six tropical American skippers that are very closely related to the *Polites* skippers. The female Little Glassy Wing and those of

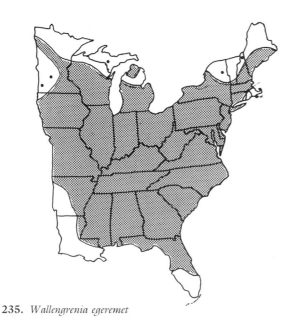

235. *Wallengrenia egeremet*

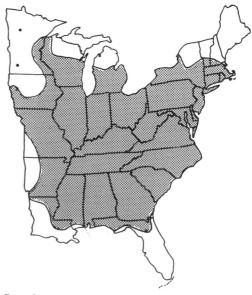

236. *Pompeius verna*

the Northern Broken Dash and the Dun Skipper share a close superficial resemblance and are sometimes referred to as the "three black witches."

Butterfly. Male forewing: $\overline{X}$ = 1.3 cm, range 1.2–1.6 cm; female forewing: $\overline{X}$ = 1.4 cm, range 1.3–1.7 cm. The males are distinguished by their predominantly black appearance and their black stigma adjoined by several white translucent patches. Ventrally, the hind wings are dark, with faint light spots and purplish overtones. The female has a glassy square spot at the end of the forewing cell that should help distinguish her from the other "black witches."

Range. The Little Glassy Wing is found from central New England west to eastern Nebraska and south to northern Florida, Mississippi, and eastern Texas.

Habitat. This dark skipper is most common near moist fields and clearings near streams, swamps, or bogs. It is usually found not far from shaded wood edges. The Little Glassy Wing occurs most often in the Upper Austral and Transition Life Zones.

Life History. *Behavior.* Although males may venture into open fields to seek nectar sources, they usually perch in small sunlit clearings on low vegetation, where they await the passage of receptive females. Courtship occurs near midday, and females exhibiting oviposition behavior were seen in late morning (1130 hr).

Broods. There is a single generation in the northern portions of this skipper's range (e.g., Michigan—June 20 to August 3), while the butterfly is bivoltine elsewhere. In northern Virginia, flights occur from early June to mid-July and from early August to early September, while in Mississippi they occur in April and May, and then again in July and August.

Early Stages. The egg is a small white or pale-green hemisphere. The mature caterpillar is yellow-green to yellow-brown and is covered with dark-brown tubercles, each bearing a light hair. The body has a dark dorsal stripe as well as three dark laterals on each side. Its head is dark red-brown with a black posterior margin.

Adult Nectar Sources. White, pink, and purple flowers are preferred by Little Glassy Wings. These include common and swamp milkweeds, dogbane, selfheal, peppermint, and joe-pye weed. Yellow flowers, such as tickseed sunflower and sneezeweed, are visited when preferred sources are rare or unavailable.

Caterpillar Host Plant. The only reported larval host is purpletop (*Tridens flavus*).

Satchem
Atalopedes campestris (Boisduval), 1852
PLATE 48 · FIGURES 285–287 · MAP 237

Etymology. The genus is derived from a combination of Atala, the daughter of an Indian chief in a novel by Chateaubriand and the Latin *pedes*, "legs." The species name is the Latin word *campestris*, "pertaining to the field."

Synopsis. The Satchem is one of the most widely distributed skippers of disturbed habitats; yet it cannot survive most winters much further north than Virginia. The only other members of this group occur in the Caribbean area.

Butterfly. Male forewing: $\overline{X}$ = 1.5 cm, range 1.5–1.6 cm; female forewing: $\overline{X}$ = 1.7 cm, range 1.6–1.8 cm. The male is easily recognized by the large quadrate black stigmatal patch on the dorsal forewing. Female Satchems are variably light to dark above, and may resemble female *Hesperia* or Fiery Skippers, but they may be reliably identified by the translucent rectangular spot at the end of their forewing cell.

Range. The Satchem is resident from Virginia west to California and south through Mexico and Central America to Brazil. In our area, the Satchem is a temporary summer visitor and sporadic breeder as far north as New York, Michigan, and Minnesota.

Habitat. This species is most common in open, sunny, disturbed areas with low weedy vegetation. Typical situations are open fields or pastures, suburban lawns, roadsides,

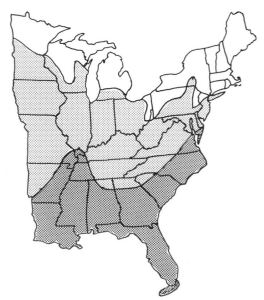

237. *Atalopedes campestris*

power-line cuts, and land-fills. In the East, the species is most at home in the Lower Austral Life Zone.

Life History.   *Behavior.* Male Satchems perch during most warm daylight hours, usually within 0.5 m of the ground, and, in many instances, on the bare ground itself. Courtship was usually observed in the morning or afternoon (1000–1629 hr) and follows a similar sequence to that of the Fiery Skipper. Mated pairs were also seen in the afternoon (1330–1530 hr), with females as the active carriers. Oviposition occurs in the afternoon (1320 hr); the female lays a single egg on dry grass blades.

Broods.   In the northern portions of its residential zone, e.g., Virginia, the Satchem has three generations: early May to mid-June, early July to mid-August, and late August to early November. In the Gulf states, such as Mississippi, where the Satchem flies from March to December, there are probably four or five flights.

Early Stages. The egg is a small white hemisphere. The mature larva has a black head and a dark olive-green body covered with dark tubercles, each of which bears a short black hair. The pupa is dark brown with a prominent white spiracle on the thorax and black points scattered on the abdomen. Much of its surface is covered with a white powder. The caterpillar lives in a tent at the base of grass clump. The mature caterpillar probably overwinters in its shelter.

Adult Nectar Sources.   Satchems favor a fairly wide array of flowers, and will visit yellow composites freely if they are available. Commonly visited plants include swamp and common milkweeds, dogbane, buttonbush, red clover, peppermint, tickseed sunflower, thistles, New York ironweed, asters, and marigold.

Caterpillar Host Plants.   Crabgrass (*Digitaria*), Bermuda grass (*Cynodon dactylon*), St. Augustine grass (*Stenotaphrum secundatum*), and goosegrass (*Eleusine*) are the known natural hosts of this common skipper.

Arogos Skipper
Atrytone arogos (Boisduval & LeConte), 1834
PLATE 48 · FIGURE 288 · MAP 238

Etymology.   The species is named after the Greek *arōgos*, "helper."

Synopsis.   The Arogos is another skipper that occurs on the prairies of the Great Plains as well as along the Atlantic coast. Its relative, the Delaware Skipper (*Atrytone logan*), is wide-ranging in disturbed as well as native situations.

Butterfly.   Male forewing: $\bar{X}$ = 1.3 cm, range 1.2–1.5 cm; female forewing: $\bar{X}$ = 1.5 cm, range 1.4–1.6 cm. Individuals from the plains populations are smaller than those from the Atlantic coast. The Arogos Skipper may be identified by its ventral hind wings, which are yellow-orange with the veins lined with paler scaling. Dorsally, the wings are broadly bordered with black and are unmarked and fulvous basally.

Range.   The Arogos Skipper occurs from Long Island south along the Piedmont and coastal plain in a very few isolated colonies to peninsular Florida and west along the Gulf to eastern Texas. A separate group of populations occurs on the prairies from southern Minnesota and adjacent Wisconsin west to eastern Wyoming and south to Missouri, Oklahoma, and northeastern Colorado.

Habitat.   The butterfly is almost always found on relatively undisturbed prairies or grasslands. In the Great Plains, it occurs on both tall and short-grass prairies; in New York and Pennsylvania, on serpentine barrens; and along the south Atlantic coastal plain on sand prairies.

Life History.   *Behavior.* Males perch on low vegetation during the afternoon. Females lay their eggs singly on the undersides of host leaves.

Broods.   In northern areas the Arogos Skipper is univoltine, usually flying during most of July, but southward it is bivoltine. In Missouri, the Arogos Skipper is found in June and September; in Georgia, during May and August; and in Florida, from mid-April to mid-May and again from early August to mid-September.

Early Stages.   The larvae live in a tent made of two leaves sewn

238. *Atrytone arogos*

together, hibernate in the fourth instar, complete development during the following spring, and pupate within an appressed leaf cocoon about a meter above the ground. The egg is a creamy white, somewhat flattened hemisphere with protruding micropyle and two faint red rings. The mature caterpillar is pale yellow-green and light green with yellow intersegmental folds. The head is gray-white with four vertical orange-brown stripes (Heitzman 1966).

Adult Nectar Sources. In the prairie habitats adults nectar at stiff coreopsis, purple coneflower, green milkweed, and ox-eyes, while in the northeast they visit purple vetch, Canada thistle, and dogbane.

Caterpillar Host Plants. In Missouri the host is big bluestem (*Andropogon gerardi*). The only eastern record, of panic grass, is based on an old painting by John Abbot and may be attributable more to artistic license than to scientific accuracy.

Delaware Skipper
Atrytone logan (Edwards), 1863
PLATE 49 · FIGURE 289 · MAP 239

Etymology. *Atrytone* is Greek epithet of Minerva. The species name *logan* is a reference to Mount Logan, Canada.

Synopsis. The Delaware Skipper is a grassland skipper that is adaptable to disturbed areas. Its proboscis is proportionately longer than that of most other skippers, being one and a half times as long as its forewing.

Butterfly. Male forewing: $\bar{X}$ = 1.4 cm, range 1.2–1.8 cm; female forewing: $\bar{X}$ = 1.6 cm, range 1.5–1.7 cm. Dorsally, the Delaware Skipper male is bright yellow-orange, with even black borders on all wings and at least faint black veining. Both sexes have a horizontal black mark at the end of the cell. Dorsally, the female is somewhat duller and has more extensive black on the basal wing areas. It is most likely to be confused with the Byssus Skipper (*Problema byssus*).

Range. This skipper occurs from southern Maine west across the Great Lakes area, Minnesota, and southern Canada to southern Alberta and central Montana. It occurs south to Florida, the Gulf states, Texas, and northeastern New Mexico.

Habitat. The Delaware Skipper may sometimes be found in dry situations, but it is more often found in damp or wet fields, marshes, or prairies. Habitats may include hayfields, roadsides, and suburban lots, as well as more native situations. It is much more tolerant of disturbance than the Arogos Skipper (*Atrytone arogos*).

Life History. *Behavior.* Males perch all day in open, grassy areas, often in flats or depressions. They usually perch and forage within a half meter of the ground. A mated pair was seen at 1530 hr.

Broods. The Delaware Skipper is bivoltine in the North (e.g., northern Michigan—July 2 to August 12), has two flights through much of the South (e.g., Georgia—May 7 to June 28 and July 10 to September 14), and possibly has three or more broods in Florida (February to October).

Early Stages. The mature larva has a blue-white body with minute black tubercles and a black crescent band on the posterior segments. The head is white with a black marginal band and three black vertical streaks on the front. The pupa is green and slender, with the head and last abdominal segment black.

Adult Nectar Sources. The Delaware Skipper will visit a variety of pink and white flowers, including common milkweed, swamp milkweed, mountain mint, shrub houstonia, sweep pepperbush, marsh fleabane, thistles, buttonbush, and pickerelweed.

Caterpillar Host Plants. Known larval hosts are all grasses, including Blue stems (*Andropogon*), switchgrass (*Panicum virgatum*), and wooly beard grass (*Erianthus divaricatus*).

239. *Atrytone logan*

Byssus Skipper
Problema byssus (Edwards), 1880

MAP 240

Etymology.   The genus name *Problema* is taken from the Latin word for 'problem.' *Byssus* is Latin for "cotton, cotton stuff."

Synopsis.   The Byssus is a large, wary skipper of two seemingly dissimilar habitats, native tall-grass prairies and coastal wetlands.

Butterfly.   Male forewing: $\overline{X}$ = 1.7 cm, range 1.5–1.8 cm; female forewing: $\overline{X}$ = 1.9 cm, range 1.8–2.0 cm. This large skipper differs from the Rare Skipper (*Problema bulenta*) by having a pale medial band on the ventral hind wing and by its antennal club, which is black below the bare sensory area. The veins of the male are often lined with black above, and the female has only limited orange areas on the dorsal wing surfaces.

Range.   The Byssus Skipper has both south Atlantic coastal plain and Midwestern prairie populations. On the coastal plain, it ranges from North Carolina south to Florida and west to Mississippi. In the Midwest, it extends from northern Indiana west to Iowa and south to Missouri and Kansas.

Habitat.   In the Midwest, the Byssus Skipper inhabits native tall-grass prairie, while on the Atlantic Coast it is associated with salt and brackish marshes.

Life History.   *Broods.* Midwestern populations have a single flight annually, during late June and July, while Atlantic Coast populations have two generations. In Georgia, the Byssus Skipper flies from May 7 to June 26 and from August 12 to September 6. In Florida, first-brood adults fly in April and May and the second brood from mid-August to mid-October.

Early Stages. Early stages have been described by Heitzman (1965a). The larvae hibernate in the fourth instar and complete their development the following spring. The pupa is located within a dense silk cocoon, in rubbish at the base of host plant. The egg is a large (1.5 mm diameter), white, somewhat flattened hemisphere. The mature caterpillar is blue-green and is densely covered with fine white hairs. The head is pale red-brown with vertical yellow-white lines and streaks. The pupa is off-white, with several small brown dots.

Adult Nectar Sources.   Pickerelweed flowers are sought by adults of the coastal populations.

Caterpillar Host Plants.   In Missouri, the Byssus Skipper uses eastern grama grass (*Tripsacum dactyloides*) for its larval host.

240. *Problema byssus*

Rare Skipper
Problema bulenta (Boisduval & LeConte), 1834

MAP 241

Etymology.   The species name is the Latin *buleuta,* "a councillor, senator"; if this is so the species name may be a misspelling.

Synopsis.   For more than 90 years after its original description, the Rare Skipper was not seen. Finally, it was rediscovered at Wilmington, North Carolina, in 1925 (Jones 1926). Apparently this butterfly is found only in the brackish marshes of large coastal rivers.

Butterfly.   Male forewing: $\overline{X}$ = 1.7 cm, range 1.6–1.8 cm; female forewing: $\overline{X}$ = 2.1 cm, range 2.0–2.3 cm. Both sexes may be distinguished from the similar Byssus Skipper (*Problema byssus*) by their immaculate yellow ventral hind wings and by the antennal club, which is orange below the bare sensory area.

Range.   To date, the Rare Skipper is known only from the lower brackish reaches of the Wicomico River (Maryland), Chickahominy River (Virginia), Cape Fear River (North Carolina), Santee River (South Carolina), and Savannah River (Georgia).

Habitat. The habitat of the rare skipper is brackish river marshes with scattered bald cypress, buttonbush, cattails, wild rice, pickerelweed, and other marsh plants. It has also been found in association with long-abandoned rice paddies.
Life History. *Broods.* The Rare Skipper may have one flight in Virginia and two southward. There is one early-May record in Georgia and two records in late June; most records are from late July through August to early September.

Adult Nectar Sources. Swamp milkweed and pickerelweed are the only two plants upon whose flowers visits by the Rare Skipper are recorded.
Caterpillar Host Plants. Distribution of the Rare Skipper is closely tied to that of marsh millet (*Zizaniopsis miliacea*) along the south Atlantic coastal plain. The skipper has been found in close association with this plant along the Santee River in South Carolina.

Mulberry Wing
Poanes massasoit (Scudder), 1864
PLATE 49 · FIGURES 290 & 291 · MAP 242

Etymology. Massasoit, a principal chief of the Wampanoag Indians, was introduced to the Puritans by Samoset at Plymouth, Massachusetts, in 1621.
Synopsis. The Mulberry Wing is the only *Poanes* that is limited to freshwater marshes. Its males use a patrolling mate-seeking behavior, as do those of the Broad-winged Skipper (*Poanes viator*); other *Poanes* males perch and are somewhat territorial.
Butterfly. Male forewing: $\overline{X} = 1.4$ cm, range 1.3–1.5 cm; female forewing: $\overline{X} = 1.5$ cm, range 1.4–1.6 cm. The Mulberry Wing is the only northern skipper that is predominantly black dorsally, and it has a yellow patch or band on the ventral hind wing. It has relatively short, rounded wings. In some individuals, the ventral hind wing is rusty brown and the yellow marks are absent.
Range. The Mulberry Wing occurs from New York and Massachusetts west across southern Ontario and the Great Lakes states to Minnesota. It ranges south to Maryland, Illinois, and Iowa.

Habitat. This dark skipper inhabits freshwater marshes or bogs with standing water and dense low vegetation. These situations are usually neutral or mildly acidic. The species is found primarily within the Upper Austral and Transition Life Zones.
Life History. *Behavior.* Male Mulberry Wings patrol through marsh vegetation with a low, weak, meandering flight. They pause frequently to rest on marsh vegetation. *Broods.* This species is strictly univoltine. The adults fly from late June or early July to the first half of August. In Michigan, some individuals may still be found during the third week of August.
Adult Nectar Sources. The Mulberry Wing rarely visits flowers, but when it does, swamp milkweed is a favorite.
Caterpillar Host Plant. Shapiro (1974) reports that the sedge *Carex stricta* is this species' host.

241. *Problema bulenta*

242. *Poanes massasoit*

Northern Golden Skipper
Poanes hobomok (Harris), 1862

PLATE 49 · FIGURES 292 & 293 · MAP 243

Etymology. Hobomok, chief of the Wampanoag Indians, helped the English upon their landing at Plymouth in 1621.

Synopsis. The Northern Golden Skipper is most unusual in having two strikingly different female forms. Both forms are found in the eastern area, but the dark form "pocahontas" is extremely rare or absent in the Midwest.

Butterfly. Male forewing: $\overline{X}$ = 1.5 cm, range 1.4–1.6 cm; female forewing: $\overline{X}$ = 1.7 cm, range 1.6–1.8 cm. The males are very similar to those of the Southern Golden Skipper (*Poanes zabulon*), but may be distinguished by their more rounded forewings and the purplish gray outer margin on both wings below. The normal females are somewhat darker than the male, but form "pocahontas" is largely purplish black above and below and has a few poorly defined white spots on the forewing.

Range. The Northern Golden Skipper occurs from Nova Scotia west across southern Canada to Saskatchewan, and south to New Jersey, Georgia (mountains), Arkansas, Kansas, and northeastern New Mexico.

Habitat. This skipper is found along the edges of damp deciduous woods, on the borders of woodland bogs, or in light gaps in forest along streams. This species is resident in the Transition and Canadian Life Zones.

Life History. *Behavior.* Males perch in the center of woodland clearings on vegetation about 2 m above the ground. Burns (1970) found that mated pairs remained attached for an average of 38 minutes (range 27–57 min).

Broods. Wherever found, the Northern Golden Skipper is univoltine. In the northern area, the Northern Golden Skipper flies from late May or early June to early or mid-July.

Further south (Georgia, Virginia), the skipper flies from late April to the end of June or early July.

Adult Nectar Sources. Adults visit flowers of blackberry, viper's bugloss, henbit, and common milkweed, and probably those of several other plants as well.

Caterpillar Host Plants. Ovipositing females select among several kinds of grasses, including panic grass (*Panicum*) and blue grass (*Poa*).

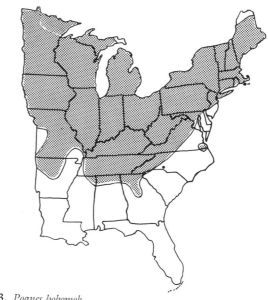

243. *Poanes hobomok*

Southern Golden Skipper
Poanes zabulon (Boisduval & LeConte), 1834

PLATE 50 · FIGURES 295–297 · MAP 244

Etymology. Zabulon, son of Jacob, is mentioned in the Old Testament.

Synopsis. The Southern Golden Skipper is strikingly dimorphic, having a light male and a dark female, unlike the Northern Golden Skipper, which has a light male, but both light *and* dark female forms.

Butterfly. Male forewing: $\overline{X}$ = 1.6 cm, range 1.4–1.7 cm; female forewing: $\overline{X}$ = 1.6 cm, range 1.5–1.6 cm. Southern Golden Skippers have more pointed forewings than do Northern Golden Skippers (*Poanes hobomok*). Males are lighter above and below, and lack the purplish gray band

along the outer margins ventrally. The females have distinct pale yellow marks on the forewings and the ventral surface of their hind wings is more two-toned red-brown and blue-grey.

Range. The Southern Golden Skipper occurs from Connecticut and southern New York west across the lower end of the Great Lakes to central Iowa. It ranges south to northern Florida, the Gulf states, and Texas. Another series of populations extends south in the mountains of Mexico and Central America to Panama.

Habitat. This skipper is found in brushy areas near moist woodland or streams. Males may wander away from these situations in search of nectar sources or perching sites. The species is generally confined to the Austral Life Zone.

Life History. *Behavior.* Males perch in sunlit clearings near streams, along wood edges, or along roads passing through woodland. They almost always perch on the edge of a broad leaf 1–1.5 m above ground. They often interact with other males, seemingly in defense of their perch areas, and may retain the same perch area for as long as a week. Perching extends through the day for as long as sunlight reaches suitable perches. Courtship has been observed from 0820–1830 hr, but is seen most often in the afternoon. Males land behind females and continually attempt to move forward to a lateral position while fluttering their wings. Unreceptive females (most observations) periodically fly away for short distances, with the male following closely.

Females are less conspicuous than males and are usually seen in more shaded brushy areas. They usually sit within a half meter of the ground.

Broods. Throughout most of the eastern United States, the Southern Golden Skipper is bivoltine, although there may be a small partial third generation in the Deep South. In the North, the first flight occurs from late May to the very beginning of July, while the second flies from early August through most of September. In Mississippi, the first brood emerges in early March. In Central America, adults are seen throughout the year.

Adult Nectar Sources. Southern Golden Skippers rely heavily upon flowers of introduced exotic plants, and seem to spend more time at flowers during the second generation. In northern Virginia, first-generation individuals visit Japanese honeysuckle (exotic), blackberry, purple vetch, red clover (exotic), everlasting pea (exotic), and common milkweed. Second-brood adults visit selfheal (exotic) in large numbers, but also visit thistles, buttonbush, New York ironweed, and joe-pye weed. At the very end of the season, when pink, purple, or white flowers are unavailable, they visit tickseed sunflower in small numbers.

Caterpillar Host Plants. Females lay single eggs on grasses, such as purpletop (*Tridens*) and lovegrass (*Eragrostis*).

Aaron's Skipper
Poanes aaroni (Skinner), 1890
PLATE 49 · FIGURE 294 · MAP 245

Etymology. This skipper was named in honor of Samual Francis Aaron (1862–1947), who published many popular articles on insects and was insect curator at the Academy of Natural Sciences in Philadelphia for a brief period.

Synopsis. This salt-marsh skipper is confined to the Atlantic Coast, and, although it is common near several metropolitan areas, its life history is essentially unknown.

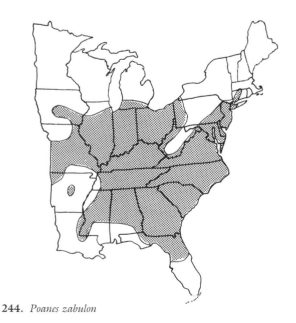

244. *Poanes zabulon*

245. *Poanes aaroni*

Butterfly. Male forewing: $\overline{X}$ = 1.5 cm, range 1.4–1.8 cm; female forewing: $\overline{X}$ = 1.6 cm, range 1.5–1.8 cm. Males have a faint forewing stigma. Both sexes have broad black borders on all wings and have an orange-tan ventral hind wing with a pale central streak. These skippers tend to be larger and more richly colored as they range southward.

Range. Aaron's Skipper is found along the immediate Atlantic coast from New Jersey discontinuously to Florida.

Habitat. This skipper is restricted to coastal salt marshes, usually in association with smooth cordgrass (*Spartina alternifolia*).

Life History. *Broods.* Aaron's Skipper is bivoltine. The first-generation adults fly progressively earlier southward, but the second flight appears in August and September at all locations. In New Jersey, the first-brood adults fly in June; in South Carolina, during May; and during late March and early April in Florida.

Early Stages. The egg is a white, finely reticulate, somewhat flattened hemisphere.

Adult Nectar Sources. Pickerelweed is favored at many localities, and salt marsh fleabane and coreopsis are also visited.

Yehl Skipper
Poanes yehl (Skinner), 1893
PLATE 50 · FIGURES 298 & 299 · MAP 246

Etymology. This skipper is named after a division of the Alaskan Tlingit Indians. Hoya is a synonym for this tribe, but it is also the name of a South Carolinian coastal settlement.

Synopsis. The Yehl Skipper is a rare denizen of wooded swamps in the coastal plain and lower Mississippi River Valley. Nothing is known of its host or early stages.

Butterfly. Male forewing: $\overline{X}$ = 1.6 cm, range 1.4–1.8 cm; female forewing: $\overline{X}$ = 1.7 cm, range 1.7–1.9 cm. Male Yehl Skippers have a strong linear stigma on the forewing and might be confused with one of the *Euphyes*. The ventral wing characters should preclude this error, particularly the mesial row of three pale spots on the hind wing. Dorsally, the female closely resembles that of the Broad-winged Skipper (*Poanes viator*), but the dark brown ventral hind wing, with its three pale dots, should dispel any doubts.

Range. Yehl's Skipper occurs along the Atlantic coastal plain from southeastern Virginia to Florida (rarely) and then west to eastern Texas. It also ranges north through the lower Mississippi Valley to Missouri and southern Illinois. It is usually found in only a few localities within each state.

Habitat. Yehl's Skipper is found within clearings or along roads through wet or swampy deciduous woodland. It is restricted to the Lower Austral Life Zone.

Life History. *Behavior.* Males perch about a meter above ground on sunlit shrubby vegetation. The males are especially wary, and are difficult to approach.

Broods. Yehl's Skipper is bivoltine, with flights in late May to mid-June and again from mid-August to mid-November.

Adult Nectar Sources. Adults have been seen visiting chinquapin, sweet pepperbush, pickerelweed, and swamp milkweed.

246. *Poanes yehl*

Broad-winged Skipper
Poanes viator (Edwards), 1865; *Poanes viator zizaniae* Shapiro, 1971
PLATE 51 · FIGURES 301 & 302 · MAP 247

Etymology. Poanes = ? Poa = see phylum Chordata (Zoology). The species name is taken from the Latin *viator*, "traveller, passenger, pedestrian." The subspecies name is taken from *zizania*, the genus name for wild rice, one of the skipper's hosts.

Synopsis. This is a large marsh Skipper that has distinctive, separate coastal and inland populations. The Broad-winged Skipper, although very local, may be incredibly abundant where found.

Butterfly.   Inland populations (*viator*)—male forewing: $\overline{X}$ = 1.6 cm; female forewing: $\overline{X}$ = 2.0 cm. Coastal populations (*zizaniae*)—male forewing: $\overline{X}$ = 1.8 cm, range 1.6–2.0; female forewing: $\overline{X}$ = 2.3 cm, range 2.1–2.5 cm. The Broad-winged Skipper has rounded wings. The forewings are mostly black with a few orange spots and the hind wings are largely dull orange with irregular black margins above and a mesial series or band of pale spots below. The male lacks a stigma, as do most other *Poanes*, and the female has one or more hyaline spots on the forewing. Careful biological and ecological studies may reveal that the inland and coastal populations consist of separate sibling species.

Range.   The coastal *zizaniae* occurs along the Atlantic coastal plain from Massachusetts south to Northern Florida and west along the Gulf Coast to Central Texas. Populations occurring along the lower Mississippi drainage north to Kentucky probably represent this subspecies. The inland *viator* ranges from Ontario, western New York, and western Pennsylvania west through the Great Lakes states to the eastern Dakotas and eastern Nebraska (Shapiro 1977).

Habitat.   Coastal plain *zizaniae* are found in or near brackish or salt-water marshes with dense stands of its host grasses. Inland *viator* is found in freshwater marshes dominated by the sedge *Carex lacustris*.

Life History.   *Behavior.* Male Broad-winged Skippers patrol with a slow, jerky flight in which the butterfly drops a bit between wing beats. The flight is usually low and meanders in and out of the reeds without regard to sun or shade. Once, at midday on a very hot, muggy day in a Maryland *viator* habitat, most individuals were found resting at the base of woody shrubs in dense shade in an apparent attempt to avoid lethal "humiditure." The skippers usually begin roosting in midafternoon (1530–1600 hr).

Broods. Inland populations are univoltine, with an annual flight from late June to early August, while coastal populations have from one to four broods, depending on latitude.

In New York, there is a single brood of the coastal population from early June to mid-August, while there are two flights in Virginia (July, August), three in Georgia (April–May, August–September, and November), and four in Texas.

Early Stages. Early stages are poorly known. The caterpillars rest in the crevice between the stem and sheath of the host and do not build a silken shelter (Shapiro 1970).

Adult Nectar Sources.   Along the coast, the Broad-winged Skipper nectars at dogbane, swamp milkweed, pickerelweed, buttonbush, salt marsh fleabane, spatterdock, thistles, and viper's bugloss. Inland populations utilize swamp milkweed, blue vervain, and purple loosestrife.

Caterpillar Host Plants.   Coastal populations have been associated with or found to feed on reed (*Phragmites communis*), wild rice (*Zizania aquatica*), and marsh millet (*Zizaniopsis miliacea*). Shapiro (1974) reported inland populations associated with the sedge *Carex lacustris*.

247. *Poanes viator zizaniae*

Palmetto Skipper
Euphyes arpa (Boisduval & LeConte), 1834
PLATE 50 · FIGURE 300 · MAP 248

Etymology.   The genus name *Euphyes* was taken from "Euphues, or the Anatomy of Wit," a work by John Lyly, published in 1578–1579. Arpa, a city in Apulia, was earlier called Argyripa (Plinius).

Synopsis.   The Palmetto is one of the few skippers in our area whose larvae feed on palms.

Butterfly.   Male forewing: $\overline{X}$ = 1.8 cm, range 1.7–1.9 cm; female forewing: $\overline{X}$ = 2.0 cm, range 1.9–2.1 cm. This large skipper is distinguished by the bright, unmarked, yellow-orange ventral hind wings, and by the restricted tawny yellow areas on the male's dorsal forewings. Moreover, the head and thoracic collar are clothed with bright orange scales.

Range.   This butterfly occurs in peninsular Florida and west along the immediate Gulf Coast to Mississippi.

Habitat.   The species is found in lowland palmetto scrub in the Lower Austral Life Zones.

Life History. *Broods*. The Palmetto Skipper flies from mid-March to mid-November in Florida and probably has three generations.

Early Stages. The caterpillars live in tube shelters at the base of palmetto fronds. The caterpillar is streaked with yellow and green and has a black collar. Its head is black with white margins and has two white streaks high on the face.

Caterpillar Host Plant.   The larval host is saw palmetto (*Serenoa repens*).

Sawgrass Skipper
Euphyes pilatka (Edwards), 1867
MAP 249

Etymology.   The species is named after a town on the banks of the St. Johns River, Florida. It is near the site of a former Seminole Village, Pilatka.

Synopsis.   The Sawgrass Skipper is one of the few butterflies found in the extensive sawgrass marshes in southern Florida.

Butterfly.   Male forewing: $\overline{X}$ = 2.0 cm, range 2.0–2.1 cm; female forewing: $\overline{X}$ = 2.2 cm, range 2.0–2.3 cm. Male Sawgrass Skippers have orange over much of the dorsal wing surfaces, and the wings have broad black outer margins and a pointed forewing apex. Ventrally, the hind wings are dull brown.

Range.   This large skipper occurs from southern Maryland (rarely) south along the Atlantic coastal marshes to Florida and west along the Gulf Coast to Louisiana.

Habitat.   This species is found in coastal brackish and freshwater marshes in the Lower Austral Life Zone.

Life History. *Broods*. In Florida, the Sawgrass Skipper has two or more (probably three) flights, extending from January through April and June through November. Farther north, there are probably two generations (late May through June and August–September).

Early Stages. The caterpillar is yellow-green with minute dark tubercles. Its head is brown, with a white upper face that has three vertical black stripes.

Adult Nectar Sources.   Pickerelweed is the only flower known to be visited.

Caterpillar Host Plant.   Sawgrass (*Cladium jamaicensis*), a sedge, is the only reported host.

248. *Euphyes arpa*

249. *Euphyes pilatka*

Dion Skipper
Euphyes dion (Edwards), 1879

PLATE 51 · FIGURE 303 · MAP 250

Etymology. The species name, *dion* is taken from Dione, a Greek nymph, consort of Zeus (Greek myth).

Synopsis. The Dion Skipper is one of the more widespread marsh skippers in the North, although it is always found in highly localized colonies. It is rare in the South.

Butterfly. Male forewing: $\overline{X}$ = 1.6 cm, range 1.5–1.7 cm; female forewing: $\overline{X}$ = 1.8 cm, range 1.6–1.9 cm. The Dion Skipper has the usual *Euphyes* pattern, with restricted orange areas on the male's forewing. Ventrally, the reddish to orange-brown hind wings have two pale yellow outward streaks, one through the cell and another above the anal fold. Southern populations have darker individuals, given the subspecies name *alabamae*, but this is only a tendency and the species is quite variable (J.M. Burns, personal communication).

Range. The Dion Skipper occurs from Connecticut and southern Ontario west to Minnesota and eastern Nebraska. It then ranges south along the Atlantic coastal plain to Florida and west along the Gulf to Texas. It also ranges north through the Mississippi Valley.

Habitat. This species is found in open marshes, bogs, swamps, and wet meadows with tussock sedges. It ranges from the Lower Austral through the Transition Life Zones.

Life History. *Behavior.* Males have strong perching behavior and an extremely rapid flight. Shapiro (1974) noted that they are highly territorial. Our observations show that the species perches primarily from midday to late afternoon. However, sometimes in late morning males have been seen to exhibit a very slow patrolling flight low through sedge vegetation.

Broods. In the northern states, the Dion Skipper is univoltine, usually flying from the first of July to the first week of August. From the lower Delaware Valley south, there are two generations. On the south Atlantic coastal plain, the Dion flies from the end of May to the first week of July and again from the last week in July through August and September.

Early Stages. The egg is light green, while young caterpillars have yellow-green bodies with yellow setae and a black collar. The head is black.

Adult Nectar Sources. This skipper visits Alsike clover, buttonbush, sneezeweed, and pickerelweed.

Caterpillar Host Plants. The Dion Skipper is known to feed on or has been closely associated with the following sedges: *Scirpus cyperinus, Carex lacustris, Carex stricta* var. *brevis,* and *Carex hyalinolepis.*

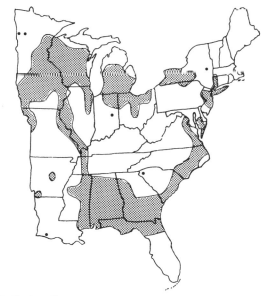

250. *Euphyes dion*

Duke's Skipper
Euphyes dukesi (Lindsey), 1923

MAP 251

Etymology. The species was named in honor of W. C. Dukes, who lived in Mobile, Alabama.

Synopsis. The rare Duke's Skipper is often found in the shade of wooded swamps. It is known in only a few widely separated localities.

Butterfly. Male forewing: $\overline{X}$ = 1.7 cm, range 1.6–1.8 cm; female forewing: $\overline{X}$ = 1.9 cm, range 1.9–2.0 cm. Dorsally, the male is entirely sooty black (dark brown after death) with a black stigma. The female is similar but has two or three pale or orange spots. Ventrally, the hind wing is pale brown with a pale-yellow streak passing through the cell.

Range. Duke's Skippers occur from southeastern Virginia south along the Atlantic coastal plain to North Carolina, along the Gulf Coast from Alabama to Louisiana, and in the Mississippi drainage from Arkansas north to Ohio and Michigan (Mather 1963).

Habitat. Duke's Skipper is found in shaded water tupelo swamps or open marshes and ditches (Irwin 1972). It occurs in the Lower and Upper Austral Life Zones.

Life History. *Broods.* There is only one generation in Michigan (August 31–September 6), but there are two flights in most areas, e.g., Virginia (June 12–July 16 and August 2–September 9), and probably three in Louisiana (May to November).

Early Stages. The caterpillars overwinter in the fourth stage. Other than this observation, the early stages are unknown.

Adult Nectar Sources. Pickerelweed, hibiscus, sneezeweed, and blue mistflower are known to be visited.

Caterpillar Host Plants. In Michigan, the butterflies select the sedge *Carex lacustris,* but they feed on *Carex hyalinolepis* in the South.

Black Dash
Euphyes conspicua (Edwards), 1863
PLATE 51—FIGURES 304–306—MAP 252

Etymology. The species name is derived from the Latin *conspicuus,* "visible, prominent."

Synopsis. The Black Dash is more of a bog species than other *Euphyes.* It is often found with the Mulberry Wing.

Butterfly. Male forewing: $\overline{X}$ = 1.5 cm, range 1.4–1.8 cm; female forewing: $\overline{X}$ = 1.7 cm, range 1.6–1.8 cm. This skipper is similar to the Dion Skipper ventrally, but its ventral hind wing is orange-red with an orange macular band.

Range. The Black Dash occurs from Massachusetts west across southern Ontario to Minnesota and eastern Nebraska. It then ranges south to Virginia, Ohio, and Illinois.

Habitat. This northern skipper occurs in boggy marshes, wet meadows, the marshy banks of slow woodland streams, and, rarely, in bogs. It is usually found in the Upper Austral and Transition Life Zones, but one isolated colony was found in the Lower Austral Zone in southeastern Virginia.

Life History. *Behavior.* Adults perch low on marsh vegetation. A mated pair was reported in early afternoon (1345 hr).

Broods. There is a single annual flight. In most areas, the species flies from early July to early August, but adults were found on June 3 in the Dismal Swamp of Virginia.

Adult Nectar Sources. Adults have been observed visiting flowers of swamp thistle, jewelweed, and buttonbush.

Caterpillar Host Plant. This species has been raised on the sedge *Carex stricta.*

251. *Euphyes dukesi*

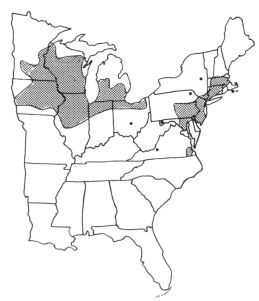

252. *Euphyes conspicua*

Berry's Skipper
Euphyes berryi (Bell), 1941
MAP 253

Etymology. The species was named after Mr. Dean Berry, a Florida collector.

Synopsis. This rare skipper has been little studied.

Butterfly. Male forewing: $\overline{X}$ = 1.6 cm, range 1.5–1.7 cm; female forewing: $\overline{X}$ = 1.8 cm, range 1.7–2.0 cm. Berry's Skipper is distinguished by the limited fulvous areas dorsally and the coffee brown ventral hind wings, which have the veins conspicuously lightened.

Range. The species is found in the coastal lowlands of North Carolina (rarely) south to southern Florida and west along the Gulf Coast to the western extremity of Florida's panhandle.

Habitat. Berry's Skipper is found in wet areas near ponds, drainage canals, and swamps in the Lower Austral Life Zone.

Life History. *Broods.* This skipper is bivoltine. In Georgia, these occur in late May and again from late July through mid-August. In Florida, adults are found from March to mid-May and again from late September to late October.

Adult Nectar Source. Berry's Skipper has been found nectaring at pickerelweed on several occasions.

253. *Euphyes berryi*

Two-spotted Skipper
Euphyes bimacula (Grote & Robinson), 1867
PLATE 52 · FIGURES 307 & 308 · MAP 254

Etymology. The species name *bimacula* is taken from the Latin *bimaculatus,* "having two spots."

Synopsis. The Two-spotted Skipper is another marsh skipper and is locally common in the North but very rare in the South.

Butterfly. Male forewing: $\overline{X}$ = 1.6 cm, range 1.5–1.7 cm; female forewing: $\overline{X}$ = 1.7 cm, range 1.6–1.8 cm. Dorsally, the male Two-spotted Skipper has a very limited, dull, tawny forewing patch and a completely dark hind wing. The female has only two light spots. Ventrally, the skipper is more distinctive, since the inner margin of the hind wing, fringes, and body are white scaled.

Range. The Two-spotted Skipper occurs from Maine west across southern Ontario, the Great Lake states, and the plains to western Nebraska and northeastern Colorado. It extends south sporadically along the Atlantic coastal plain to Georgia and has been found on the Gulf Coast.

Habitat. This skipper is found in sedgy wet meadows, marshes, and bogs. In the South it is found in open sedgy areas near swamp woods.

Life History. *Behavior.* Males perch within 1 m of the ground atop low stalks or on the ground in windy areas. They

254. *Euphyes bimacula*

interact strongly with adjacent males and have been seen in lateral chases lasting at least 60 m. One mated pair was seen in early afternoon (1300 hr).

Broods. In the North, the Two-spotted Skipper is univoltine, with a flight from late June to the middle or end of July. Along the southeastern coastal plain, the skipper is probably bivoltine, with one flight in mid-May to early June and another from late July to mid-August.

Adult Nectar Sources. In the North, the skippers are exceptionally fond of larger blue flag, but also nectar at common milkweed, selfheal, and spiraea. In the South, they nectar at pickerelweed, zenobia, and sweet pepperbush (Dorr 1980).

Caterpillar Host Plant. Shapiro (1974) reports the host to be the sedge *Carex trichocarpa.*

Dun Skipper
Euphyes ruricola metacomet (Harris), 1862
PLATE 52 · FIGURES 309 & 310 · MAP 255

Etymology. The species name is derived from the Latin *ruricola,* "peasant, person working in the country," while the subspecies was named after Metacomet, a Wampanoag Indian chief, also called King Philip. He lead a bloody uprising against the New England colonists (1675–76).

Synopsis. This is the smallest *Euphyes* in our area. It is also the most widespread and most adaptable to altered habitats.

Butterfly. Male forewing: $\bar{X}$ = 1.4 cm, range 1.3–1.5 cm; female forewing: $\bar{X}$ = 1.3 cm, range 1.2–1.5 cm. The male's wings are entirely black, although there is a sheen to the wing's ventral surface. Dorsally, the head and thorax are clothed with yellow-orange scales. The females have very small white forewing spots.

Range. The Dun Skipper extends from Nova Scotia and southern Quebec west across southern Canada to Manitoba and south to Florida, the Gulf Coast, and eastern Texas. Separate populations occur in the high plains and Rocky Mountains, as well as Arizona, California, and Oregon.

Habitat. This common skipper is associated with wet areas in or near deciduous woodland. Specific situations include meadows, roadsides, edges of swamps and marshes, power-line cuts, and seeps.

Life History. *Behavior.* Males perch in open areas about 1 m above the ground, especially in late afternoon. Mated pairs have been seen in the afternoon. Females are often seen in semishaded habitats nearby. Ovipositing females have been observed fluttering through sedges in late morning (1130 hr). Freshly emerged males may visit muddy spots along roads or streams.

Broods. In the North, the Dun Skipper is univoltine (e.g., Maine—June 29–August 9), while southward it is bivoltine (e.g., Delaware Valley—Mid-May–early July, mid-August–late September). In Georgia, the first brood emerges in early March.

Early Stages. The newly laid egg is a pale-green hemisphere. The final-instar caterpillar has a pale, translucent green body with a white overcast caused by many white horizontal dashes. The head is tricolored, with black in back and an oval spot centered high on the face. The remainder of the face is caramel brown with two vertical cream bands. The pupa is pale white-green, pale yellow-green, and pale brown; it also has a slight whitish dusting. The life cycle from egg to adult requires 2 months in the summer. Pupation occurs in a white silk-lined tube near the base of the host.

Adult Nectar Sources. Adults favor white, pink, or purple flowers. Those visited with greatest frequency are common milkweed, dogbane, purple vetch, New Jersey tea, selfheal, viper's bugloss, and peppermint.

Caterpillar Host Plant. Sedges, including *Cyperus esculentus,* are probably the only hosts (Heitzman 1964a), although grasses have been reported by other observers.

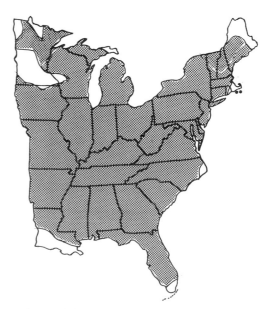

255. *Euphyes ruricola metacomet*

Monk

Asbolis capucinus (Lucas), 1857

PLATE 53 · FIGURE 313 · MAP 256

Etymology. *Asbolus* was a black, shaggy dog of Actaeon (Ovidius). *Capucinus* is a Latin word for "member of the Franciscan order," or "wearing a hooded cape."

Synopsis. The Monk is a Cuban species now naturalized in Florida, where it feeds on various native and introduced palms.

Butterfly. Male forewing: $\overline{X} = 2.1$ cm, range 2.2–2.3 cm; female forewing: $\overline{X} = 2.6$ cm, range 2.3–2.7 cm. Dorsally, the Monk male is black, with distinct stigma, while females are a paler brown-black. Ventrally, this large skipper has mahogany-red and black hind wings.

Range. This Cuban species successfully colonized southern Florida in the late 1940s.

Habitat. The Monk frequents both disturbed and natural habitats where appropriate palms occur.

Life History. *Broods.* There are three or four flights in southern Florida, January–April, July–August, and November–December.

Adult Nectar Sources. The Monk visits hibiscus as well as other large flowers.

Caterpillar Host Plants. In the larval stage, this large skipper feeds on various palms, including coconut (*Cocos nucifera*), palmetto (*Sabal*), date palm (*Phoenix dactylifera*), and Everglades palm (*Acoelorrhaphe wrightii*).

Dusted Skipper

Atrytonopsis hianna (Scudder), 1868

PLATE 52 · FIGURES 311 & 312 · MAP 257

Etymology. *Atrytonopsis* is derived from *Atrytōně*, "epithet of Minerva," and *opsis*, "appearance." The Hianna are a tribe living in Morocco.

Synopsis. This is the only eastern representative of a genus, most of whose species occur in Mexico and the southwestern United States. The Dusted Skipper is usually found in localized colonies.

Butterfly. Male forewing: $\overline{X} = 1.7$ cm, range 1.5–1.8 cm; female forewing: $\overline{X} = 1.7$ cm, range 1.6–1.8 cm. The Dusted Skipper is a medium-sized, gray-black butterfly with pointed forewings. Ventrally, the wings are dusted outwardly with gray and there is at least a single white spot at the base of the hindwing. There may be a postbasal and postmedian series of white spots on the ventral hind wings. White-spotted indi-

256. *Asbolis capucinus*

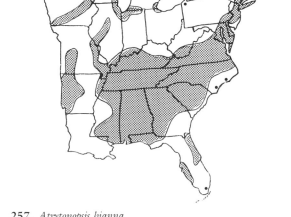

257. *Atrytonopsis hianna*

viduals increase southward and have had the name *loammi* applied to them. We consider these to represent a clinal subspecies, rather than a distinct species.

Range. The Dusted Skipper occurs from southern New England west across the Great Lakes states and southern Manitoba to the western Dakotas, eastern Wyoming, and central Colorado. It ranges southward to southern Florida, the Gulf states, and Oklahoma.

Habitat. This skipper is found in bluestem grasslands, and often on acid pine or pine–oak barrens or prairies. Old fields, woodland clearings, and power-line swaths are also utilized. These habitats are subject to fire, and the butterfly must either survive burning or be a good colonist (Shapiro 1965).

Life History. *Behavior.* Males perch on the ground or slightly above it on grass stems on slopes or flats during most sunny daylight hours.

Broods. In most of the Dusted Skipper's range, it is univoltine with a flight during May and the first half of June. In Florida there are two flights, in March–April and again in October.

Early Stages. The newly laid egg is bright lemon yellow and hemispherical. The mature caterpillar is pale pink-lavender dorsally, with the prothorax and lateral portions of the abdomen pale gray. The anal segment is pale brown and the prothoracic shield is dark brown. The body is also covered with long yellow-white hair. The head is deep red-purple. The pupa is dark brown with the wing cases light brown. The thoracic spiracles are ruby red. The caterpillar lives in a tent of several leaves sewn together, and when mature, lives in a sealed nest at the base of the host over the winter. Pupation occurs in a sealed case at the base of the grass clump 1–3 inches above ground (Heitzman 1974).

Adult Nectar Sources. The Dusted Skipper favors Japanese honeysuckle, blackberry, wild strawberry, vervain, red clover, phlox, and wild hyacinth.

Caterpillar Host Plants. The larvae of this distinctive skipper feed on big bluestem (*Andropogon gerardi*) and little bluestem (*A. scoparius*).

Linda's Roadside Skipper
Amblyscirtes linda Freeman, 1943

MAP 258

Etymology. *Amblyscirtes* could possibly be derived from the Greek *amblyskō*, "misbeget." This species was named for H. A. Freeman's daughter Linda.

Synopsis. The Linda's Roadside Skipper is a close relative of the more western *Amblyscirtes aenus* (Bronze Roadside Skipper) and, in fact, some consider it no more than a subspecies.

Butterfly. Male forewing: $\overline{X}$ = 1.2 cm, range 1.2–1.3 cm; female forewing: $\overline{X}$ = 1.4 cm, range 1.3–1.4 cm. Fresh individuals have bright orange-brown overscaling above and below, but when worn might be confused with either the Roadside Skipper (*Amblyscirtes vialis*) or Bell's Roadside Skipper (*A. belli*). There is an obscure postmedian spot band on the ventral hind wing.

Range. This small skipper occurs from western Tennessee west through southern Illinois, Missouri, and Arkansas to eastern Oklahoma.

Habitat. Linda's Roadside Skipper is found along or near streams through undisturbed woodland.

Life History. *Behavior.* Males of the related Bronze Roadside Skipper (*Amblyscirtes aenus*) perch in gully bottoms, especially on rocks, during most warm daylight hours.

Broods. The skipper is bivoltine, with flights in late April to early May and in late June to early July.

Early Stages. The eggs, which are shiny white and hemispherical, are laid singly near the edge on the lower surface of host leaves. The larvae live in tents of folded and sealed host leaves. Pupation occurs in a leaf that is rolled and sealed at both ends. The mature caterpillar is pale blue-white with a faint blue dorsal line. The body is covered with snow-white setae. The caterpillar's head is white with lateral and median black bands on the front. The pupa is creamy yellow with scattered red-orange setae (Heitzman and Heitzman 1969).

258. *Amblyscirtes linda*

Adult Nectar Sources. Individuals of the spring brood visit blackberry flowers.

Caterpillar Host Plant. Broadleaf uniola (*Uniola latifolia*) is the only reported host.

Pepper and Salt Skipper
Amblyscirtes hegon (Scudder), 1864
MAP 259

Etymology. "Hegans" was a name applied to an Indian tribe near Kittery, Maine, by the English victims of an attack.
Synopsis. The Pepper and Salt Skipper is a widespread forest species usually found in northern areas but also ranging southward in the mountains. It may be found together with the Roadside Skipper on some occasions.
Butterfly. Male forewing: $\overline{X}$ = 1.2 cm, range 1.1–1.2 cm; female forewing: $\overline{X}$ = 1.2 cm, range 1.1–1.3 cm. This small skipper has a more spotted forewing than the Roadside Skipper (*Amblyscirtes vialis*). The ventral hind wing is dusted with light gray-green scales and there is a poorly defined postmedian row of buff or gray spots.
Range. The Pepper and Salt Skipper occurs from Nova Scotia and Maine west across the Great Lakes area to Minnesota and Manitoba, thence southward to Georgia (mountains), northern Florida (rarely), Louisiana, northern Mississippi and Texas.
Habitat. This skipper is found in glades or at the edges of mixed or coniferous forest as well as at the edges of bogs or boggy streams.
Life History. *Broods.* The Pepper and Salt Skipper is univoltine throughout its range, but rare, late-emerging adults are occasionally found in late July. The brief flight is normally in the spring, although in northern New York the skipper flies from mid-June to mid-July. Michigan flight dates extend from May 29 to June 14. In Georgia, at its southern terminus, the Pepper and Salt Skipper flies from April 7 to May 28. *Early Stages.* The mature caterpillar is pale white-green and has three narrow, dark-green dorsal stripes and a white stripe along each side. The head is dark brown with two pale-brown

vertical stripes on the front as well as light-brown bands on the lateral portions. The pupa is pale yellow with green on the wing pads and dull orange on the proboscis case.
Adult Nectar Sources. The butterfly visits viburnum and blackberry for its nectar supplies.
Caterpillar Host Plants. The caterpillars feed on Kentucky blue grass (*Poa pratensis*), Indian grass (*Sorghastrum nutans*), *S. secundum,* and broadleaf uniola (*Uniola latifolia*).

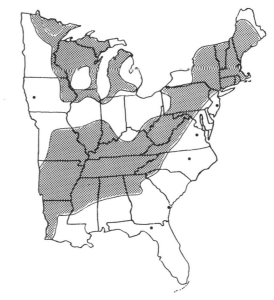

259. *Amblyscirtes hegon*

Lace-winged Roadside Skipper
Amblyscirtes aesculapius (Fabricius), 1793
PLATE 53 · FIGURE 314 · MAP 260

Etymology. The species is named after *Aesculapius,* the Latin god of medicine.
Synopsis. Although the Lace-winged Roadside Skipper has been described for almost 200 years, the life history of this southeastern swamp forest butterfly is virtually unknown.
Butterfly. Male forewing: $\overline{X}$ = 1.4 cm, range 1.3–1.5 cm; female forewing: $\overline{X}$ = 1.5 cm, range 1.4–1.6 cm. The bold

"cobweb" pattern on this black skipper's ventral hind wings is distinctive. The buff or pale yellow outlined veins criss-cross with two irregular macular bands to give the effect. The fringes are strongly checkered.
Range. The Lace-winged Roadside Skipper is found from southeastern Virginia south to Florida, then west to eastern Texas, Louisiana, and Missouri.

Habitat. This species occurs in dense wet woods with growths of cane (*Arundinaria*). Adults are most often seen in glades or along forest roads. The species is found only in the Lower and Upper Austral Life Zones.

Life History. *Broods.* The Lace-winged Roadside Skipper is bivoltine. In Georgia, adults are found from April 8 to June 5 and again from July 24 to September 15. In Florida and Mississippi the species if first found in early March.

Adult Nectar Sources. Adults are found visiting selfheal, dogbane, white clover, blackberry, sweet pepperbush, and elephant's-foot.

Caterpillar Host Plant. It is likely that the Lace-winged Roadside Skipper feeds on cane (*Arundinaria*) in the larval stage.

Carolina Roadside Skipper
Amblyscirtes carolina (Skinner), 1892
PLATE 53 · FIGURES 315–317 · MAP 261

Etymology. The species name is derived from the Carolinas, where it was first collected.

Synopsis. The Carolina Roadside Skipper has been confused with the Reversed Roadside Skipper. The two have somewhat different, but overlapping distributions and flight periods (Mather 1975).

Butterfly. Male forewing: $\overline{X}$ = 1.3 cm, range 1.2–1.3 cm; female forewing: $\overline{X}$ = 1.5 cm, range 1.4–1.6 cm. The Carolina Roadside Skipper's ventral hind wing is rusty brown, but it is so extensively overscaled with dull yellow that only a few spots of rusty brown show through.

Range. This skipper is found from southeastern Virginia south to Georgia, then east to southern Illinois, Arkansas, Mississippi, and northeastern Louisiana. It is more often found on the Piedmont than is the Reversed Roadside Skipper (*Amblyscirtes reversa*).

Habitat. The Carolina Roadside Skipper occurs in or near wet woods usually near streams or swamps. These habitats always have an abundant undergrowth of cane. This is a species of the Lower and Upper Austral Life Zones.

Life History. *Behavior.* Males have been seen perching in sunlit openings during the afternoon.

Broods. The Carolina Skipper has three yearly generations. In Virginia, they are in April, June, and mid-August to early September. These dates appear to apply generally through its range.

Adult Nectar Sources. This skipper visits flowers of ironweed, wild strawberry, blackberry, cinquefoil, swamp milkweed, and sweet pepperbush.

Caterpillar Host Plants. Larval hosts are not known. Switch cane (*Arundinaria tecta*) is the most obvious candidate.

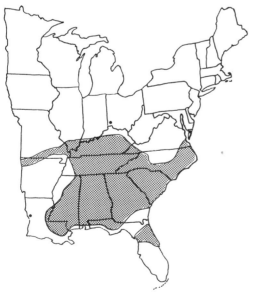

260. *Amblyscirtes aesculapius*

261. *Amblyscirtes carolina*

Reversed Roadside Skipper
Amblyscirtes reversa F. M. Jones, 1926

MAP 262

Etymology.   The species name refers to the reversed pattern on the ventral hind wings.

Synopsis.   Although considered by many to be a form of the Carolina Roadside Skipper, the Reversed Roadside Skipper merits distinction as a separate species. It breeds true, has different seasonality, and tends to be more coastal in its distribution (Mather 1975).

Butterfly.   Male forewing: $\overline{X}$ = 1.3 cm, range 1.2–1.5 cm; female forewing: $\overline{X}$ = 1.4 cm, range 1.4–1.5 cm. The ventral hind wings of the Reversed Roadside Skipper lack the extensive yellow overscaling of the Carolina Roadside Skipper (*Amblyscirtes carolina*). Instead, in this species the ventral hind wing is rusty brown with a submarginal row of small yellow spots and a yellow streak through the cell. The male's forewing apex seems somewhat more acute.

Range.   The Reversed Roadside Skipper occurs from southeastern Virginia south along the Atlantic coastal plain and occasionally the Piedmont to Georgia, and then again in Mississippi and southern Illinois along the Mississippi River flood plain.

Habitat.   This skipper prefers swamp woods or wet woods near sluggish streams. In any case, there must be an undergrowth of switch cane.

Life History.   *Broods.* There seem to be two annual generations along the Atlantic coastal plain (May to mid-June and July to early August) and three along the Gulf Coast (April, June, and August).

Early Stages. The mature caterpillar almost severs the leaf in which its cocoon is encased so that the cocoon and leaf rotate freely in the slightest breeze.

Caterpillar Host Plant.   The Reversed Roadside Skipper was reared from switch cane (*Arundinaria tecta*) in Georgia, but the early stages were not detailed (Harris 1972).

Nysa Roadside Skipper
Amblyscirtes nysa Edwards, 1877

MAP 263

Etymology.   The species name may be based on a town of that name in Missouri.

Synopsis.   Found primarily in Texas, the Nysa Roadside Skipper enters only the far western portion of our area.

Butterfly.   Male forewing: $\overline{X}$ = 1.0 cm, range 1.0–1.1 cm; female forewing: $\overline{X}$ = 1.2 cm, range 1.2–1.3 cm. Nysa is a small black roadside skipper with several small white spots in the forewing apical area. The ventral hind-wing pattern is

262. *Amblyscirtes reversa*

263. *Amblyscirtes nysa*

diagnostic, with several broad dark areas together with small patches of light gray and yellow-brown.

Range. The Nysa Roadside Skipper occurs from western Missouri and western Arkansas west to Kansas and southern Arizona as well as northern Mexico. It is most widespread in Texas.

Habitat. This skipper prefers open areas, including dry ravines, yards, and gardens.

Life History. *Behavior.* The adults often rest on bare ground. They fly very low and are difficult to follow. Oviposition takes place in late afternoon (1500–1700 hr). Females deposit eggs randomly on leaves and stems.

Broods. There are four generations in Missouri, with adults from May to October.

Early Stages. The egg is a shiny white hemisphere. The caterpillar lives in a tent made of a rolled and sewn grass blade, midway from the base. The mature caterpillar is pale green with a dark-green dorsal stripe. There are minute green blotches over much of its body. The head is creamy white with vertical, bright, orange-brown streaks and bands. The pupa is cream with orange-brown shading on the head and numerous orange-brown setae over the head and abdomen. The larva pupates in a silk-lined case made from a host-plant leaf, and it lies amid rubbish at the base of the host grass clump (Heitzman 1964b).

Adult Nectar Sources. A variety of low flowers are visited, including marigold, blue spiraea, lima bean, and canteloupe.

Caterpillar Host Plants. A number of grasses can be selected as hosts, including crab grass (*Digitaria sanguinalis*), barnyard grass (*Echinochloa crusgalli*), yellow foxtail (*Setaria glauca*), and St. Augustine grass (*Stenotaphrum secundatum*).

Roadside Skipper
Amblyscirtes vialis (Edwards), 1862
PLATE 53 · FIGURE 318 · MAP 264

Etymology. The species name *vialis* is derived from the Latin *via*, "road."

Synopsis. There are about twenty-four species of *Ambyscirtes* skippers. Most occur in the Southwest and northern Mexico. The Roadside Skipper is the most widespread skipper in North America.

Butterfly. Male forewing: $\overline{X}$ = 1.2 cm, range 1.1–1.2 cm; female forewing: $\overline{X}$ = 1.2 cm, range 1.1–1.3 cm. Ventrally, the Roadside Skipper may be distinguished by the violet-gray scaling on the outer half of the hind wing and forewing apex. Dorsally, the wings are black except for a few small white spots at the forewing apex.

Range. This skipper occurs from Nova Scotia and Maine west across southern Canada to British Columbia and then south to northern Florida, the Gulf states, Texas, northern New Mexico, and central California.

Habitat. The Roadside Skipper is most often found in open areas near or within woodland, often in the vicinity of streams. It occurs in the Lower Austral through the Canadian Life Zones.

Life History. *Behavior.* Males perch on the ground or low vegetation in glades or wood edges. They usually wave their antennae alternately in small circles.

Broods. In the North, the Roadside Skipper is univoltine, usually with adults found from mid-May to late June. Two flights are usual from Illinois and the Delaware Valley southward. They commonly occur from mid-April to late May or early June and again from mid-July to near the end of August. In Mississippi and Florida, the skipper has been found in March.

Early Stages. The egg is pale green and hemispherical. The mature larva is pale green and is covered with small, green, setiferous tubercles. The caterpillar's head is dull white with several vertical red-brown stripes on the face. The pupa is green with small areas of red or yellow around the head.

Adult Nectar Sources. Low blue flowers, such as selfheal and verbena, are preferred nectar sources.

Caterpillar Host Plants. Various grasses are selected as larval hosts, including blue grass (*Poa*), wild oats (*Avena*), bent grass (*Agrostis*), Bermuda grass (*Cynodon dactylon*), and broadleaf uniola (*Uniola latifolia*).

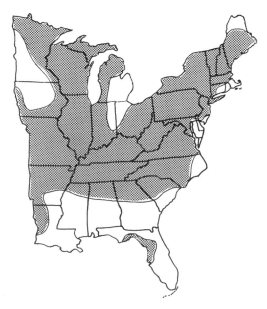

264. *Amblyscirtes vialis*

Celia's Roadside Skipper
Amblyscirtes celia (Skinner), 1895

Synopsis. A skipper of western Texas and northern Mexico, Celia's Roadside Skipper has been found once or twice in Louisiana. Celia's Roadside Skipper is sometimes considered to be the same species as Bell's Roadside Skipper.

Bell's Roadside Skipper
Amblyscirtes belli Freeman, 1941

MAP 265

Etymology. The species was named in honor of Ernest Layton Bell (1876–1964), a banker by profession and a lepidopterist by avocation. He was a Research Associate at the Museum of Natural History in New York and described more than 200 species, predominantly skippers.

Synopsis. This is a woodland skipper of the mid-South that is closely related to the more southwestern Celia's Roadside Skipper.

Butterfly. Male forewing: $\overline{X}$ = 1.3 cm, range 1.3–1.4 cm; female forewing: $\overline{X}$ = 1.4 cm, range 1.3–1.4 cm. Bell's Roadside Skipper has a linear series of white spots running through the forewing cell almost to the apex. The ventral hind wing is hoary gray-brown with a pattern of dull gray spots.

Range. Bell's Roadside Skipper occurs from Georgia west to Missouri, eastern Kansas, Oklahoma, and Texas.

Habitat. This skipper is found in grassy areas in open woods, by woodland creeks, or along wood edges. It inhabits the Lower and Upper Austral Life Zones.

Life History. *Broods.* There are three flights in Missouri (May, July, and September).

Early Stages. The egg is shiny white and hemispherical. The mature caterpillar's body is pale green with a white overcast. The body is also covered with short black setae along the back and pale orange ones on the side. The head is creamy white with vertical orange-brown bands. The pupa has a pale cream head and wing cases together with light-brown horns and a pale-yellow abdomen ringed with pale orange. The eye cases are bright red. The caterpillar lives in a folded leaf tent and feeds on the leaf from the tip down. The pupa is in a sealed leaf-case cocoon that is thinly lined with powdery silk. Larvae of the last yearly brood overwinter in the fourth instar (Heitzman 1965b).

Caterpillar Host Plant. Broadleaf uniola (*Uniola latifolia*) is the only reported host.

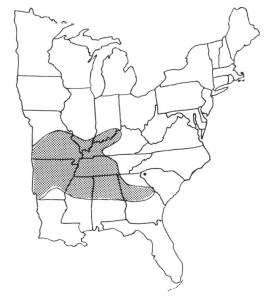

265. *Amblyscirtes belli*

Least Florida Skipper
Amblyscirtes alternata (Grote & Robinson), 1867

MAP 266

Etymology. The species name is derived from the Latin *alternus,* "reciprocal."

Synopsis. The Least Florida Skipper is a rarely encountered small skipper of pine flats on the southeastern coastal plain.

Butterfly. Male forewing: $\overline{X}$ = 1.1 cm, range 1.0–1.2 cm; female forewing: $\overline{X}$ = 1.2 cm, range 1.1–1.2 cm. This species is similar to the Roadside Skipper (*Amblyscirtes vialis*), but has a more pointed forewing, several small white spots in the center of the forewing, widely scattered gray (rather than violet-gray) scales over the entire ventral hind wing surface, and white, rather than buff, checked fringes.

Range. The Least Florida Skipper occurs from southeastern Virginia south along the coastal plain to Florida, and west along the Gulf Coast to Arkansas and eastern Texas.

Habitat. This skipper has been found in open pine woods in the Lower and Upper Austral Life Zones.

Life History. *Broods.* This small skipper is bivoltine, with emergence occurring earlier in more southern portions of its range. Flight dates in Virginia are April 26 to June 5 and July 15 to August 15, while in Mississippi, adults may be found by late March. There may be three broods in Florida, with the third appearing in November.

Eufala Skipper
Lerodea eufala (Edwards), 1869
PLATE 54 · FIGURE 319 · MAP 267

Etymology. The genus name may be derived from a combination of the Greek *leros,* "silly, foolish," and the Latin *dea,* "goddess." Eufala is the name of two cities, one in southeastern Alabama (formerly Irvington) and one in east Oklahoma.

Synopsis. The Eufala Skipper is another butterfly unable to survive freezing winters but that frequently expands its range northward during the summer. It is similar to several other small dark skippers.

Butterfly. Male forewing: $\overline{X}$ = 1.3 cm, range 1.2–1.3 cm; female forewing: $\overline{X}$ = 1.4 cm, range 1.2–1.5 cm. The Eufala Skipper has a distinct gray appearance below due to heavy overscaling on the ventral hind wings. There are usually several hyaline spots on the forewing, including a costal series, as on the Three Spot Skipper (*Cymaenes tripunctis*).

Range. The Eufala Skipper is resident from coastal Georgia and possibly South Carolina south through Florida and east across the southern United States to central California. It expands its range through emigration in late summer and

periodically reaches Michigan and Minnesota. It ranges south through Central and South America to Patagonia and is also found on Cuba.

Habitat. The Eufala Skipper is found in open, sunny areas, including the inner edges of salt marshes, road edges, vacant lots, and agricultural areas.

Life History. *Behavior.* Males perch on low vegetation in flat grassy areas.

Broods. In Georgia, there are two annual flights (February to late May and August to early October); in Florida the Eufala Skipper flies virtually all year and may have four or more generations.

Early Stages. The egg is a glistening pale green when first laid and is hemispherical. The mature caterpillar is bright green with a dark dorsal stripe, faint white lateral lines, and several yellow longitudinal stripes. Its head is dull white with orange-brown blotches. The pupa is slender, has a pointed head projection, and is green with a prominent dark-green dorsal line (Comstock 1929).

266. *Amblyscirtes alternata*

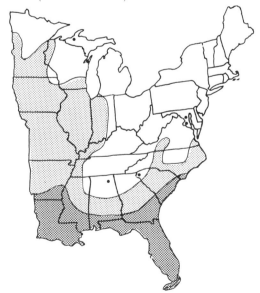

267. *Lerodea eufala*

Adult Nectar Sources. Adults visit the flowers of croton, lippia, alfalfa, composites, and probably several others.
Caterpillar Host Plants. In California, the Eufala Skipper feeds on exotic Johnson grass (*Sorghum halepense*) and occa-

sionally on milo (*Sorghum vulgare*). In Cuba it feeds on sugarcane. It will feed on St. Augustine grass (*Stenotaphrum secundatum*) in the laboratory.

Twin Spot Skipper
Oligoria maculata (Edwards), 1865
PLATE 54 · FIGURE 320 · MAP 268

Etymology. The genus name is taken from the Greek *oligōria,* "light esteem, contempt." *Maculata* is derived from the Latin *maculatus,* "having spots."
Synopsis. Although the Twin Spot Skipper is common in several southeastern states, its biology is little known. It is the only species in its genus.
Butterfly. Male forewing: $\overline{X}$ = 1.6 cm, range 1.5–1.7 cm; female forewing: $\overline{X}$ = 1.7 cm, range 1.6–1.8 cm. This is a brown-black butterfly with relatively rounded wings and four hyaline forewing spots. The ventral hind wing is most distinctive, with its pale reddish overlay and three white spots, two of which are closely spaced.
Range. The Twin Spot Skipper is resident on the southeastern coastal plain from North Carolina south through Florida and east along the Gulf Coast to Texas. This skipper rarely strays north to New Jersey, Pennsylvania, and Maryland.
Habitat. The Twin Spot Skipper has been variously reported in pine flatwoods and in coastal swamps. It is confined to the Lower Austral Life Zone.
Life History. *Broods.* The Twin Spot Skipper is bivoltine everywhere except Florida, where there are probably three flights. In Georgia and South Carolina, there are flights in April and May and then again in August and September; in Mississippi, the first brood occurs earlier (April–May).
Early Stages. The mature caterpillar is pale green with the last

two segments darker. The prothoracic shield and head are light brown. The pupa is dull green.
Adult Nectar Sources. Adults were found visiting pickerelweed flowers in Georgia.

268. *Oligoria maculata*

Brazilian Skipper
Calpodes ethlius (Stoll), 1782
MAP 269

Etymology. The genus name is derived from the Greek *Kalpē,* a proper name, and *ōdes,* a suffix formed from *eidos,* "like."
Synopsis. The presence of the Brazilian Skipper (also known as the Canna Skipper) is most often revealed by the sight of its caterpillars on ornamental cannas. It has tremendous dispersal capability.
Butterfly. Male forewing: $\overline{X}$ = 2.3 cm, range 2.2–2.4 cm; female forewing: $\overline{X}$ = 2.4 cm, range 2.2–2.7 cm. This large,

dark skipper has large translucent spots on both forewings and hind wings. Its forewings are pointed, and the ventral surface of the wings is reddish brown.
Range. The Brazilian Skipper is resident in the United States only in southern Florida and southern Texas, and ranges south through the Caribbean and Latin America to Argentina. In the United States, it periodically colonizes areas far to the north, particularly along the Atlantic and Gulf Coasts, as well as the Mississippi Valley.

Habitat. In the United States, the Brazilian Skipper is found in residential areas and garden plots where cannas are grown.

Life History. *Behavior.* The mate-seeking behavior has not been observed, but oviposition has been seen in the afternoon (1300–1445 hr). The adults are strong, high fliers. In Costa Rica, the butterflies will visit flowers from first light until sunset, and may utilize "trap-lining," a method of returning to the small individual flowers along a given route periodically through the day.

Broods. In Florida, adult Brazilian Skippers are found all year, but are more abundant in the summer. They usually appear elsewhere in the eastern states during late summer.

Early Stages. The egg, blue-green when first laid, is a low hemisphere. The mature larva is pale green and translucent, with a dark-orange head. The pupa is pale green and has a pointed head with a long tongue case.

Adult Nectar Sources. Nectar sources have not been reported in the United States, but, in Costa Rica, the skippers visit large white or pale yellow flowers of shrubs, trees, and woody lianas.

Caterpillar Host Plants. Various plants of the canna family (Marantaceae) are utilized as larval hosts. These are primarily various cannas in our area, especially the green-leaved varieties. Arrowroot (*Thalia geniculata*) has been fed upon in Florida.

Salt Marsh Skipper
Panoquina panoquin (Scudder), 1864
PLATE 54 · FIGURES 321 & 322 · MAP 270

Etymology. The genus and species names are probably derived from an American aboriginal name.

Synopsis. The Salt Marsh Skipper is a common skipper of the Atlantic and Gulf Coast salt marshes. Although it has several close relatives, it is distinctly marked.

Butterfly. Male forewing: $\overline{X}$ = 1.4 cm, range 1.3–1.5 cm; female forewing: $\overline{X}$ = 1.6 cm, range 1.6–1.7 cm. The ventral pattern of yellow veins on the dark brown hind wing, together with a short white dash at the end of the cell, should easily distinguish the Salt Marsh Skipper from all others.

Range. The Salt Marsh Skipper is confined to the immediate coast from Long Island, south to Florida and east along the Gulf Coast to Louisiana.

Habitat. This species is found within and adjacent to ex-

tensive coastal salt or brackish marshes. It occasionally wanders to nearby open fields and wood edges.

Life History. *Broods.* The Salt Marsh Skipper is bivoltine in the northern part of its range (May and July–August), but there seem to be three flights along the Gulf Coast (e.g., Mississippi—April, June, and October). In Florida, there may be four flights, since the species is on the wing from mid-February to December.

Adult Nectar Sources. The Salt Marsh Skipper nectars at verbena, thistle, blue mistflower, salt marsh fleabane, lippia, gumweed, red clover, sweet pepperbush, and privet.

Caterpillar Host Plants. Seashore saltgrass (*Distichlis spicata*) is suspected to be the host.

269. *Calpodes ethlius*

270. *Panoquina panoquin*

Obscure Skipper
Panoquina panoquinoides (Skinner), 1891

MAP 271

Etymology. The species name means "similar to panoquin."

Synopsis. The Obscure Skipper is closely related to the Salt Marsh Skipper of the Atlantic coast and largely replaces it on the Gulf Coast and southward.

Butterfly. Male forewing: $\bar{X}$ = 1.3 cm, range 1.2–1.4 cm; female forewing: $\bar{X}$ = 1.4 cm, range 1.3–1.4 cm. The Obscure Skipper is like a small, dull version of the Salt Marsh Skipper. On the ventral surface of the hind wings, the veins are pale and there are three pale spots in the postmedian area.

Range. The Obscure Skipper occurs along the coastal areas of Florida and west along the Gulf Coast to Texas. In addition, this skipper is found in the Caribbean as well as South America.

Habitat. This skipper is found in salt marshes and adjacent open fields and dunes.

Life History. *Broods.* Adult Obscure Skippers are found from February through December in Florida, but are most common from March to May and September to October. There are probably two major generations.

Early Stages. The egg is white and hemispherical. Mature larvae are green with four green-white dorsal lines and a pale-yellow stripe along each side. The caterpillar's head is light to dark brown. The pupa is translucent green with longitudinal stripes.

Adult Nectar Sources. Adults visit rattlebox and probably other flowers as well.

Caterpillar Host Plants. Bermuda grass (*Cynodon dactylon*) and sugarcane are the only two reported hosts.

Ocola Skipper
Panoquina ocola (Edwards), 1863

PLATE 54 · FIGURE 323 · MAP 272

Etymology. Ocala was the name of an Indian tribe of north central Florida, possibly located near present-day Ocala, Marion County.

Synopsis. The Ocola Skipper is one of the few of its family known to partake in mass emigrations. In addition, the species is likely to appear in areas far to the north of its native haunts.

Butterfly. Male forewing: $\bar{X}$ = 1.6 cm, range 1.5–1.7 cm;

271. Panoquina panoquinoides

272. Panoquina ocola

female forewing: $\overline{X}$ = 1.8 cm, range 1.7–1.9 cm. This species has elongated forewings relative to the hind wings, and they project far beyond the hind wings when the skipper is at rest. The ventral wings are usually devoid of markings, but fresh females have a distinct purplish iridescence.

Range. In our area the Ocola Skipper is probably resident only in the Deep South, although it is found frequently in Georgia and along the Gulf Coast. It wanders north irregularly as far as New Jersey, Ohio, and Indiana. Beyond our area, it extends south through the Caribbean area, Mexico, and Central America to Argentina.

Habitat. The Ocola Skipper is found in low, damp fields and pastures. It also occurs in gardens and along forest edges. It is resident in the Tropical Life Zone.

Life History. *Behavior.* This species rests with forewings pulled down, so that only their outer portions project from between the hind wings. Thousands of these skippers were seen flying northeast near Lake Pontchartrain, Louisiana, on October 15, 1950 (Penn 1955). This is one of the few times that skippers have been observed in such large movements. *Broods.* The Ocola Skipper is found year round in Florida, but only during much briefer periods in other eastern states. *Early Stages.* The egg is white and hemispherical, while the mature caterpillar's first two body segments are blue-green and the remainder are gray-green. There is a dark dorsal line and a green-white stripe along each side. The head is light green. The chrysalis is green with four yellow stripes along the abdomen. The pupal head is pointed and the tongue case extends just barely beyond the wing pads.

Adult Nectar Sources. These butterflies take nectar at flowers of pickerelweed, buttonbush, swamp milkweed, shepherd's needle, and lantana.

Caterpillar Host Plants. Sugarcane and the grass *Hymenachne amplexicaule* are fed upon in Puerto Rico.

Subfamily Megathyminae: Giant Yucca Skippers

We follow Stanford (1981) in considering this group of specialized skippers to be a subfamily. The adults are medium to large robust insects with predominantly black wings, usually with white or yellow bands and spots. The adult's head is distinctly narrower than the thorax, and its antennae are clubbed, unlike those of other skippers. The Giant Yucca Skippers are strictly a New World group, centered in our arid Southwest and northern Mexico. Adult reproductive behavior is unknown. The skippers are extremely rapid flyers, and have been seen taking moisture at sand. They are thought not to take food as adults. Most species are univoltine. The large, hemispherical eggs are laid on host leaves. Upon hatching, the young larva burrows down the leaf to the caudex of the host and proceeds to burrow into the growing point or roots, wherein it constructs a silk-lined tunnel. Frass and other debris are incorporated into a silken turret constructed before pupation. The caterpillars are plain creamy white and grublike. The pupae are brown with a chalky bloom. Unlike those of all other butterflies, the pupae move actively up and down their burrows.

Giant Yucca Skipper
Megathymus yuccae (Boisduval & LeConte), 1834
PLATE 54 · FIGURE 324 · MAP 273

Etymology. *Megathymus* is derived from the Greek *mega,* "big," and *thymus,* an aromatic plant (mint family) or the name of a gland. The species name, *yuccae,* is taken from the host-plant name.

Synopsis. The robust, large-bodied yucca skippers belong to a subfamily separate from other skippers. Our two species both bore into the stems of yuccas.

Butterfly. Male forewing: $\overline{X}$ = 2.8 cm, range 2.6–3.0 cm; female forewing: $\overline{X}$ = 3.3 cm, range 3.1–3.5 cm. In males, the submarginal yellow forewing band is clear-cut in the Giant Yucca Skipper, while it is a much broader blurred patch in the Cofaqui Skipper (*Megathymus cofaqui*). In the Giant Yucca Skipper female, the submarginal band is offset and separated from the yellow costal bar, while in the Cofaqui Skipper female these two markings are merged to form a broad arching patch. Several subspecies are said to occur in our area, but they are not currently recognized.

Range. The Giant Yucca Skipper occurs from southeastern Virginia south along the Atlantic coastal plain and Piedmont to Florida, thence west in the Gulf states and from Arkansas through Colorado and the Southwest to southern California. The species also occurs south into northern Mexico and Baja California.

Habitat. In our area, this skipper inhabits coastal dunes, pine flatwoods, old fields, bottomlands, and large granite outcroppings primarily in the Lower Austral Life Zone.

Life History. *Behavior.* The Giant Yucca Skipper is active in late afternoon and early evening. They are extremely rapid fliers. Males of western subspecies are perchers, selecting a high vantage point. If a male is removed, his territory will be reoccupied within an hour.

Broods. There is a single spring flight each season, lasting no more than a month at any particular locality. Adults are usually active in March and April, but may be found from mid-February to mid-May in Florida.

Early Stages. Eggs are laid on yucca leaves and the young larvae feed near the leaf apex. Later they bore into the growing point (caudex) and feed within the root, sometimes forming a tunnel as much as half a meter deep. Their presence is indicated by a silk tent or chimney projecting from the caudex. Small host individuals are usually selected (Butler and Covell 1957). The mature larvae, which are white with a black head and cervical shield, overwinter in the tents, and pupate in late winter or early spring.

Adult Food. Although freshly emerged males take moisture at wet spots, no flower visitation or other substantive feeding has been reported.

Caterpillar Host Plants. The Giant Yucca Skipper feeds only on yuccas, including Small's yucca (*Yucca smalliana*), bear grass (*Yucca filamentosa*), Spanish dagger (*Yucca gloriosa*), and Spanish bayonet (*Yucca aloifolia*) (Remington 1958).

Cofaqui Skipper
Megathymus cofaqui (Strecker), 1876
MAP 274

Etymology. Cofaqui was a settlement in eastern Georgia through which the explorer De Soto passed in April 1540.

Synopsis. The Cofaqui Skipper is another yucca feeder. Its larvae construct their tents at the base of the plant, not projecting from the growing point like those of the Giant Yucca Skipper.

Butterfly. Male forewing: $\overline{X} = 2.4$ cm, range 2.1–2.7 cm; female forewing: $\overline{X} = 2.6$ cm, range 2.5–2.7 cm. The features that separate the Cofaqui Skipper from the Giant Yucca Skipper are pointed out in the account of the latter species. Male Cofaqui Skippers have erect hairlike scales on the basal portion of the dorsal hind wing. The subspecies *harrisi* is found in the Georgia Piedmont and probably in Tennessee as well.

Range. The Cofaqui Skipper occurs from central Georgia and possibly southeastern Tennessee south through much of peninsular Florida.

Habitat. This skipper is found on coastal sand dunes, pine flatwoods, or other situations where its host plants are found.

Life History. *Behavior.* Adults emerge from their tents in midmorning and require 3 hours before they are ready to fly. The butterfly has a strong, rapid flight, although each sortie is fairly brief. On hot afternoons, or perhaps at other times when they are not active, the adults rest on the shady side of

273. *Megathymus yuccae*

274. *Megathymus cofaqui*

tree trunks, 1 or 2 meters above the ground. A female was observed ovipositing an hour before dusk.

Broods. In Georgia, there is a single prolonged flight from early July to mid-August, with a few late individuals reported in mid-September. In Florida, there are two generations, with adults in March–April and then again from mid-August to mid-November.

Early Stages. The eggs are laid singly on host leaves. Upon hatching, the young caterpillar bores into the stem and forms a silk-lined tunnel in the lower stem and roots. When fully developed, the caterpillar constructs a silken "tent" projecting from the ground near the base of the host or from the stalk. When fresh, the tent is sticky and becomes covered with soil particles and leaf fragments. The pupa is found in the tent and when disturbed it actively wriggles downward into the tunnel. Emergence of the adult occurs 1 or 2 weeks after pupation (Harris 1954).

Caterpillar Host Plants.   The Cofaqui Skipper selects Spanish bayonet (*Yucca aloifolia*) and bear grass (*Yucca filamentosa*) as its caterpillar host plants.

Species Unreliably Reported in the Eastern United States

(arranged phylogenetically)

SCIENTIFIC NAME	COMMON NAME
Battus devilliers	Devilliers Swallowtail
Papilio rutulus	Western Tiger Swallowtail
Eurytides celadon	Cuban Kite Swallowtail
Colias eurydice	California Dog Face
Eumaeus minijas	Cycad Butterfly
Agriades franklinii	Arctic Blue
Anartia fatima	Fatima
Diaethria clymena	Eighty-eight Butterfly
Mestra cana	St. Lucia Mestra
Historis odius	Orion
Historis acheronta	Cadmus
Marpesia coresia	Waiter
Marpesia chiron	Many-banded Dagger Wing
Polygonus manueli	Manuel's Skipper
Hesperia viridis	Green Skipper
Choranthus radians	Rayed Skipper
Choranthus haitensis	Haitian V-Mark Skipper
Panoquina sylvicola	Sugar Cane Skipper
Panoquina hecebolus	Hecebolus Skipper

Glossary

Androconia: Specialized wing scales, often in patches or along veins, which may release chemical sex attractants (pheromones) during courtship or mating. Limited to males.

Antennae: The long, segmented, paired sensory appendages projecting from the top of a butterfly's head.

Aposematic: Pertaining to brightly colored, distasteful insects, usually protected from predators.

Batesian mimicry: The close resemblance of an edible insect to a distasteful one, which confers protection from predators.

Bergmann's Rule: A phenomenon displayed by animals (usually vertebrates) in which individuals tend to be larger at higher latitudes and elevations.

Bivoltinism: Having two generations or flights each year.

Brood: In this book, a more or less synchronous generation of a butterfly species.

Bursa copulatrix: A membranous pouch of the female internal reproductive system. The sperm sacs (spermatophores) are stored within this structure.

Chalaza: On larvae, sclerotized (hardened) patches bearing more than one seta.

Chitin: The hard substance of which an insect's external skeleton is composed.

Chorion: The outer shell of an insect egg.

Chrysalis: see Pupa.

Corolla: The combined structure made up of the petals of a flower, sometimes fused.

Cremaster: The hooked projections at the posterior terminus of a Lepidopteran pupa.

Crypsis: The phenomenon of background resemblance in animals.

Deciduous: Capable of falling off. In this book, pertaining to trees or shrubs that shed their leaves.

Diapause: A common obligatory process of developmental cessation, usually triggered by a particular day-length (photoperiod) regime.

Dicotyledon: One of two divisions of flowering plants (Angiospermae). There are two seedling leaves.

Dimorphism: Having two forms, usually differing by color, size, and/or shape in butterflies.

Disturbed habitats: Sites that are periodically or permanently altered by man's activities.

Ecdysis: The process by which a caterpillar sheds its 'skin.'

Emigration: A one-way movement outward from an area of residency.

Estivation: A process of summer dormancy engaged in by the Mourning Cloak, tortoise shells and angle wings.

Embryogenesis: The process of development of the young organism within the egg following fertilization.

Falcate: Hooked. Here, pertaining to the shape of the forewing apex of some butterflies, e.g., Florida Leaf Wing.

Flash coloration: The sudden display of bright colors, for example the brightly colored dorsal wing surfaces of some butterflies.

Flavonoids: A class of chemical compounds responsible for some yellow, orange, and brown pigments.

Genitalia: The adult structures involved with mating and reproduction.

Genus: A taxonomic grouping that includes one or more related species. *Plural:* genera.

Ground litter: The dead, loose organic matter on the forest floor or around the base of plants.

Hammock: An 'island' of dense, evergreen woody vegetation, usually surrounded by more open habitats.

Hemolymph: The internal body fluid of insects, functionally equivalent to the blood of vertebrates, but usually lacking hemoglobin.

Hibernation: A process of winter dormancy that involves either diapause or physiological slowdown due to cold temperatures.

Holarctic: A faunal region including the supratropical areas of the Northern Hemisphere.

Instar: The stage of a caterpillar or larva.

Integument: The skin or outer covering of animals.

Iridescence: The metallic or opalescent reflection of structural color.

Labrum: A plate of an insect head above the oral area.

Larva: The immature feeding stage of holometabolous insects. *Plural:* larvae.

Life Zones: Geographic areas divided on the basis of climatic features. The zones are predictive of the presence or absence of animal or plant species.

Meconium: A structure that blocks the larval gut during the pupal stage. It is usually broken and voided upon emergence of the adult.

Melanins: Pigments responsible for some dark or black colors.

Mesic: Moist or wet habitats.

Micropyle: The small area atop an insect egg through which the sperm cells pass to achieve fertilization.

Migration: A regular two-way movement of animal populations.

Mimicry: The resemblance of a harmless, relatively palatable animal (mimic) to a harmful, poisonous, or distasteful animal (model).

Monocotyledon: One or two groupings of flowering plants (Angiospermae). The young seedlings have a single cotyledon. Major leaf veins are parallel and unconnected.

Monophagous: Dietary specialization. In butterflies, species whose caterpillars feed on a single host.

Ocelli: The small simple eyes of insect larvae, or eyelike markings on butterfly wings. *Singular:* ocellus.

Oligophagous: Dietary specialization. In butterflies, species whose caterpillars eat several plants in one or a few closely related genera or families.

Ommatidia: The separate visual units of which an insect's compound eye are composed. *Singular:* ommatidium.

Overwinter: The process of passing the cold winter months.

Oviposition: The act of laying or depositing eggs.

Palpi: The appendages adjacent to the oral region. *Singular:* palpus.

Patrolling: A mate-location system in which the males fly about in search of receptive females.

Perching: A mate-location system in which the males take up perches where they await the passage of receptive mates.

Pheromone: A chemical involved in communication between individuals, usually for sex attraction or courtship.

Photoperiod: The period of continuous light, usually the length of the daylight period.

Piedmont: Literally translated, "the foothills." In the eastern United States, the region between the coastal plain and mountains.

Poikilothermic: Referring to cold-blooded organisms that cannot internally regulate their body temperatures.

Polymorphism: Having two or more forms. In butterflies, usually involving color, size, or wing shape.

Polyphagous: Dietary specialization. In butterflies, species whose caterpillars feed on many hosts in several to many unrelated plant families.

Polyphenism: Having two or more seasonal forms.

Population: A group of individuals of the same species occurring in the same geographic area.

Proboscis: The coiled feeding apparatus of adult Lepidoptera.

Prolegs: The false legs of caterpillars, found on the third, fourth, fifth, sixth, and tenth segments in butterflies.

Pteridine: A group of chemical compounds responsible for white, yellow, or orange pigments in some butterflies, especially Pieridae.

Pupa: The transformation stage of advanced insects within which the larva metamorphoses to the adult. *Plural:* pupae.

Riparian: Relating to rivers, as in the animal and plant communities nearby.

Scrub: A plant formation composed of low, dense, woody, vegetation.

Setae: The hairs of insects, found on larvae, pupae, or adults. The wing scales are modified setae. *Singular:* seta.

Species: A taxonomic level or basic biological unit, all of whose individuals potentially can breed with each other and produce fertile offspring.

Spermatophore: The membranous structure produced by male butterflies within which are the spermatozoa. The spermatophore is passed to the female during mating.

Spiracles: The openings along the lateral areas of caterpillars through which they breathe.

Stadium: The time period of a larval instar. *Plural:* stadia.

Subspecies: A taxonomic subdivision of a species, consisting of a geographic unit, most of whose individuals are separable from those of other subspecies.

Switch gene: A single locus on a chromosome that controls a major change in appearance or color pattern.

Taxonomy: The science of classification and naming.

Territoriality: The defense of a defined area containing resources important for feeding or reproduction. The term is dubiously applied to butterfly behavior.

Trap-lining: A method of foraging wherein an animal periodically visits the same potential resource sites in approximately the same sequence. In butterflies, it is the sequential visitation of the same nectar sources at periodic intervals.

Ultraviolet: Colors of extremely short wave-lengths at the far end of the spectrum, not usually visible to the human eye.

Univoltine: Having a single generation or flight each year.

Voltinism: The number of generations or flights each year.

Bibliography

Angevine, M. W., and P. F. Brussard. 1979. Population structure and gene frequency analysis of sibling species of *Lethe. J Lepid Soc* 33:29–36.

Anthony, G. D. 1970. Population structure of *Oenois melissa semidea. J Res Lepid* 7:133–48.

Arms, K., P. Feeny, and R. C. Lederhouse. 1974. Sodium: stimulus for puddling behavior by Tiger Swallowtail butterflies. *Papilio glaucus. Science* 185:373–74.

Badger, F. S., Jr. 1958. *Euptychia mitchellii* (Satyridae) in Michigan and Indiana tamarack bogs. *Lepid News* 12:41–46.

Baggett, H. D. 1982. In *Rare and Endangered Biota of Florida,* Vol. 6, Invertebrates. R. Franz (ed.), pp. 75–77. Gainesville. University of Florida Press.

Bates, M. 1923. Notes on Florida Lepidoptera. *Fla Entomol* 7:42–43.

Beck, A. F. 1983. Notes on the Great Dismal Swamp population of *Mitoura hesseli* Rawson and Ziegler. *J Lepid Soc* In press.

Beebe, W., J. Crane and H. Fleming. 1960. A comparison of the eggs, larvae and pupae in fourteen species of heliconiine butterflies from Trinidad, W.I. *Zoologica* 45:111–54.

Benson, W. W., and T. C. Emmel. 1973. Demography of gregariously roosting populations of the nymphaline butterfly *Marpesia berania* in Costa Rica. *Ecology* 54:326–35.

Bitzer, R. J., and K. C. Shaw. 1981. Territorial behavior of the Red Admiral, *Vanessa atalanta* (L.) (Lepidoptera: Nymphalidae). *J Res Lepid* 18:36–49.

Boisduval, J. A., and J. E. LeConte. 1833. *Histoire Générale et Iconographie des Lepidopteres et des Chenilles de l'Amérique Septentrionale,* Volume 1. Paris. 228 pp.

Bowers, D. 1978. Observations on *Erora laeta* (Lycaenidae) in New Hampshire. *J Lepid Soc* 32:140–44.

Bowers, M. D. 1978. Overwintering behavior in *Euphydryas phaeton* (Nymphalidae). *J Lepid Soc* 32:282–88.

Bowers, M. D., 1980 Unpalatability as a defense strategy of *Euphydryas phaeton* (Lepidoptera: Nymphalidae). *Evolution* 34:586–600.

Brimley, C. S. 1938. Lepidoptera. In *The Insects of North Carolina.* Raleigh:Division of Entomology, North Carolina Department of Agriculture. pp. 255–313.

Brooks, J. C. 1962. Food plants of *Papilio palamedes* in Georgia. *J Lepid Soc* 16:198.

Brower, A. E. 1974. A list of the Lepidoptera of Maine—Part 1. The Macrolepidoptera. University of Maine (Orono), Life Sciences and Agricultural Station, *Tech Bull* 66, 136 pp.

Brower, L. P. 1958. Larval foodplants in the *Papilio glaucus* group. *Lepid News* 12:103–14.

Brower, L. P. 1962. Evidence for interspecific competition in natural populations of the Monarch and Queen butterflies, *Danaus plexippus* and *D. gilippus berenice* in south central Florida. *Ecology* 43:549–52.

Brower, L. P., F. M. Pough, and H. R. Meck, 1970. Theoretical investigations of automimicry. 1. Single trial learning. *Proc Natl Acad Sci* 66:1059–66.

Brower, J. V. Z. 1958. Experimental studies of mimicry in some North American butterflies. Part II. *Battus philenor* and *Papilio troilus, P. polyxenes* and *P. glaucus. Evolution* 12:123–36.

Brown, F. M. 1955. Studies of Nearctic *Coenonympha tullia* (Rhopalocera: Satyridae). *Bull Amer Mus Nat Hist* 105:359–410.

Brown, F. M. 1961. *Coenonympha tullia* on islands in the St. Lawrence River. *Can Ent* 93:107–17.

Burns, J. M. 1964. Evolution of skipper butterflies in the genus *Erynnis. Univ Calif Publ Ent* 37:1–216.

Burns, J. M. 1966a. Expanding distribution and evolutionary potential of *Thymelicus lineola* (Lepidoptera: Hesperiidae), an introduced skipper, with special reference to its appearance in British Columbia. *Can Ent* 98:859–66.

Burns, J. M. 1966b. Preferential mating versus mimicry: disruptive selection and sex-limited dimorphism in *Papilio glaucus. Science* 153:551–53.

Burns, J. M. 1968. Mating frequency in natural populations of skippers and butterflies as determined by spermatophore counts. *Proc Natl Acad Sci* 61:852–59.

Burns, J. M. 1969. Cryptic sleeping posture of a skipper butterfly, *Erynnis brizo. Psyche* 76:382–86.

Burns, J. M. 1970. Duration of copulation in *Poanes hobomok* (Lepidoptera: Hesperiidae) and some broader speculations. *Psyche* 77:127–30.

Burns, J. M., and F. M. Johnson. 1967. Esterase polymorphism in natural populations of a sulphur butterfly, *Colias eurytheme. Science* 156:93–96.

Burns, J. M., and R. O. Kendall. 1969. Ecologic and spatial distribution of *Pyrgus oileus* and *Pyrgus philetas* (Lepidoptera: Hesperiidae) at their northern distributional limits. *Psyche* 76:41–53.

Butler, R. B., and C. V. Covell, Jr. 1957. *Megathymus yuccae* in North Carolina. *Lepid News* 11:137–41.

Cardé, R. T., A. M. Shapiro, and H. K. Clench. 1970. Sibling species in the *eurydice* group of *Lethe* (Lepidoptera: Satyridae). *Psyche* 77:70–103.

Chermock, R. L., and O. D. Chermock. 1947. Notes on the life histories of three Floridian butterflies. *Can Ent* 79:142–44.

Chew, F. S. 1977. The effects of introduced mustards (Cruciferae) on some native North American cabbage butterflies (Lepidoptera: Pieridae). *Atala* 5:13–19.

Chew, F. S. 1980. Natural interspecific pairing between *Pieris virginiensis* and *P. napi oleracea* (Pieridae). *J Lepid Soc* 34:259–61.

Chew, F. S., P. Mathews, and J. Douglas. 1977. Of cabbages and caterpillars: ecological studies of two *Pieris* species in northern Vermont. *The Northern Raven* 5:4–9.

Clark, A. H. 1926. Carnivorous butterflies. *Smithsonian Institution Publ* 2856:439–508.

Clark, A. H. 1932. The butterflies of the District of Columbia and vicinity. Washington, DC: United States National Museum, *Bulletin* 157:1–337.

Clark, A. H. 1936. The Gold-Banded Skipper (*Rhabdoides cellus*). *Smithsonian Miscellaneous Collection* 95(7):1–50.

Clark, A. H., and L. F. Clark. 1951. The butterflies of Virginia. *Smithsonian Miscellaneous Collection* 116:1–239.

Clark, S. H., and A. P. Platt. 1969. Influence of photoperiod on development and larval diapause in the viceroy butterfly. *Limenitis archippus. J Insect Physiol* 15:1951–57.

Clench, H. K. 1961. *Panthiades m-album* (Lycaenidae): remarks on its early stages and on its occurrence in Pennsylvania. *J Lepid Soc* 15:226–32.

Clench, H. K. 1965. The beginning of the butterfly season. *J Lepid Soc* 19:239–41.

Clench, H. K. 1970. Communal roosting in *Colias* and *Phoebis* (Pieridae). *J Lepid Soc* 24:117–20.

Clench, H. K. 1972. *Celastrina ebenina*, a new species of Lycaenidae (Lepidoptera) from the eastern United States. *Ann Carnegie Mus* 44:33–44.

Clench, H. K. 1976a. Fugitive color in the males of certain Pieridae. *J Lepid Soc* 30:88–90.

Clench, H. K. 1976b. *Nathalis iole* (Pieridae) in the southeastern United States and the Bahamas. *J Lepid Soc* 30:121–26.

Clench, H. K, and P. A. Opler. 1983. Studies on Nearctic *Euchloe*. Part 8. Distribution, ecology, and variation of *Euchloe olympia* populations. *Ann Carnegie Mus* 55:xx–xx.

Comstock, J. A. 1929. Studies in Pacific Coast Lepidoptera (continued). *Bull South Calif Acad Sci* 28:22–32.

Comstock, J. A. 1954. Life history notes on *Ascia monuste crameri. Bull South Calif Acad Sci* 53:46–49.

Comstock, J. A., and C. M. Dammers. 1934. Additional notes on the early stages of California Lepidoptera. *Bull South Calif Acad Sci* 33:25–34.

Comstock, J. A., and C. M. Dammers. 1935. Notes on the early stages of two butterflies and one moth. *Bull South Calif Acad Sci* 34:81–87.

Comstock, W. P. 1940. Butterflies of New Jersey. *J NY Ent Soc* 48:47–84.

Cook, J. H. 1906–7. Studies in the genus *Incisalia*. II. *Incisalia augustus. Can Ent* 38:214–17, 39:145–49.

Cook, J. H. 1907. Studies in the genus *Incisalia*. III. *Incisalia henrici. Can Ent* 39:229–34.

Cook, L. M., E. W. Thomason, and A. M. Young. 1976. Population structure, dynamics and dispersal of the tropical butterfly *Heliconius charitonius. J Anim Ecol* 45:851–63.

Coolidge, K. R. 1924. Life history of *Heodes helloides* Bdv. (Lepid.: Lycaenidae). *Ent News* 35:306–12.

Covell, C. V., Jr. 1977. Project Ponceanus and the status of the Schaus' Swallowtail (*Papilio aristodemus ponceanus*) in the Florida Keys. *Atala* 5:4–6.

Covell, C. V., Jr., and G. B. Straley. 1973. Notes on Virginia butterflies, with two new state records. *J Lepid Soc* 27:144–54.

Dorr, L. J. 1980. *Euphyes bimacula* (Hesperiidae) in the southeastern coastal plain. *J Lepid Soc* 34:373–74.

dos Passos, C. F. 1936. The life history of *Calephelis borealis. Can Entomol* 68:167–70.

dos Passos, C. F., and A. B. Klots. 1969. The systematics of *Anthocharis midea* Hübner (Lepidoptera: Pieridae). *Entomol Amer* 45:1–34.

Douglas, M. M., and J. W. Grula. 1978. Thermoregulatory adaptations allowing ecological range expansion by the pierid butterfly, *Nathalis iole* Boisduval. *Evolution* 32:776–83.

Downes, J. A. 1973. Lepidoptera feeding at puddle-margins, dung, and carrion. *J Lepid Soc* 27:89–99.

Downey, J. C., and A. C. Allyn. 1975. Wing-scale morphology and nomenclature. *Bull Allyn Mus* 31:1–32.

Downey, J. C., and A. C. Allyn. 1979. Morphology and biology of the immature stages of *Leptotes cassius theonus* (Lucas) (Lepid.: Lycaenidae). *Bull Allyn Mus* 55:1–27.

Drees, B. M, and L. Butler. 1978. Rhopalocera of West Virginia. *J Lepid Soc* 32:198–206.

Dyar, H. G. 1900. Life history of *Callidryas agarithe. Ent News* 11:618–19.

Ebner, J. A. 1970. *The Butterflies of Wisconsin*. Milwaukee Public Museum Popular Science Handbook 12. 205 pp.

Edwards, W. H. 1868–1897. *The Butterflies of North America,* 3 vols. American Entomological Society. Boston: Houghton-Mifflin.

Ehle, G. 1957. Unusual occurrence of *Melitaea nycteis* (Nymphalidae) in Lancaster County, Pennsylvania. *Lepid News* 11:38–41.

Ehrlich, P. R. 1958. The comparative morphology, phylogeny, and higher classification of the butterflies. Lawrence, Kansas: University of Kansas Science Bulletin, Vol. 38.

Ehrlich, P. R., and S. E. Davidson. 1960. Techniques for capture-recapture studies of Lepidoptera populations. *J Lepid Soc* 14:227–29.

Ehrlich, P. R., and S. E. Davidson. 1961. The internal anatomy of the Monarch butterfly, *Danaus plexippus* L. (Lepidoptera: Nymphalidae). *Microentomology* 24:85–133.

Ehrlich, P. R., and P. H. Raven. 1965. Butterflies and plants: a study in coevolution. *Evolution* 18:586–608.

Emmel, T. C. 1969. Taxonomy, distribution and biology of the genus *Cercyonis* (Satyridae). I. Characteristics of the genus. *J Lepid Soc* 23:165–75.

Emmel, T. C. 1970. Population biology of the neotropical satyrid butterfly, *Euptychia hermes*. *J Res Lepid* 7:153–65.

Emmel, T. C. 1972. Dispersal in a cosmopolitan butterfly species (*Pieris rapae*) having open population structure. *J Res Lepid* 11:95–98.

Evans, W. H. 1958. A breeding experiment with pupal coloration of *Eurema nicippe* (Pieridae). *Lepid News* 12:95.

Fales, J. H. 1959. A field study of the flight behavior of the Tiger Swallowtail butterfly. *Ann Ent Soc Amer* 52:486–87.

Ferge, L. A., and R. M. Kuehn. 1976. First records of *Boloria frigga* (Nymphalidae) in Wisconsin. *J Lepid Soc* 30:233–34.

Ferris, C. D. 1977. Taxonomic revision of the species *dorcas* Kirby and *helloides* Boisduval in the genus *Epidemia* Scudder (Lycaenidae: Lycaeninae). *Bull Allyn Mus* 45:1–42.

Ferris, C. D., and M. S. Fisher. 1973. *Callophrys (Incisalia) polios* (Lycaenidae): distribution in North America and description of a new subspecies. *J Lepid Soc* 27:112–18.

Fiske, W. F. 1901. An annotated catalogue of the butterflies of New Hampshire. Durham, New Hampshire: College Agricultural Experiment Station, *Technical Bulletin* 1:3–80.

Gatrelle, R. R. 1971. Notes on the confusion between *Lethe creola* and *Lethe portlandia* (satyridae). *J Lepid Soc* 25:145–46.

Graham, S. M., W. B. Watt, and L. F. Gall. 1980. Metabolic resource allocation versus mating attractiveness: adaptive pressures on the 'alba' polymorphism of *Colias* butterflies. *Proc Natl Acad Sci* 77:3615–19.

Grey, L. P. 1965. The flight period of *Boloria eunomia*. *J Lepid Soc* 19:184–85.

Hafernik, J. A., Jr. 1982. Phenetics and ecology of hybridization in buckeye butterflies. *Univ. California Publ. Entomol.*

Harris, L., Jr. 1954. An account of the unusual life history of a rare yucca skipper (Megathymidae). *Lepid News* 8:153–62.

Harris, L., Jr. 1972. *Butterflies of Georgia*. Norman, Oklahoma: University of Oklahoma Press, 326 pp.

Harvey, D. T., and T. A. Webb. 1981. Ants associated with *Harkenclenus titus, Glaucopsyche lygdamus* and *Celastrina argiolus* (Lycaenidae). *J Lepid Soc* 34:372.

Haskin, J. R. 1933. The life histories of *Eurema demoditas, Lycaena theonus* and *L. hanno*. *Ent News* 44:153–56.

Heitzman, J. R. 1963. The complete life history of *Staphylus hayhursti*. *J Res Lepid* 2:170–72.

Heitzman, J. R. 1964a. The early stages of *Euphyes vestris*. *J Res Lepid* 3:151–53.

Heitzman, J. R. 1964b. The habits and life history of *Amblyscirtes nysa* (Hesperiidae) in Missouri. *J Res Lepid* 3:154–56.

Heitzman, J. R. 1965a. The life history of *Problema byssus* (Hesperiidae). *J Lepid Soc* 19:77–81.

Heitzman, J. R. 1965b. The life history of *Amblyscirtes belli* in Missouri. *J Res Lepid* 4:75–78.

Heitzman, J. R. 1966. The life history of *Atrytone arogos* (Hesperiidae). *J Lepid Soc* 20:177–81.

Heitzman, J. R. 1973. A new species of *Papilio* from the eastern United States. *J Res Lepid* 12:1–10.

Heitzman, J. R. 1974. *Atrytonopsis hianna* biology and life history in the Ozarks. *J Res Lepid* 13:239–45.

Heitzman, J. R., and C. F. dos Passos. 1974. *Lethe portlandia* (Fabricius) and *L. anthedon* (Clark), sibling species, with descriptions of new subspecies of the former (Lepidoptera: Satyridae). *Trans Amer Ent Soc* 100:52–99.

Heitzman, J. R., and R. Heitzman. 1969. The life history of *Amblyscirtes linda* (Hesperiidae). *J Res Lepid* 8:99–104.

Hendricks, D. P. 1974. "Attacks" by *Polygonia interrogationis* (Nymphalidae) on chimney swifts and insects. *J Lepid Soc* 28:236.

Heppner, J. B. 1974. Habitat: *Brephidium pseudofea* (Lycaenidae). *J Res Lepid* 13:99–100.

Hessel, S. A. 1956. *Eurema nicippe* (Pieridae) breeding in Connecticut. *Lepid News* 10:200.

Hoffman, R. J. 1973. Environmental control of seasonal variation in the butterfly *Colias eurytheme*. I. Adaptive aspects of photoperiodic response. *Evolution* 27:387–97.

Hovanitz, W. H. 1963. Geographic distribution and variation of the genus *Argynnis*. II. *Argynnis idalia*. *J Res Lepid* 1:117–23.

Howe, W. H. (ed.) 1974. *The Butterflies of North America*. New York: Doubleday & Company, Inc., 633 pp.

Hubbell, S. P. 1957. *Boloria frigga* (Nymphalidae) in Michigan. *Lepid News* 11:37–38.

Irwin, R. R. 1970. Notes on *Lethe creola* (Satyridae), with designation of lectotype. *J Lepid Soc* 24:143–51.

Irwin, R. R. 1972. Further notes on *Euphyes dukesi* (Hesperiidae). *J Res Lepid* 10:185–88.

Irwin, R. R., and J. C. Downey. 1973. Annotated checklist of the butterflies of Illinois. Champaign, Illinois: Natural History Survey. *Biological Notes* 81:1–60.

Jeffords, M. R., J. G. Sternburg, and G. P. Waldbauer. 1979. Batesian mimicry: field demonstration of the survival of pipevine swallowtail and monarch color patterns. *Evolution* 33:275–86.

Jones, F. M. 1926. The rediscovery of *Hesperia bulenta* Bdv. and LeC., with notes on other species (Lepidoptera: Hesperiidae). *Ent News* 37:194–98.

Jones, F. M., and C. P. Kimball. 1943. The Lepidoptera of Nantucket and Martha's Vineyard Islands, Massachusetts. Nantucket, Massachusetts: Publications of the Nantucket Maria Mitchell Association, 4:1–217.

Kendall, R. O. 1964. Larval foodplants for twenty-six species of Rhopalocera (Papilionoidea) from Texas. *J Lepid Soc* 18:129–57.

Kendall, R. O. 1976. Larval foodplants and life history notes for some metalmarks (Lepidoptera: Riodinidae) from Mexico and Texas. *Bull Allyn Mus* 32:1–12.

Kilduff, T. S. 1972. A population study of *Euptychia hermes* in northern Florida. *J Res Lepid* 11:219–28.

Kimball, C. P. 1965. *The Lepidoptera of Florida.* Gainesville, Florida: Florida Department of Agriculture, 363 pp.

Klots, A. B. 1951. *A Field Guide to the Butterflies of North America, East of the Great Plains.* Boston: Houghton Mifflin Company, 349 pp.

Klots, A. B. 1960. Notes on *Strymon caryaevorus* McDunnough (Lepidoptera, Lycaenidae). *J NY Ent Soc* 68:190–98.

Klots, A. B., and H. K. Clench. 1952. A new species of *Strymon* Huebner from Georgia (Lepidoptera, Lycaenidae). *Amer Mus Nov* 1600:1–19.

Klots, A. B., and C. F. dos Passos. 1982. Studies of North American *Erora* (Scudder) (Lepidoptera, Lycaenidae). *J NY Ent Soc* 89:295–331.

Knudson, E. C. 1974. *Urbanus dorantes dorantes* Stoll (Hesperiidae): another example of Florida's population explosion.

Kohler, S. 1977. Revision of North American *Boloria selene* (Nymphalidae) with description of a new subspecies. *J Lepid Soc* 31:243–68.

Lambremont, E. N. 1954. The butterflies and skippers of Louisiana. *Tulane Studies Zool* 1:127–64.

Lawrence, D. A., and J. C. Downey, 1966. Morphology of the immature stages of *Everes comyntas* Godart. *J Res Lepid* 5:61–96.

Lenczewski, B. 1980. Butterflies of Everglades National Park. Report T-588. Homestead, Florida: National Park Service, 110 pp.

MacNeill, C. D. 1964. The skippers of the genus *Hesperia* in western North America with special reference to California. *Univ Calif Publ Ent* 35:1–130.

MacNeill, D. C. 1974. Hesperiidae. In *The Butterflies of North America,* W. H. Howe (ed). New York: Doubleday & Company, Inc., pp. 423–578.

Macy, R. W., and H. H. Shepard. 1941. *Butterflies: A Handbook of the Butterflies of the United States.* Minneapolis: University of Minnesota Press, 247 pp.

Masters, J. H. 1968a. *Euphydryas phaeton* in the Ozarks. *Ent News* 79:85–91.

Masters, J. H. 1968b. First records of two butterflies in Wisconsin (Nymphalidae, Pieridae). *J Lepid Soc* 22:252.

Masters, J. H. 1970a. Distributional notes on the genus *Mestra* (Nymphalidae) in North America. *J Lepid Soc* 24:203–8.

Masters, J. H. 1970b. Ecological and distributional notes on *Erebia discoidalis* (Satyridae) in the north central States. *J Res Lepid* 9:11–16.

Masters, J. H. 1970c. Record of *Colias gigantea* from southwest Manitoba and Minnesota. *J Res Lepid* 8:129–32.

Masters, J. H. 1972a. Habitat: *Oeneis macounii* Edwards. *J Res Lepid* 10:301–2.

Masters, J. H. 1972b. A new subspecies of *Lycaeides argyrognomon* (Lycaenidae) from the eastern Canadian forest zone. *J Lepid Soc* 26:150–54.

Masters, J. H. 1974. Biennialism in *Oeneis macounii* (Satyridae). *J Lepid Soc* 28:237–42.

Masters, J. H., and J. T. Sorensen. 1968. Bionomic notes on the satyrid butterfly *Oeneis macounii* at Riding Mountain, Manitoba. *Blue Jay* 1968:38–40.

Masters, J. H., and J. T. Sorensen. 1969. Field observations on forest *Oeneis* (Satyridae). *J Lepid Soc* 23:155–61.

Masters, J. H., and J. T. Sorensen. 1973. Habitat: *Oeneis jutta ascerta. J Res Lepid* 11:94.

Mather, B. 1954. Size of *Papilio glaucus* in Mississippi *Lepid News* 8:131–34.

Mather, B. 1955. Data on *Danaus plexippus* in the Gulf States. *Lepid News* 9:119–24.

Mather, B. 1956. *Eurema daira daira* in Mississippi. *Lepid News* 10:204–6.

Mather, B. 1963. *Euphyes dukesi. J Res Lepid* 2:161–69.

Mather, B. 1970. Variation of *Graphium marcellus* in Mississippi (Papilionidae). *J Lepid Soc* 24:176–89.

Mather, B. 1974. Size variation in *Euptoieta claudia* in Mississippi. *J Lepid Soc* 28:220–23.

Mather, B. 1975. *Amblyscirtes carolina* and *A. reversa* (Hesperiidae) in Mississippi and Georgia. *J Lepid Soc* 29:177–79.

Mather, B., and K. Mather. 1958. The butterflies of Mississippi. *Tulane Studies Zool* 6:63–109.

Mather, B., and K. Mather. 1960. The butterflies of Mississippi— Supplement No. 1. *J Lepid Soc* 13:71–72.

Matteson, J. H. 1930. *Anaea portia*—the leaf-wing, and a list of the Rhopalocera of Miami, Florida. Privately printed. 16 pp.

McAlpine, W. S. 1938. Life history of *Calephelis muticum* (McAlpine). *Bull Brooklyn Ent Soc* 33:111–20.

McAlpine, W. S. 1973. Observations on the life history of *Oarisma powesheik* (Parker) 1870. *J Res Lepid* 11:83–93.

McAlpine, W. S., S. P. Hubbell, and T. E. Pliske. 1960. Distribution, habits, and life history of *Euptychia mitchellii. J Lepid Soc* 14:209–26.

McCabe, T. L. 1981. The Dakota Skipper, *Hesperia dacotae* (Skinner): range and biology, with special reference to North Dakota. *J Lepid Soc* 35:179–93.

McDunnough, J. 1920. Notes on the life history of *Phyciodes batesii* Reak. (Lepid.). *Canad Entomol* 52:56–59.

McGuire, W. W. 1982a. New oviposition and larval hostplant records for North American *Hesperia* (Rhopalocera: Hesperiidae). *Bull Allyn Mus* 72:1–6.

McGuire, W. W. 1982b. Notes on the genus *Hesperia* in Texas: temporal and spatial relationships. *Bull Allyn Mus* 73:1–21.

Merritt, J. R. 1952. Butterflies and hilltops. *Lepid News* 6:101–2.

Miller, L. D., and F. M. Brown. 1981. A catalogue/checklist of the butterflies of America north of Mexico. Sarasota, Florida: Lepidopterist's Society, *Memoir* 2, 280 pp.

Miller, L. D., and H. K. Clench. 1968. Some aspects of mating behavior in butterflies. *J Lepid Soc* 22:125–32.

Moore, S. 1960. A revised annotated list of the butterflies of Michigan. *Occas Papers Mus Zool Univ Michigan* 617:1–39.

Muyshondt, A. 1973. Some observations on *Dryas iulia iulia* (Heliconiidae). *J Lepid Soc* 27:302–3.

Newcomb, W. W. 1909. The life history of *Chrysophanus dorcas* Kirby. *Canad Ent* 43:160–68.

Nielsen, E. T., and A. T. Nielsen. 1950. Contributions toward the knowledge of the migrations of butterflies. *Amer Mus Nov* 1471:1–29.

Nielsen, M. C. 1958. Observations on *Hesperia pawnee* [sic] in Michigan. *Lepid News* 12:37–40.

Nielsen, M. C. 1964. Discovery and observations of *Boloria eunomia* (Nymphalidae) in Michigan. *J Lepid Soc* 18:233–37.

Nielsen, M. C. 1970. Distributional maps for Michigan butterflies, Part I: skippers. *Mid-continent Lepid Ser* 1:1–10.

Nielsen, M. C., and L. A. Ferge. 1982. Observations of *Lycaeides agryrognomon nabokovi* in the Great Lakes region (Lycaeniade). *J Lepid Soc* 36:233–34.

Oliver, C. G. 1970. The environmental regulation of seasonal dimorphism in *Pieris napi oleracea* (Pieridae). *J Lepid Soc* 24:77–81.

Oliver, C. G. 1979. Experimental hybridization between *Phyciodes tharos* and *P. batesii* (Nymphalidae). *J Lepid Soc* 33:6–20.

Oliver, C. G. 1980. Phenotypic differentiation and hybrid breakdown within *Phyciodes "tharos"* (Lepidoptera: Nymphalidae) in the northeastern United States. *Ann Ent Soc Amer* 73:715–21.

Oliver, C. G. 1982. Distinctiveness of *Megisto c. cymela* and *M. c. viola* (Satyridae). *J Lepid Soc* 36:153.

Opler, P. A. 1974. Studies on Nearctic *Euchloe*. Part 7. Comparative life histories, hosts and the morphology of immature stages. *J Res Lepid* 13:1–20.

Pease, R. W., Jr. 1962. Factors causing seasonal forms in *Ascia monuste* (Lepid.). *Science* 137:987–88.

Penn, G. H. 1955. Mass flight of Ocola skippers. *Lepid News* 9:79.

Platt, A. P., and L. P. Brower. 1968. Mimetic versus disruptive coloration in integrading populations of *Limenitis arthemis* and *astyanax* butterflies. *Evolution* 22:699–718.

Platt, A. P., R. P. Coppinger, and L. P. Brower. 1971. Demonstration of the selective advantage of mimetic *Limenitis* butterflies presented to caged avian predators. *Evolution* 25:692–701.

Pollard, E. 1977. A method for assessing changes in the abundance of butterflies. *Biol Conserv* 12:115–33.

Powell, J. A. 1968. A study of area occupation and mating behavior in *Incisalia iroides* (Lepidoptera: Lycaenidae). *J NY Ent Soc* 76:47–57.

Proctor, N. S. 1976. Mass hibernation site for *Nymphalis vau-album* (Nymphalidae). *J Lepid Soc* 30:126.

Pyle, R. M. 1981. *The Audubon Society Field Guide to North American Butterflies*. New York: Alfred A. Knopf, 916 pp.

Randle, W. S. 1953. Observations on the life history of *Calephelis borealis*. *Lepid News* 7:119–22, 133–38.

Rausher, M. D. 1980. Host abundance, juvenile survival, and oviposition preference in *Battus philenor*. *Evolution* 34:342–55.

Rawlins, J. E. 1980. Thermoregulation by the black swallowtail butterfly, *Papilio polyxenes* (Lepidoptera: Papilionidae). *Ecology* 61:345–57.

Rawlins, J. E., and R. C. Lederhouse. 1978. The influence of environmental factors on roosting in the black swallowtail, *Papilio polyxenes asterius* Stoll (Papilionidae). *J Lepid Soc* 32:145–59.

Rawson, G. W. 1961. The recent rediscovery of *Eumaeus atala* (Lycaenidae) in Florida. *J Lepid Soc* 15:237–44.

Rawson, G. W. 1976. Notes on the biology and immature stages of the white peacock butterfly, *Anartia jatrophae guantanamo* (Nymphalidae). *J Lepid Soc* 30:207–10.

Rawson, G. W., and J. B. Ziegler. 1950. A new species of *Mitoura* Scudder from the pine barrens of New Jersey (Lepidoptera, Lycaenidae). *J NY Ent Soc* 58:69–82.

Rawson, G. W., J. B. Ziegler, and S. A. Hessel. 1951. The immature stages of *Mitoura hesseli* Rawson and Ziegler (Lepidoptera, Lycaenidae). *Bull Brooklyn Ent Soc* 46:123–34.

Reist, J. D. 1979. *Callphrys niphon* (Lycaendae) in Alberta with notes on the identification of *C. niphon* and *C. eryphon*. *J Lepid Soc* 33:248–53.

Remington, C. L. 1942. The distribution of *Hemiargus isola* (Reakirt) east of the Mississippi River. *Bull Brooklyn Ent Soc* 37:6–8.

Remington, C. L. 1952. Biology of Nearctic Lepidoptera. II. Foodplant and pupa of *Hemiargus isolus*. *Psyche* 59:129–30.

Remington, C. L. 1958. Autecology of *Megathymus yuccae*, with notes on foodplant specificity. *Lepid News* 12:175–85.

Remington, C. L., and R. W. Pease, Jr. 1955. Studies in foodplant specificity. I. The suitability of swamp white cedar for *Mitoura gryneus* (Lycaenidae) *Lepid News* 9.4–6.

Rickard, M. A. 1968. Life history of *Dryas julia delia* (Heliconiidae). *J Lepid Soc* 22:75–76.

Riley, N. D. 1975. *A Field Guide to the Butterflies of the West Indies*. London: W. Collins Sons & Co. 224 pp.

Riley, T. J. 1980. Effects of long and short day photoperiods on the seasonal dimorphism of *Anaea andria* (Nymphalidae) from central Missouri. *J Lepid Soc* 34:330–37.

Robbins, R. K. 1981. The "falsehead" hypothesis: predation and wing pattern variation of lycaenid butterflies. *Amer Nat* 118:770–75.

Ross, G. N. 1964. Life history studies on Mexican butterflies. I. Notes on the early stages of four papilionids from Catemaco, Veracruz. *J Res Lepid* 3:9–18.

Rutkowski, F. 1966. Rediscovery of *Euptychia mitchellii* (Satyridae) in New Jersey. *J Lepid Soc* 20:43–44.

Rutkowski, F. 1971. Observations on *Papilio aristodemus ponceanus* (Papilionidae). *J Lepid Soc* 25:126–36.

Rutowski, R. L. 1979. Courtship behavior of the Checkered White, *Pieris protodice* (Pieridae). *J Lepid Soc* 33:42–49.

Rutowski, R. L. 1980. Courtship solicitation by females of the Checkered White butterfly, *Pieris protodice*. *Behav Ecol Sociobiol* 7:113–17.

Saunders, A. A. 1932. *The Butterflies of Allegany State Park*. New York State Museum Handbook 13. Albany, New York: The University of the State of New York. 270 pp.

Saverner, P. A. 1908. Migratory butterflies. *Ent News* 19:218–20.

Scott, J. A. 1973. Life span of butterflies. *J Res Lepid* 12:225–30.

Scott, J. A. 1975a. Mate-locating behavior in western North American butterflies. *J Res Lepid* 14:1–40.

Scott, J. A. 1975b. Movements of *Euchloe ausonides* (Pieridae). *J Lepid Soc* 29:24–31.

Scott, J. A. 1975c. Variability of courtship of the buckeye butterfly, *Precis coenia* (Nymphalidae). *J Res Lepid* 14:142–47.

Scott, J. A., and P. A. Opler. 1975. Population biology and adult behavior of *Lycaena xanthoides* (Lycaenidae). *J Lepid Soc* 29:63–66.

Scriber, J. M., and P. P. Feeny. 1976. New foodplant and oviposition records for *Battus philenor* (Papilionidae). *J Lepid Soc* 30:70–71.

Scriber, J. M., G. L. Lintereur, and M. H. Evans. 1982. Foodplant suitabilities and a new oviposition record for *Papilio glaucus canadensis* (Lepidoptera: Papilionidae) in northern Wisconsin and Michigan. *Great Lakes Ent* 15:39–46.

Scudder, S. H. 1887. The introduction and spread of *Pieris rapae* in North America, 1860–1885. *Mem Boston Soc Nat Hist* 4:53–69.

Scudder, S. H. 1889. *The Butterflies of the Eastern United States and Canada with Special Reference to New England*, 3 Vols. Cambridge, Massachusetts: publ. by author, 1958 pp.

Shapiro, A. M. 1965. Ecological and behavioral notes on *Hesperia metea* and *Atrytonopsis hianna* (Hesperiidae). *J Lepid Soc* 19:215–21.

Shapiro, A. M. 1966. *Butterflies of the Delaware Valley*. Philadelphia: American Entomological Society, Special Publication, 79 pp.

Shapiro, A. M. 1968. Photoperiodic induction of vernal phenotype in *Pieris protodice* Boisduval & LeConte (Lepidoptera: Pieridae). *Wasmann J Biol.* 26:137–49.

Shapiro, A. M. 1970. The biology of *Poanes viator*. *J Res Lepid* 9:109–23.

Shapiro, A. M. 1973. The ecological associations of the butterflies of Staten Island. *J Res Lepid* 12:65–126.

Shapiro, A. M. 1974. Butterflies and skippers of New York State. *Search* 4:1–60.

Shapiro, A. M. 1976. Seasonal polyphenism. In *Evolutionary Biology*, Vol. 9, M. Hecht, W. Steere, and B. Wallace (eds). New York: Plenum Press. pp. 259–333.

Shapiro, A. M. 1977. Evidence for two routes of post-Pleistocene dispersal in *Poanes viator* (Hesperiidae). *J Res Lepid* 16:173–75.

Shapiro, A. M. 1979. *Erynnis baptisiae* (Hesperiidae) on crown vetch (Leguminosae). *J Lepid Soc* 33:258.

Shapiro, I. 1975. Courtship and mating behavior of the Fiery Skipper, *Hylephila phyleus* (Hesperiidae). *J Res Lepid* 14:125–41.

Sheppard, P. M. 1965. The monarch butterfly and mimicry. *J. Lepid Soc* 19:227–30.

Shull, C. A. 1907. Life history and habits of *Anthocharis (Synchloe) olympia* Edw. *Ent News* 18:73–82.

Shull, E. M. 1979. Mating behavior of butterflies (Papilionoidea) and skippers (Hesperioidea) in Indiana. *Proc Indiana Acad Sci* 88:200–208.

Silberglied, R. E., and O. R. Taylor, Jr. 1973. Ultraviolet differences between the sulphur butterflies, *Colias eurytheme* and *C. philodice*, and a possible isolating mechanism. *Nature* 241:406–8.

Silberglied, R. E., and O. R. Taylor, Jr. 1978. Ultraviolet reflection and its behavioral role in the courtship of the sulfur butterflies *Colias eurytheme* and *C. philodice* (Lepidoptera: Pieridae). *Behav Ecol Sociobiol* 3:203–43.

Smith, D. S., D. Leston, and B. Lenczewski. 1982. Variation in *Eurema daira* (Lepidoptera: Pieridae) and the status of *palmira* in southern Florida. *Bull Allyn Mus* 70:1–8.

Smith, J. E. 1797. The *Natural History of the Rarer Lepidopterous Insects of Georgia*, 2 Vols. London: J. Edwards; Cadell and Davies; J. White. 214 pp.

Stamp, N. E. 1979. New oviposition plant for *Euphydryas phaeton* (Nymphalidae). *J Lepid Soc* 33:203–4.

Stanford, R. E. 1981. Hesperioidea. In *Butterflies of the Rocky Mountain States*, C. D. Ferris and F. M. Brown (eds). Norman, Oklahoma: University of Oklahoma Press. pp. 67–144.

Strohecker, H. F. 1938. The larval and pupal stages of two tropical American butterflies. *Ohio J Sci* 38:294–95.

Tamburo, S. E., and F. G. Butcher. 1955. Biological studies of the Florida dusky wing skipper, and a preliminary study of other insects of Barbados cherry. *Fla Ent* 38:65–69.

Teale, E. W. 1955. An apparent migration of the mourning cloak. *Lepid News* 9:143.

Tietz, H. M. 1952. *The Lepidoptera of Pennsylvania: A Manual*. University Park: Pennsylvania Agricultural Experiment Station, Pennsylvania State College School of Agriculture.

Urquhart, F. A. 1960. The *Monarch Butterfly*. Toronto: University of Toronto Press.

Urquhart, F. A., and N. R. Urquhart. 1976a. The overwintering site of the eastern population of the monarch butterfly (*Danaus p. plexippus;* Danaidae) in southern Mexico. *J Lepid Soc* 30:153–58.

Urquhart, F. A., and N. R. Urquhart. 1976b. A study of the peninsular Florida populations of the monarch butterfly (*Danaus p. plexippus:* Danaidae). *J Lepid Soc* 30:73–87.

Vawter, A. T., and P. F. Brussard. 1975. Genetic stability of populations of *Phyciodes tharos* (Nymphalidae: Melitaeinae). *J Lepid Soc* 29:15–23.

Wagner, W. H., Jr. 1977. A distinctive dune form of the marbled white butterfly, *Euchloe olympia* (Lepidoptera: Pieridae) in the Great Lakes Area. *Great Lakes Entomol* 10:107–12.

Wagner, W. H., Jr. 1978. The northern Great Lakes white, *Pieris virginiensis* (Lepidoptera: Pieridae) in comparison with its southern Appalachian counterpart. *Great Lakes Entomol* 11:53–57.

Wagner, W. H., Jr., and M. K. Hansen. 1980. Size reduction southward in Michigan's mustard white butterfly, *Pieris napi* (Lepidoptera: Pieridae). *Great Lakes Entomol* 13:77–80.

Wagner, W. H., Jr., and T. L. Mellichamp. 1978. Foodplant, habitat, and range of *Celastrina ebenina* (Lycaenidae). *J Lepid Soc* 32:20–36.

Watt, W. B., D. Han, and B. E. Tabashnik. 1979. Population structure of pierid butterflies. II. A "native" population of *Colias philodice eriphyle* in Colorado. *Oecologia* (Berl) 44:44–52.

West, D. A., W. M. Snellings, and T. A Herbek. 1972. Pupal color dimorphism and its environmental control in *Papilio polyxenes asterius* Stoll (Lepidoptera: Papilionidae). *J NY Ent Soc* 80:205–11.

Williams, C. B. 1937. Butterfly travelers. *Nat Geogr* 71:568–85.

Williams, C. B. 1970. The migrations of the Painted Lady butterfly, *Vanessa cardui* (Nymphalidae), with special reference to North America. *J Lepid Soc* 24:157–75.

Wolcott, G. N. 1927. Entomologie d'Haiti. Haiti: Port au Prince. Department of Agriculture.

Young, A. M. 1980. Some observations on the natural history and behaviour of the Camberwell Beauty (Mourning Cloak) butterfly *Nymphalis antiopa* (Linnaeus) (Lepidoptera: Nymphalidae) in the United States. *Ent Gazette* 31:7–18.

Young, A. M., and J. H. Thomason. 1975. Notes on communal roosting of *Heliconius charitonius* (Nymphalidae) in Costa Rica. *J Lepid Soc* 29:243–55.

Index of Butterfly Names

Index of Nectar Source and Plant Names